B/NB

# ADVANCES IN ELECTRONICS AND ELECTRON PHYSICS

## VOLUME 69

EDITOR-IN-CHIEF

# PETER W. HAWKES

*Laboratoire d'Optique Electronique*
*du Centre National*
*de la Recherche Scientifique*
*Toulouse, France*


ASSOCIATE EDITOR—IMAGE PICK-UP AND DISPLAY

# BENJAMIN KAZAN

*Xerox Corporation*
*Palo Alto Research Center*
*Palo Alto, California*

# Advances in
# Electronics and
# Electron Physics

EDITED BY
## PETER W. HAWKES

*Laboratoire d'Optique Electronique*
*du Centre National*
*de la Recherche Scientifique*
*Toulouse, France*

**VOLUME 69**

ACADEMIC PRESS, INC.
**Harcourt Brace Jovanovich, Publishers**
San Diego   New York   Berkeley   Boston
London   Sydney   Tokyo   Toronto

ACADEMIC PRESS, INC.
1250 Sixth Avenue, San Diego, California 92101


United Kingdom Edition published by
ACADEMIC PRESS INC. (LONDON) LTD.
24–28 Oval Road, London NW1 7DX


LIBRARY OF CONGRESS CATALOG CARD NUMBER: 49-7504

ISBN 0–12–014669–X (alk. paper)

PRINTED IN THE UNITED STATES OF AMERICA

87 88 89 90     9 8 7 6 5 4 3 2 1

# CONTENTS

### Voltage Measurement in the Scanning
### Electron Microscope
A. GOPINATH

### New Experiments and Theoretical Development
### of the Quantum Modulation of Electrons
### (Schwarz–Hora Effect)
HEINRICH HORA and PETER H. HANDEL

# CONTRIBUTORS TO VOLUME 69

Numbers in parentheses indicate the pages on which the authors' contributions begin.

ARTHUR D. FISHER, Naval Research Laboratory, Washington, D.C. 20375-5000 (115)

A. GOPINATH,[1] Chelsea College, University of London, London, England (1)

PETER H. HANDEL, Department of Physics, University of Missouri at St. Louis, St. Louis, Missouri 63121 (55)

HEINRICH HORA, Department of Theoretical Physics, University of New South Wales, Kensington 2033, Australia (55)

DAVID F. KYSER, Philips Research Laboratories Sunnyvale, Signetics Corporation, Sunnyvale, California 94088 (175)

JOHN N. LEE, Naval Research Laboratory, Washington, D.C. 20375-5000 (115)

KENJI MURATA, Electronics Department, College of Engineering, University of Osaka Prefecture, Sakai, Osaka 591, Japan (175)

HARRY WECHSLER, Department of Electrical Engineering, University of Minnesota, Minneapolis, Minnesota 55455 (261)

[1]Present address: Department of Electrical Engineering, University of Minnesota, Minneapolis, Minnesota 55455.

# PREFACE

The spread of subjects in this volume is wide, though no wider than is reasonable in a series devoted to electron physics and electronics. We open with a detailed account by A. Gopinath of a family of modes of operation of the scanning microscope that are of considerable scientific and commercial interest, since they enable us to study the local behavior of integrated circuits and related semiconductor devices in a dynamic fashion.

Electron beams are also central to the second chapter, in which H. Hora and P. Handel describe the present state of knowledge concerning the quantum modulation of electrons. The first experiments in this field were much contested and for some years there was no clear understanding of the effect named after the first author of this chapter and H. Schwarz. Theoreticians and experimentalists have subsequently shed much light on the situation, and a second effect (named after the second author of this chapter) has emerged. Although the effects are no longer in doubt and the theory is now reasonably well understood, this is clearly a subject in which we can expect further developments.

Optical processing has long been regarded as very attractive, especially when large volumes of information are to be manipulated, and it is no secret that major research efforts are being devoted to the development of suitable systems. In the third chapter, J. N. Lee and A. D. Fisher present the types of devices that are needed in this field and give an extremely useful survey of the material available and the characteristics of the various elements. We hope to complement this device-oriented survey with an account of the techniques that are being implemented with the aid of such elements in a future volume.

In the next chapter, by K. Murata and D. F. Kyser, we return to microcircuit engineering, but to a very different aspect from that discussed by A. Gopinath. In order to understand the limits and problems of both x-ray and electron-beam microlithography, it is necessary to simulate the beam–specimen interactions and this has become an activity in itself, with highly specialized programs capable of simulating the scattering processes involved. The authors cover the whole subject, from elementary scattering and the Monte Carlo technique to all the different kinds of targets currently employed.

Finally, we have a chapter on pattern recognition, which reflects my efforts to expand the coverage of digital image processing in this series. H. Wechsler examines the question of invariance in pattern recognition, a most important problem for it is extremely difficult to devise an automatic system that is as tolerant to changes of size, shape, and contrast range as are the human eye and brain. The author describes and analyzes systematically the various methods that are known for tolerating such changes, while warning us that the arduous task of invariant pattern recognition is very far from completed: artificial intelligence is still far behind the dreams of science fiction!

The following contributions will be included in future volumes.

| | |
|---|---|
| Image Processing with Signal-Dependent Noise | H. H. Arsenault |
| Scanning Electron Acoustic Microscopy | L. J. Balk |
| Electronic and Optical Properties of Two-Dimensional Semiconductor Heterostructures | G. Bastard *et al.* |
| Pattern Recognition and Line Drawings | H. Bley |
| Magnetic Reconnection | A. Bratenahl and P. J. Baum |
| Sampling Theory | J. L. Brown |
| Dimensional Analysis | J. F. Cariñena and M. Santander |
| Electrons in a Periodic Lattice Potential | J. M. Churchill and F. E. Holmstrom |
| The Artificial Visual System Concept | J. M Coggins |
| Accelerator Physics | F. T. Cole and F. Mills |
| High-Resolution Electron-Beam Lithography | H. G. Craighead |
| Corrected Lenses for Charged Particles | R. L. Dalglish |
| Enviromental Scanning Electron Microscopy | D. G. Danilatos |
| The Development of Electron Microscopy in Italy | G. Donelli |
| Amorphous Semiconductors | W. Fuhs |
| Median Filters | N. C. Gallagher and E. Coyle |
| Bayesian Image Analysis | S. and D. Geman |
| Vector Quantization and Lattices | J. D. Gibson and K. Sayood |
| Aberration Theory | E. Hahn |
| Low-Temperature Scanning Electron Microscopy | R. P. Huebner |
| Ion Optics | D. Ioanoviciu |
| Systems Theory and Electromagnetic Waves | M. Kaiser |

# Voltage Measurement in the Scanning Electron Microscope

## A. GOPINATH*

*Chelsea College*
*University of London*
*London, England*

## I. INTRODUCTION

Voltage contrast in the scanning electron microscope (SEM) is widely used for semiconductor device and integrated circuit investigations. This contrast is the basis of currently available electron-beam testing systems. The contrast signal is small compared to the video signal, and its magnitude varies nonlinearly with specimen potential. By means of electron energy analysis and

* Present address: Department of Electrical Engineering, University of Minnesota, Minneapolis, Minnesota 55455.

suitable circuitry, the contrast is linearized, and voltage changes as low as 0.5 mV have been measured. Despite the limited bandwidth of such measurement systems, voltage waveforms and dynamic potential distributions have been obtained in devices operating up to frequencies of 9.1 GHz, using sampling techniques.

This article discusses the mechanism of voltage contrast, describes the contrast linearization schemes for measuring voltage in the SEM, and also outlines the techniques used for recording periodic waveforms at the higher frequencies using the sampling methods.

A large number of researchers have contributed to this area. However, only papers which are relevant to the development of this article are cited, and any omissions are not intentional.

## II. Historical Background

The first published work on voltage contrast was by Oatley and Everhart (1957), and detailed studies led to some understanding of the contrast mechanism. Well and Bremner (1968) used the 127° sectoral analyzer to determine the shift in the secondary electron energy distribution to linearize the contrast, and subsequently in 1969 used a 63.5° analyzer for the same experiment. Driver (1969) used a hemispherical retarding potential analyzer with a floating specimen and measured the shift in energy distribution using a sine wave and a tuned amplifier. Fleming and Ward (1970) used a feedback loop in which the potential of the floating specimen was varied with respect to a fixed 1 kV collector grid to keep the collected current constant. Gopinath and Sanger (1971a) used a rectangular cage to deflect the secondaries to a planar retarding potential analyzer in front of the scintillator detector, in combination with a feedback loop. Beaulieu et al. (1972) also used a retarding potential analyzer in front of the scintillator detector and a feedback loop, in combination with periodic pulsing of the bias and a lock-in amplifier to detect 10 mV voltage changes. Hannah (1974) used a 63.5° sectoral analyzer with automatic detection of the peak of the secondary electron distribution, and also resolved 10 mV. Fentem and Gopinath (1974) used a hemispherical retarding potential analyzer with a feedback loop, and obtained a resolution of 80 mV. Hardy et al. (1975) used a hemispherical retarding potential analyzer to detect the peak of the secondary distribution with a resolution of 0.5 V. Balk et al. (1976) used a 300 V attraction grid together with a 127° sectoral analyzer set for 300 eV electrons, a planar retarding grid analyzer, and a feedback loop to measure voltages of 25 mV. Tee and Gopinath (1976) improved on the work of Fentem to detect a 10 mV peak value sine wave at 1 Hz frequency. Gopinath et al. (1978) reported using a planar four-grid analyzer placed just above the specimen and also obtained 10 mV resolution. Duykov *et al.* (1978) also used a planar analyzer to obtain a resolution of 50 mV. Rau and Spivak (1978) used a planar analyzer and a feedback loop to obtain a resolution of 20 mV. Feuerbaum (1979) used a planar analyzer together with a deflection scheme to collect the filtered secondaries by the scintillator detector, a feedback loop, and a lock-in amplifier to obtain a resolution of 1 mV. Menzel and Kubalek (1979a) improved on the work of Balk to obtain a resolution of 1 mV, and 0.5 mV has also been resolved by Menzel (1981). Plows (1981) has modified the planar analyzer to have a ring scintillator detector with fiber-optical pipes, and obtained a resolution of 1 mV. Goto *et al.* (1981) have used a hemispherical analyzer with an extraction electrode close to the specimen to counter the effect of local fields.

The bandwidth of the SEM video display system is several megahertz, and this determines the maximum frequency of any waveform and hence any voltage contrast waveforms that may be recorded with the beam held at a single spot on the specimen. For video display, the frame time determines the voltage changes frame to frame that may be recorded. However, specimen voltage waveforms that are harmonic or near harmonic of the display line frequency, up to the maximum of the video bandwidth, may be recorded on the video display in the form of bars or slanted lines, and this is termed "voltage coding" by Lukianoff and Touw (1975).

To record waveforms above the maximum video frequency, sampling systems have to be used. Plows and Nixon (1968) demonstrated stroboscopic voltage contrast studies of a shift register clocked at 7 MHz. Robinson *et al.* (1968) showed dynamic voltage contrast distributions across a transferred electron device oscillating at 85 MHz. Hill and Gopinath (1973) examined dynamic voltage distributions across a transferred electron device oscillating at 9.1 GHz, and showed the movement of dipole domains. The sampling mode operation, recording high-frequency waveforms on a copper stub, was initially demonstrated by Thomas *et al.* (1976), and observations on an integrated circuit (IC) and switching transistor delays were published by Gopinathan *et al.* (1978). Quantitative voltage contrast micrographs were published by Balk *et al.* (1976). Other sampling mode and stroboscopic mode studies on clocked ICs and other devices operating up to 1 GHz frequencies have been demonstrated by other groups.

### III. INTRODUCTION TO THE SEM

The scanning electron microscope scans a fine electron-beam probe over the surface of the specimen, and in synchronism with this, the electron beam of a cathode-ray tube is scanned over its display face, as shown schematically in Fig. 1. The specimen, under beam bombardment, emits electrons whose number varies according to the local emissive properties of the specimen. This emissive current is collected and amplified to intensity modulate the display tube beam, and the resultant image on the display tube face is an emissive micrograph of the specimen. If the linear dimensions of the specimen area scanned by the beam probe are the same as those scanned by the beam of the display tube, the image on the display tube has a magnification of unity. Reducing the linear dimensions of the area scanned by the beam probe, while retaining the scan over the entire face of the display tube, produces magnification. Thus, if the beam probe scans a rectangular region $l_{1b}l_{2b}$ and the corresponding display scan is over the area given by $L_{1d}L_{2d}$, then the linear magnification in each direction is

$$M_i = L_{id}/l_{ib}, \qquad i = 1, 2 \tag{1}$$

By progressively reducing the scan area of the beam probe by means of attenuators, magnifications of up to 100,000 times or larger may be obtained, limited only by the size of the beam-probe spot. Details smaller than the probe cross section cannot be resolved, and hence it is necessary to use a very fine beam probe when high resolution is required. The purpose of the electron optical column is therefore to reduce the beam-probe spot size, by progressively demagnifying the electon source size. Currently, the use of tungsten filaments, lanthanum hexaboride emitters, and tungsten field-emission sources result in ultimate spot sizes of the order of 50 Å, 25 Å, and a few angstroms, respectively. At low demagnifications, the spot size is determined

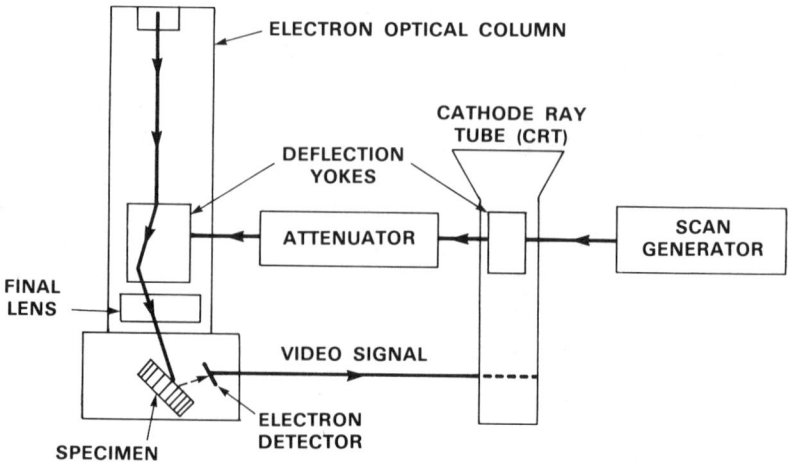

FIG. 1.   Schematic diagram of the scanning electron microscope.

by the demagnification of the column lenses and the source size. However, for larger demagnifications, the spot size is determined by the final lens aberration coefficients. Oatley (1972) has given the spot size in this limit by the expression

$$d_s^2 = \left(d_{co}\frac{\alpha_c}{\alpha_s}\right)^2 + \left(\frac{1}{2}C_s\alpha_s\right)^3 + \left(C_c\alpha_s\frac{\delta E_0}{E_0}\right)^2 + \left(\frac{1.22\lambda}{\alpha_s}\right)^2 \tag{2}$$

where $d_{co}$ is the diameter of the first crossover of the beam, $\alpha_c$ is the semiangle of convergence of the beam at the first crossover, $\alpha_s$ is the semiangle of convergence of the beam at the specimen, $C_s$ is the spherical aberration coefficient of the final lens, $C_c$ is the chromatic aberration coefficient of the final lens, $\delta E_0/E_0$ is the energy spread of the beam, and $1.22\lambda/\alpha_s$ is the diameter of the Airy disk. For large currents and large spot sizes, only the demagnification given by the first term in the above expression needs to be taken into account.

The secondary electrons, together with the backscattered electrons and any tertiaries generated by the backscattered electrons, constitute the emissive mode signal. These electrons are attracted to a collector cage, with the face toward the specimen made of coarse mesh, held at a potential of 250 V. The electrons, passing through the mesh, are attracted to a scintillator with a thin aluminum film coating, which is held at 10–12 kV. The resultant light flashes from the scintillator are conveyed to a photomultiplier by means of a light guide made of Perspex or other plastic, glass, or quartz. These light flashes constitute the detected electrons and are amplified by the photomultiplier and video amplifier to intensity modulate the display tube. The bandwidth of the video amplifier is usually in the range 2–3 MHz, and that of the photomultiplier is about 80 MHz. The scintillator response varies depending on the material used and may be as low as 10 ns.

The geometric contrasts in the emissive mode are to a large measure identical to those obtained with reflected light images. Specimen features seen in a reflection light microscope may be examined magnified many times in the SEM, with a much larger depth of field. Thus the SEM has proved to be an extremely valuable instrument in areas where the geometric structure of the specimen is of great interest. In the semiconductor device area, these images provide information in identifying visual processing defects or design shortcomings. Other modes of operation, including voltage contrast or electron-beam-induced conductivity, or the charge-collection, cathodoluminescence, and x-ray emission modes, are often used for complete device investigations. In this article, the discussion is confined to the voltage contrast mode and associated studies of fast voltage waveforms in the SEM.

For a detailed introduction to the SEM, readers are referred to the book by Oatley (1972).

## IV. Qualitative Voltage Contrast

The primary beam in the SEM impinging on the specimen produces slow secondary electrons (energies less than, say, 15 eV) with some particular energy distribution. If the specimen emission point potential is lowered below or raised above ground potential, every emitted electron, assuming the emission process is not perturbed, gains or loses a proportional amount of energy. A determination of this gain or loss of energy of all emitted electrons by the change in the origin of the energy distribution curve using electron energy analysis measures the change in emission point potential. Figure 2 shows a

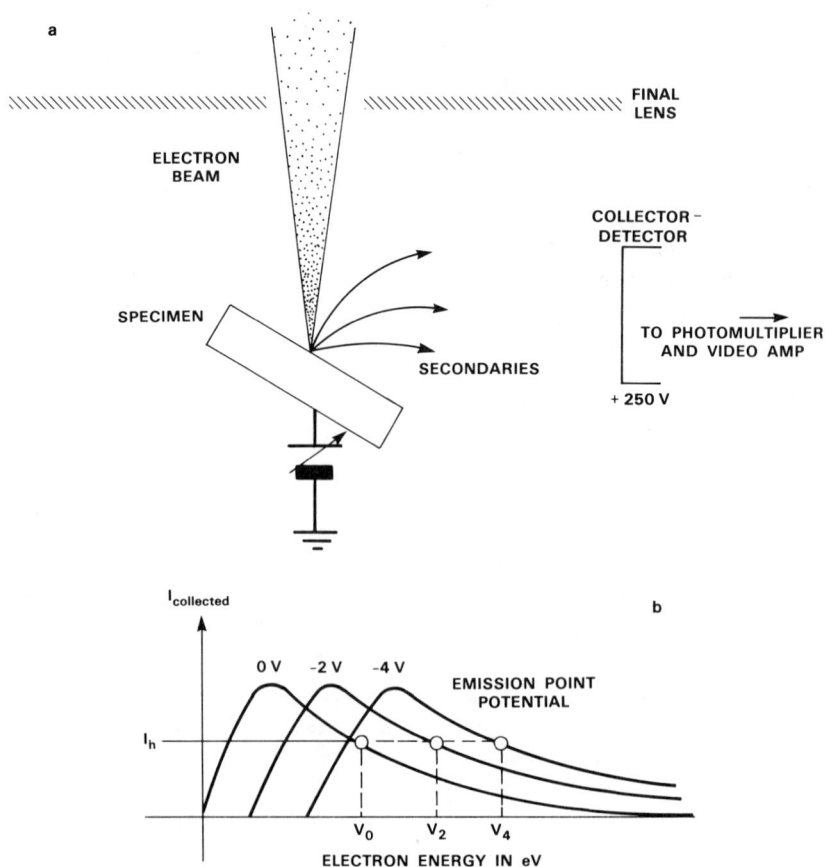

FIG. 2(a) Schematic diagram of geometry of specimen and collector in the SEM. (b) Schematic diagram of energy distribution of secondary electrons and its movement with specimen bias for a 0 V collector.

schematic diagram of the energy distribution curve of secondaries and its movement with the change in emission point potential, with respect to a zero potential collector. In the usual emissive mode voltage contrast in the SEM, some crude form of energy selection of the number of electrons reaching the collector occurs. This energy selection is caused by the disposition of various electrodes and their potentials in the SEM chamber, and include the collector cage of the scintillator detector and its potential, the specimen stage, the final lens and the chamber walls, the specimen geometry and the potentials of the various electrodes distributed on it, and especially those adjacent to the emission point. Thus, conventional contrast shows negative regions (below ground potential) bright, which implies a larger number of secondaries are collected as they are more energetic, and positive regions appear dark. Addition of other electrodes and shields have demonstrated reversal of this contrast, but these results are confusing and remain unexplained except in some unsatisfactory manner. The work of Everhart *et al.* (1959), plotting electron trajectories in two dimensions, supports this simple explanation of voltage contrast.

For most visual inspection of semiconductor devices and integrated circuits, this qualitative contrast appears to be adequate. This is especially the case with digital integrated circuits, which currently are a large fraction of all manufactured ICs. However, visually differentiating between different levels of voltages on a circuit when, for example, all potentials are negative or positive, proves problematic especially if the levels are small, less than 1 V, as the contrast is nonlinear. In this situation, one of the several measurement schemes needs to be used, these being discussed later.

The voltage contrast signal level is low and is superimposed on the emissive-mode topography signal, and separation of the contrast signal may be advantageous. Subtraction of the topography-only signal, obtained from a zero-bias specimen, from the topography-with-contrast signal, obtained from the biased specimen, provides the contrast-only signal. Several methods are possible: the simplest is to pulse the bias, and then perform the subtraction with analog circuitry. Oatley (1969) was the first to demonstrate this scheme, and in his method, the primary beam was also pulsed, but at twice the bias pulse rate. Figure 3 shows the schematic diagram of his system as used to isolate the contrast. While contrast isolation is accomplished, the contrast signal is also dependent on topography, and therefore, fortunately, some topography is retained. However, linearity is not apparent in these contrast-only signals. Other methods of contrast isolation include specimen bias pulsing followed by phase-sensitive detection or a lock-in amplification–detection scheme without beam pulsing (Gopinath and Sanger, 1971b) and digital storage and subtraction schemes (Fujioka *et al.*, 1982). Note that the lock-in detection scheme is now widely used.

With digital circuits examined in the SEM, clocking at SEM video-display frame rates (25, 30, 50, or 60 Hz) proves to be unacceptably slow. One would never get through all the functions in a complex chip in a reasonable time! Thus, clocking at a harmonic of the line rate (note that modern SEM displays have the line frequency as a harmonic of the frame frequency, and also synchronized) results in much greater device or circuit information, with some very interesting display patterns. If this clock is shifted to a frequency slightly above or below this harmonic of the line frequency, these patterns move across the display in diagonals. This scheme was given the name voltage coding by Lukianoff and Touw (1975), who proposed and demonstrated it, and has found wide use in testing digital circuits in the SEM. Other schemes which require beam pulsing, known as logic mapping, and frequency tracing are discussed later.

We have stated earlier that the qualitative contrast signal is small, and it is often in the range of a few percent of the typical topography signal. Thus, an incident beam of 10 kV, current $I_b$, generates approximately $0.1I_b$ slow secondaries on a flat semiconductor specimen, although this may be a little higher in the metallized regions. Tilting the specimen may raise this to $0.2I_b$,

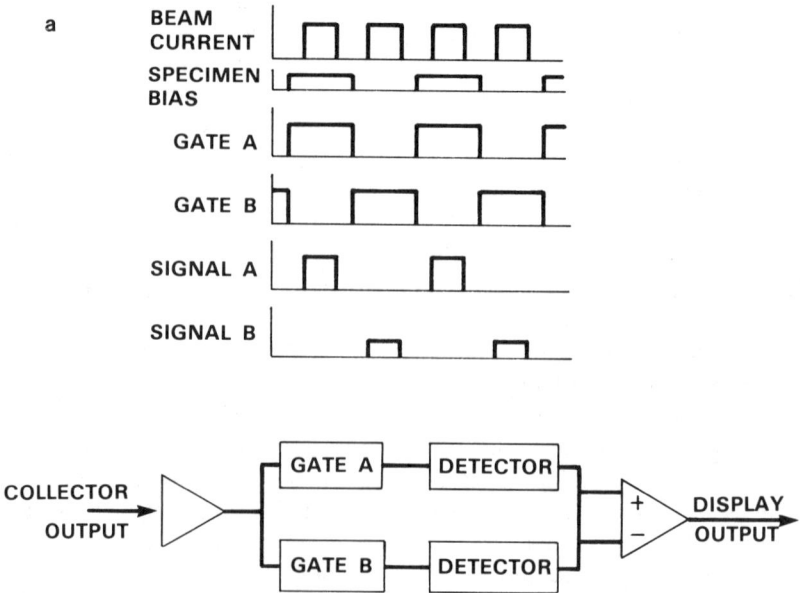

FIG. 3(a)  Schematic diagram of voltage-contrast isolation as suggestion by Oatley (1969). (b) Shift register using Oatley's scheme for contrast isolation showing the outputs of (i) one channel only; (ii) two channels with 0 V bias on specimen; and (iii) two channels with 5 V bias on specimen (reproduced with permission from Oatley, 1959, and the Institute of Physics, London).

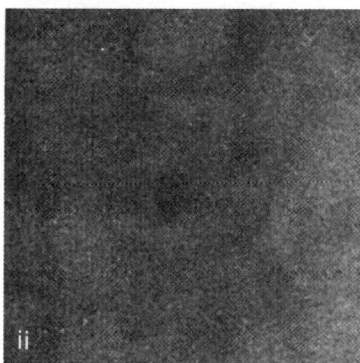

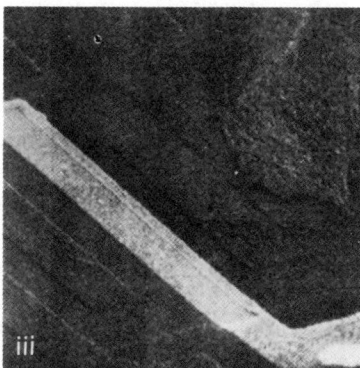

**b**

Figure 3 (*continued*)

according to Lambert's cosine rule (Gopinath, 1974). Everhart (1968) has suggested that slow secondaries constitute 65% of the emissive-mode signal, and the remainder is from backscattered primaries and tertiaries. A rough guess is that the video signal is about $0.2I_b$ with the slow secondaries contributing about 65% of this signal, with the rest of the secondaries not being detected by the collector detector. The voltage contrast information is carried in the usual emissive mode by the low-energy fraction, with energies 3 eV or less, and these are approximately 20% of all emitted slow secondaries. Thus, the contrast signal is of the order of $0.026I_b$, which is about 13% of the video signal. In practice, this estimate is thought to be high, typical values being about 5% of the video signal or less. Voltage resolution in the qualitative contrast mode is at best 100 mV, but is usually about 0.5 V. Note that these estimates are lower than those suggested by Nakamae *et al.* (1981), who use a scintillator detector with an open cyclinder as the cage. Their results are based on computer trajectory plots and experiments, but since these are related largely to quantitative results, the paper and results are discussed in Section V.

## V. Voltage Contrast Linearization for Potential Measurements

As stated earlier in Section IV, when the emission-point potential is changed, by lowering it below or raising it above ground potential, all emitted electrons gain or lose a proportionate amount of energy with respect to a fixed potential collector. This is shown in Fig. 2, where the electron energy distribution curves are shifted linearly with emission-point potential. Hence, energy analysis of the emitted electrons determines this shift in the secondary energy distribution curve and therefore measures the change in potential. Thus, all voltage measurement schemes in the SEM measure change in potential, not the absolute potential. This method requires all secondary electrons to be energy analyzed, and therefore errors arise when this is not the case, for example when local fields on the specimen prevent complete collection. This simple scheme of energy analysis of the secondaries with and without bias requires further modification to make it practical, by means of a feedback loop. The different measurement schemes differ from each other by the type of energy analyzer and the feedback loop scheme used. Currently, the retarding potential analyzer is widely used in one form or another for voltage linearization, and the feedback loop controls the potential of the retarding grid or electrode. We will, however, discuss other schemes of linearization to illustrate its development.

## A. Floating-Specimen Feedback Scheme

Fleming and Ward (1970) first proposed using a feedback scheme to measure voltages and voltage distributions without any reference to secondary energy distribution curves. It was suggested that the nonlinearity in voltage contrast may be linearized by varying the potential difference between the specimen and collector to keep the collected current constant. The collector was a piece of plane grid, held at a fixed potential of 1 kV, placed close to the specimen. The specimen and its bias supplies were floated, and the feedback loop adjusted the specimen–bias floating potential to keep the collected current, and hence the video signal constant, as shown schematically in Fig. 4. The change in the specimen–floating potential is the change in the specimen emission-point potential. This is, of course, the usual "null method" of measurement, and it proved to be very good for voltage measurements in this context, and some excellent results were obtained. Varying the specimen-to-collector potential difference to keep the collected current constant, in effect, moves the appropriate emission-point energy distribution curve to the origin. Thus, the above explanation of voltage contrast linearization still holds except that the feedback loop modifies the scheme. Floating the specimen and its bias supplies proves to be cumbersome, and placing a 1 kV collector grid close to the specimen, which on occasion may be very valuable, was not a welcome idea. Thus, the scheme without any modifications has not been adopted.

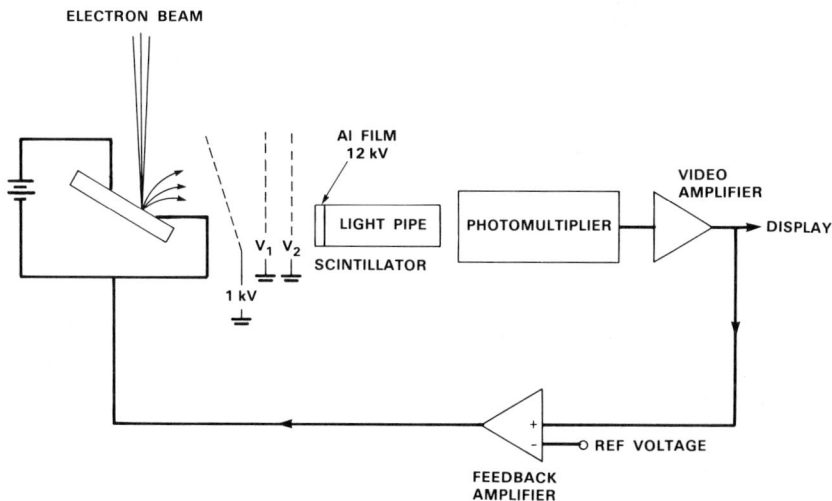

FIG. 4. Schematic diagram of floating-specimen scheme for linearization of voltage contrast (after Fleming and Ward, 1970).

Balk *et al.* (1976) modified the collection scheme of this method. A planar attraction plate, held at 300 V, was placed over the specimen, and the secondaries, thus accelerated, were fed into a 127° sectoral energy analyzer set to collect 300 eV electrons, with a wide slit and hence poor resolution, also placed over the specimen. The electrons from this analyzer collector are detected by the usual scintillator detector. The schematic diagram in Fig. 5 shows the collection system and that was used with a feedback loop which varied the floating potential of the specimen to keep the collected current constant. The bias to the specimen was pulsed, and a lock-in amplifier was used to isolate the contrast. Results obtained from the system were excellent, despite the primary beam having to pass through the sectoral analyzer and the long working distance. Subsequently, the scheme was modified with a planar retarding potential analyzer placed at the outlet of the sectoral analyzer collector, and the feedback loop was connected to the retarding grid of the planar analyzer (see next section), so that the specimen was not floated. Using this modified scheme with further changes to reduce the system noise by reducing the loop bandwidth, Menzel (1981) has measured a 1 Hz sine wave, with a 0.5 mV peak value, which is currently the lowest value of voltage measured in the SEM.

## B. Retarding Potential Analyzer Feedback Scheme

The retarding potential analyzer uses a retarding-potential grid through which all electrons being analyzed have to pass. Assuming a perfectly permeable retarding grid, held at, for example, −3 V, and secondaries with energies greater than 3 eV will pass through and be collected. Therefore, a

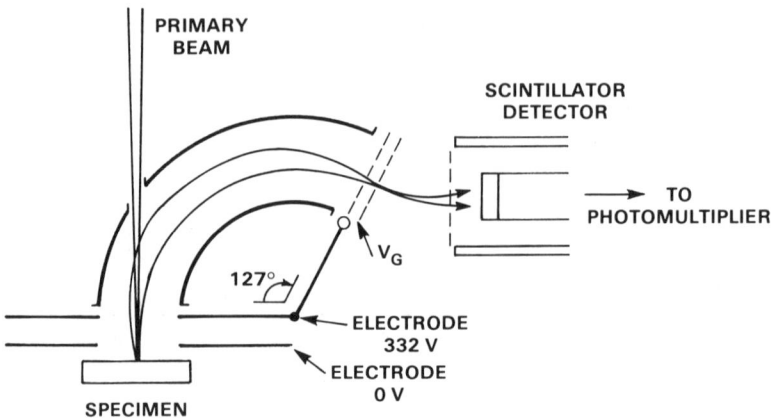

FIG. 5.    Schematic diagram of sectoral-collector scheme (after Balk *et al.*, 1976).

curve of the collected current against retarding potential on this control grid results in an integrated energy distribution curve of the emitted electrons. It follows that linear shifts in the integrated energy distribution curves with emission-point potential are also obtained, and these are shown schematically in Fig. 6. A feedback loop which maintains the collected current constant by varying the potential of the retarding grid measures the potential change of the emission point. The potential of the retarding grid plus or minus a constant is the potential of the emission point. The schematic diagram of the measurement system is shown in Fig. 7.

The three-hemisphere grid system of Fentem and Gopinath (1974) demonstrated the operation of the scheme, and this was subsequently replaced

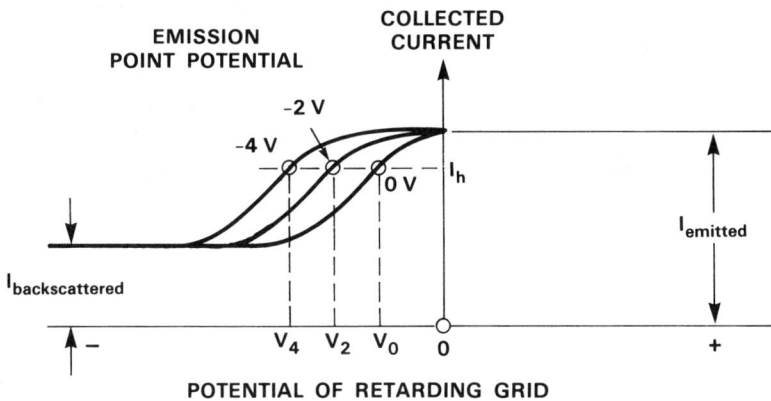

FIG. 6.   Integrated energy distribution curves from a retarding potential analyzer.

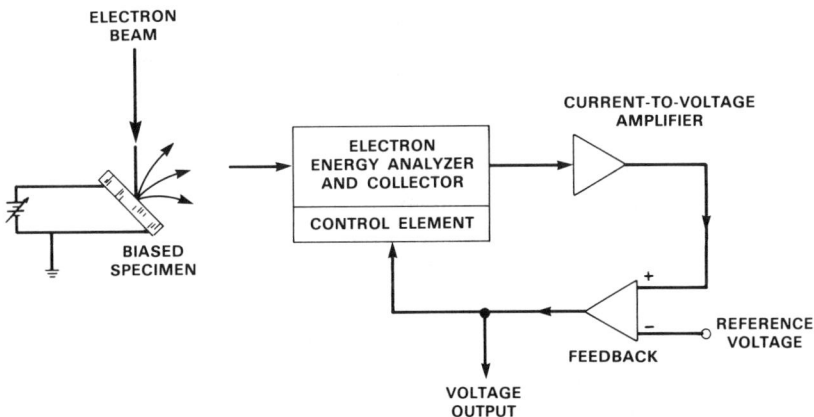

FIG. 7.   Schematic diagram of the feedback loop for voltage measurement.

by a four-hemisphere grid version (Tee and Gopinath 1976, 1977). Despite its limitations, the scheme is discussed in some detail, because current systems are basically modified versions of it. A schematic diagram of the system is shown in Fig. 8. The inner most hemispherical grid, held at 60 V, attracts the secondary electrons, the second grid is the retarding grid, the third is the collector grid, also held at 60 V, and the fourth grid is held at some negative potential to repel the slow tertiaries created by the backscattered electrons, and also to improve collector efficiency. Excellent measurement capabilities were reported, with the measurement of a 10 mV peak value, 1 Hz sine wave. Voltage distributions were obtained across 10 $\mu$m long mesa-structure Gunn devices despite the very high surface fields, after correcting for the errors due to the high fields by means of trajectory plots (Tee et al., 1978).

There are several problems with this scheme. The hemispherical grids limit the room available for the specimen, although this may be overcome by careful design. The collected current is measured by an electrometer amplifier, and therefore the open-loop bandwidth of the system is low, less than 1 kHz. Attempts to improve the closed-loop bandwidth by increasing the open-loop gain resulted in loop oscillation due to the positive feedback from the stray capacitance between the retarding grid and the collection grid. This was

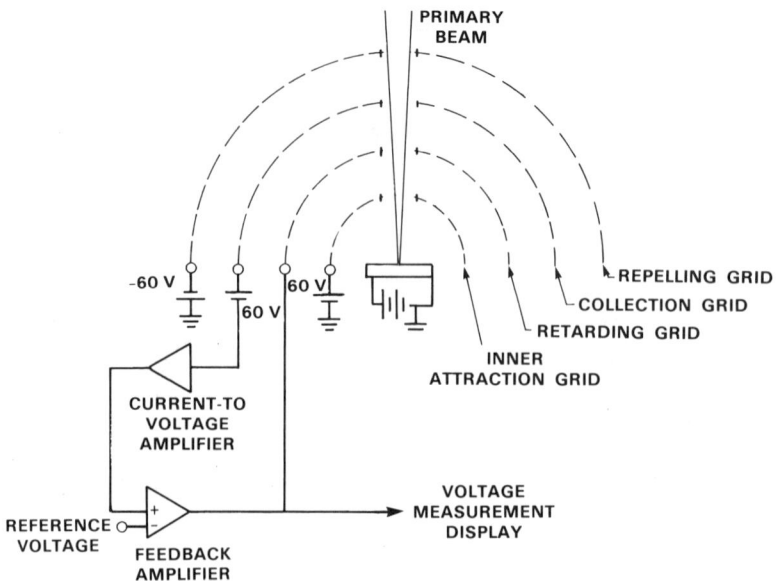

FIG. 8.   Schematic diagram of the four-hemispherical retarding potential analyzer scheme (after Tee and Gopinath, 1976).

compensated to some extent by feeding an equal but opposite current, using a unity gain inverting amplifier which feeds a lumped capacitor of similar value to the stray capacitor. Thus, bandwidths of a few kilohertz have been obtained and were considered reasonable. A detailed discussion of these problems has been given by Thomas *et al.* (1980).

A method of eliminating the slow electrometer amplifier was suggested by Fentem (1974). The hemispherical collector was originally solid, and a large hole was opened in the direction of the scintillator detector. Thus, some of the filtered electrons pass through the hole and are collected and detected by the scintillator. The amplified version of this electron signal from the video amplifier is fed back to the retarding grid, at the correct polarity, to keep the collected current constant. Another version used the third hemisphere, in grid form, as a means of accelerating the electrons away from the retarding grid, with a bare scintillator as collector detector. A similar version of this scheme has been used by Goto *et al.* (1981) with an electrode close to the specimen to attract all the slow secondaries into the analyzer, as shown in Fig. 9. Thus, the essential symmetry of the hemispherical analyzer is preserved, the use of the scintillator detector improves the bandwidth, and the stray capacitance feedback problems are eliminated in the loop. In practice, only a fraction of the filtered electrons are collected, leading to poorer voltage resolution. The space problem was solved by Gopinathan (1978) by using circular planar grids, mounted normal to the plane of the primary beam, fixed permanently just below the final lens—in effect, a planar retarding potential analyzer was used, again with a grided collector. The problem of bandwidth and stray feedback remain with the grided collector.

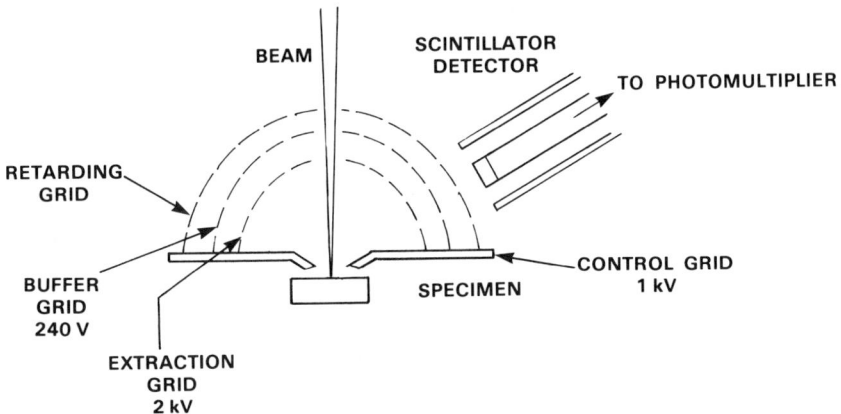

FIG. 9. Hemispherical retarding potential analyzer scheme with extraction electrode (after Goto *et al.*, 1981).

## 1. Feurbaum Analyzer

In this version (Feuerbaum, 1979) of the planar retarding analyzer, planes horizontal, the filtered electrons, emerging from the retarding grid, pass through a planar attraction grid, through a second vertically placed attraction grid, and are collected by the scintillator, which is close to this second grid. Figure 10 shows a sketch of this analyzer, which is used with a feedback loop, a pulsed bias, and a lock-in amplifier, and has resolved voltage changes of 1 mV. The advantage of this over the Kubalek analyzer discussed above is the short working distance of 10 mm, the large scan area, and the lack of large potentials to defocus the beam spot. The advantage over the four-grid planar analyzer discussed above is the use of the scintillator detector, and consequent improvement in bandwidth, and the elimination of the stray feed-back effects. Various versions of this analyzer together with its modified forms are currently being used in several operating systems and are also used in commercial systems.

## 2. Plows Analyzer

The planar retarding potential analyzer is used here, with the collector in the form of a ring scintillator, held at 10 kV. The light flashes from this scintillator are conveyed to the photomultiplier by means of a bundle of fiber-optic light guides, as shown in Fig. 11. Careful design and shielding have been necessary to eliminate the effect of the scintillator potential affecting the primary beam, which in some instances may be with energies as low as 1 keV. Voltage resolution with these analyzers is reported to be 1 mV, and these are currently being built commercially.

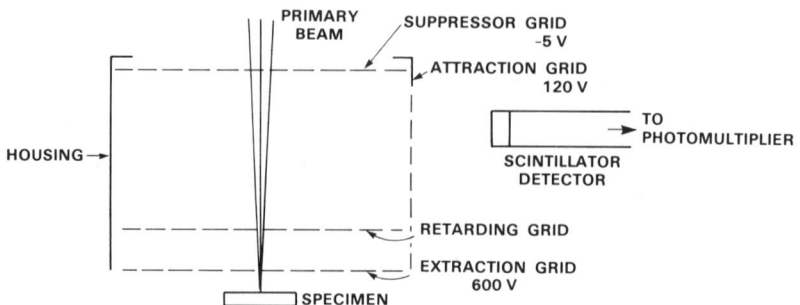

FIG. 10.    Planar retarding potential analyzer (after Feuerbaum, 1979).

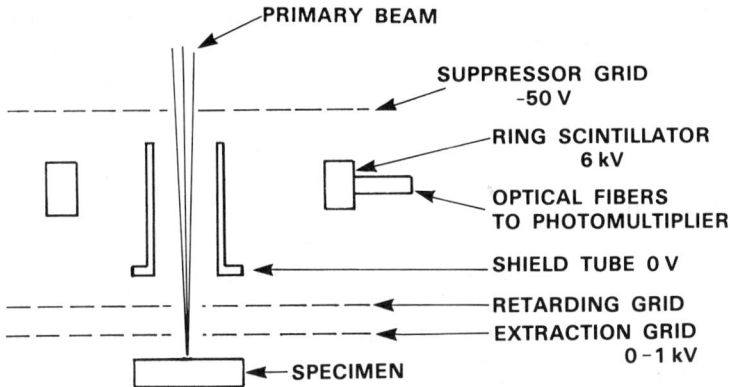

FIG. 11.   Planar analyzer with ring scintillator (after Plows, 1981).

## C. Errors in Voltage Measurement

Several sources of error are present in the measurement system discussed above. In this section, systematic errors are considered; those errors related to noise and other such phenomena are not discussed here. The major systematic errors are due to performance inadequacies in the electron analyzers, the imperfect collection of secondary electrons, and the effect of different emissivities at different positions on the specimen.

### 1. Performance of Retarding Potential Analyzers

The retarding potential analyzer with planar grids horizontal or with concentric hemispherical grids is currently used in a majority of the systems in use (the Feuerbaum version or its modifications that have altered the collection and detection of the filtered secondaries). As discussed earlier, the ideal analyzer of this type would have a sharp cutoff of electrons with energies below the retarding grid potential, and allow all electrons at or above this energy to pass through the grid. However, the finite size of the grid wire diameter $d$, the mesh opening size, and the distance $D$ of the retarding grid to the nearest upstream attraction grid determine the performance of the analyzer. This aspect of these analyzers has been discussed in detail by Huchital and Rigden (1972), and the effect of their performance in the SEM voltage measurement context has been outlined by Menzel and Kubalek (1983a). The passband of electron energies, $\Delta V$, is defined as the change in grid potential to change the collected current from cutoff condition to full transmission of a monoenergetic normal incidence beam of voltage $V_0$.

Menzel and Kubalek (1983a), however, define this quantity as the 10–90%
transmission of the monoenergetic beam. The effect of nonzero passband of
energies, $\Delta V$, results in an analyzer resolution defined as $\Delta V/V_0$, and the
integrated energy distribution curves are therefore smeared out, and the slopes
of the curves decrease with increasing values of $\Delta V$. Huchital and Rigden
(1972) show that the value of the resolution is given by the ratio $d/D$ for planar
analyzers, with a normal incidence beam. For hemispherical analyzers with
large radii compared to $D$, this relationship also holds, but the expression
becomes complex in form when the radii are comparable to $D$. The effect of
this finite resolution results in the slope of the integrated distribution curve,
given by $\partial I_c/\partial V = g_m$, becoming smaller, and the voltage resolution of the
voltage measurement system also reduces, as will be shown later, in
Section VII.

The second problem with these analyzers is the off-normal electron incidence at the retarding grid. The effect of this is to increase the resolution, as
defined above, by the factor $\sin^2 \alpha$, and thus the corresponding passband $\Delta V$
becomes $(\sin^2 \alpha)V_0$, where $\alpha$ is the angle of incidence to the normal (Huchital
and Rigden, 1972), if the finite size of the mesh is ignored.

In the hemispherical analyzer, due to its finite size, the edges of the
specimen are off-axis, and therefore produce off-normal electron incidence at
the retarding grid. Since the electron emission varies as the cosine of the angle
to the normal at the plane of the specimen, the planar analyzer (Fig. 12) has
electrons with incidence angles varying from the normal to a maximum value
$\alpha_{max}$, determined by the position of the emission point and the size of the
analyzer. Menzel and Kubalek (1983a) suggest that the effect of the off-normal
incidence, in this case, may be approximated by a transmission coefficient
given by $T_0 \cos \alpha$. Electrons emitted from the specimen at angle $\alpha$ will cross the
retarding barrier of $V_G$, provided only that the normal component of velocity

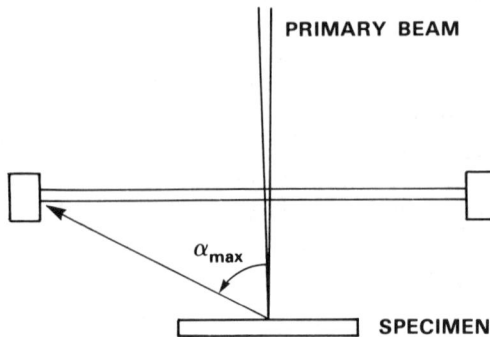

FIG. 12.    Incidence angle in the planar analyzer (after Menzel and Kubalek, 1983a).

is large enough. In effect, this requires that $eV_{em} \cos^2 \alpha > eV_G$, where $eV_{em}$ is the emitted electron energy and $V_G$ is the retarding grid potential.

The collected current in the planar analyzer is therefore given by

$$I_c = I_{c\,max} T_0 \int_{d\alpha}^{\alpha_{max}} \int_{eV_G/\cos^2 \alpha}^{20eV} 2N(eV) \cos^2 \alpha \, d(eV) \qquad (3)$$

where

$$I_{c\,max} = \int_0^\pi d\alpha \int_0^{20eV} 2N(eV) \cos \alpha \, d(eV) \qquad (4)$$

Using the above equation, the curves in Fig. 13 are plotted to show that the maximum value is reduced and also that the slope is reduced when compared to the ideal hemispherical analyzer, where the angle $\alpha_{max}$ is equal to $\pi$. Note that the mesh effects further degrade this curve. With this equation as formulated by Menzel and Kubalek (1983a), all plots have been obtained for two-dimensional Cartesian coordinates.

## 2. Local (Specimen) Electric Fields

The effect of local electric fields on the surface of the specimen is to prevent complete collection of the emitted secondaries. Since the fields vary spatially, the various energy fractions that enter the analyzer also vary spatially, and in turn give rise to measurement errors.

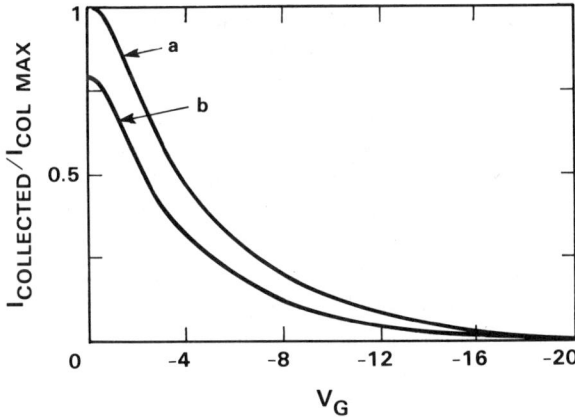

FIG. 13.  Collected current normalized to the hemispherical analyzer maximum value as a function of retarding grid potential for (a) hemispherical analyzer and (b) planar retarding potential analyzer (reproduced with permission of Menzel and Kubalek (1981a), and FACM Publishers, Inc., Mahwah, NJ).

While it is possible to measure negative specimen voltages with reasonable accuracy, positive voltages show larger errors. The positive electrode, its adjacent electrodes on the specimen and its mounting package cause retarding fields which selectively prevent the escape of low-energy secondaries. Thus, the measurement schemes in the SEM, have often used an accelerating electrode in the form of a mesh or a ring, held close to the specimen to pull out these secondaries. While considerable mitigation of this effect is obtained by this method, it is not likely to disappear entirely. The work of Nakame *et al.* (1981) has attempted to resolve this problem.

This paper has computed electron trajectories in two dimensions with a known energy distribution in three cases: the simple voltage contrast detector, which is the scintillator shielded by a cylindrical cage without a grid at its front face, the planar retarding analyzer, and a concentric spherical analyzer. The specimen in all cases is a metal strip 8–10 $\mu$m wide, with two additional symmetrically placed strips also 8–10 $\mu$m wide, separated from the central strip by 10–12 $\mu$m. The effect of a planar extraction grid held above the specimen in the latter two cases, to provide fields of 0.625 and 6.25 kV/cm in each case, are also examined. Two types of local field effects are noticed, type I, which arises from a nonzero potential at the measurement point on the specimen, and type II, which is from nonzero potentials on neighboring electrodes. These two types of local field contrast have been previously described as trajectory contrast and local field contrast by Gopinath (1974), but the above notation is retained here for clarity. The specimen used in this work allows the separation of these two types of contrast.

Figure 14 shows a section of the simple contrast detector with the scintillator unscreened, together with the equipotential distributions assuming two dimensional geometry. The results of trajectory plots translated into collected current are summarized in Fig. 15, where the computed curent is compared with experiment, and shows reasonable agreement for a variety of bias conditions. Unlike the earlier estimate of contrast magnitude, these experiments show up to 30% change of the collected current, neglecting tertiaries and backscattered primaries. Of course, the unscreened scintillator is the main cause of this difference and improvement. Perhaps all voltage contrast observations in the SEM ought to be made with this type of scintillator and cage without screening!

The second set of results of interest are for the planar analyzer, shown in Fig. 16, which is somewhat difficult to analyze. The results of positive voltage on the strip specimen, with zero volts for $V_1$ and $V_2$, on the adjacent left and right strips, respectively, are in good agreement with experiment for the lower extraction field, but poor for the higher extraction field. Since these plots are of collected current as a function of $V_{\text{specimen}} - V_R$, the curves should coincide for zero errors in the measurement, and thus substantial errors will

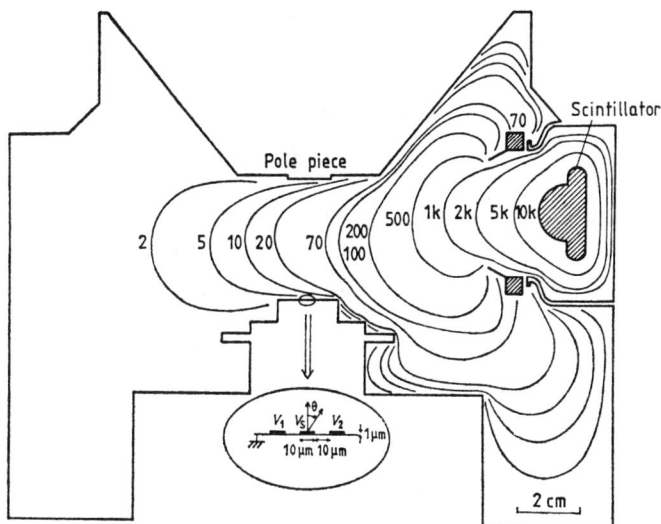

FIG. 14. Equipotential distributions in the normal SEM used for qualitative voltage contrast with an unscreened scintillator (reproduced with permission from Nakamae *et al.*, 1981, and the Institute of Physics, London).

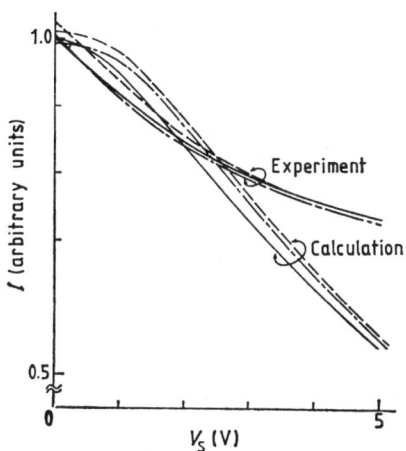

FIG. 15. Calculated and measured collected current of the simple voltage-contrast detector in Fig. 14 as a function of central electrode potential $V_s$ of the specimen, also shown in Fig. 14 (reproduced with permission from Nakamae *et al.*, 1981, and the Institute of Physics, London). Values of $V_1$ and $V_2$ are plotted, respectively, as ——, 0, 0; — · —, 5, 0; and ---, 0, 5 V.

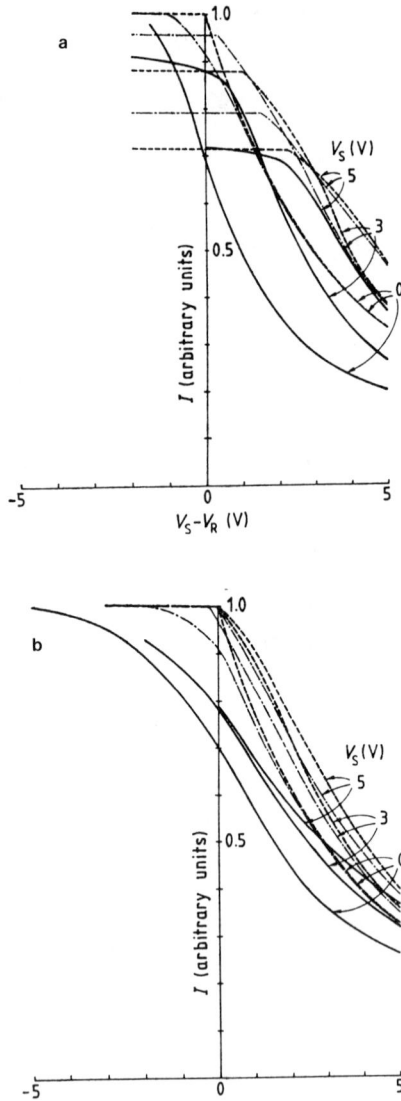

FIG. 16. Calculated and measured current as a function of $V_s - V_R$, the difference between the specimen central electrode potential and the retarding grid potential with different potentials on the other specimen electrodes, for the planar analyzer, for extraction voltages of (a) 100 V and (b) 1000 V (reproduced with permission of Nakamae et al., 1981, and the Institute of Physics, London) Values of $V_1$ and $V_2$ are plotted, respectively, as (experiment)——, 0, 0; (calculation)---, 0, 0; and —·—, 0.5 V.

probably occur in any linearization. The expected errors due to the two types of field effects for different extraction fields were calculated. It was shown that the errors are substantial for type-I (almost 50%), and much smaller for type-II local field effects (about 10%), with low field, and reduce considerably with the higher extraction field.

Khurshed and Dinnis (1983) have examined the effect of local fields in the collection efficiencies for the Feuerbaum analyzer, the Plows analyzer, and the hemispherical analyzer without the extraction grid. The geometry of the specimen is similar to that of Nakamae et al. (1981) discussed above. Their results show that the two versions of the Feuerbaum analyzer do very well, while the hemispherical analyzer does very poorly indeed without the extraction grid.

A similar but more specific calculation was performed by Tee et al. (1978), who measured voltage distributions along the edge of a 10 $\mu$m long mesa transferred electron device, where surface fields of 20 kV/cm or higher may be present, depending on the bias. By plotting trajectories at different points along the surface, the collection error was obtained and the linearization error estimated, and the voltage distribution plots were corrected.

## 3. Specimen Topography and Compositional Effects

When the composition or the topography of the specimen changes as the beam scans, the emissivity also changes, resulting in a spurious voltage measurement error. A simple method of minimizing this error is to perform contrast separation by subtracting the topography signal, which may have been previously stored or alternatively by means of pulsing the bias and using a lock-in system. An alternative is to use normalization, as proposed by Tee and Gopinath (1976), where the total emitted current entering the analyzer is used for this purpose. In the retarding potential analyzer, the total current is obtained by measuring the current collected at the inner attraction or extraction grid, and summing to the collector current. Thus, normalization may be performed during linearization without the need for pulsed bias and lock-in amplifier.

## 4. Other Errors

Beam current drift may be corrected by normalization as above. Other errors may arise due to bandwidth limitations and amplifier effects. These have been discussed in detail with regard to the retarding potential analyzer with the grid collector by Thomas et al. (1980). The extension of this work to other systems should in principle present no difficulties. Another method of dealing with this problem is to use a digital computer to perform the normalization, as discussed by Nye and Dinnis (1985a).

The measurement of voltages on semiconductors passivated with various dielectrics is fraught with difficulties. The preferred method is to strip this passivation using various solvents. If this cannot be done, measurement at a suitable beam voltage, which would allow the beam-generated conductive region to reach the metallization or the semiconductor, may have to be performed (see Taylor, 1978; Thomas et al., 1980). The problem with this scheme is the residual damage and charge deposition in the dielectric.

The alternative to the above scheme is to capacitively couple out the voltage without penetrating the dielectric. This requires normalization of the surface in the first instance to some steady uniform potential by scanning the beam at a specific accelerating potential, so that the emissivity $\delta$ (ratio of emitted to incident electrons) is greater than unity, with a grid held close to the specimen, at either 100 or $-5$ V. In the former case the surface reaches some positive potential at which $\delta$ becomes unity, and in the latter, some negative potential for which $\delta$ also becomes unity. Fujioka et al. (1983) show that it is then possible to measure the potential of a buried electrode, although with some error. Other work in this area has been reported by Gorlich et al. (1984), Todokoro et al. (1983), and Nye and Dinnis (1985b).

## VI. AUGER ELECTRON VOLTAGE MEASUREMENT SCHEMES

Auger electron emission occurs with characteristic electron energies. In the normal SEM with poor vacuum conditions, the carbon electrons may be readily seen with a suitable analyzer. Thus, the movement of this Auger carbon electron peak with specimen emission-point potential was used by MacDonald (1970) to measure voltages in the SEM. Other groups have used this and other methods for voltage measurement, and some discussion of these follow.

The more recent work of measuring voltages using Auger electrons have used specially constructed ultra-high-vacuum Auger electron machines built for compositional analysis, with high-resolution electron spectrometers, and with electrostatic scanning of the beam probe. Such systems have spatial resolution of the order of 0.5 $\mu$m with a thermionic source, and of the order of 50 nm with a field emission source. Since beam stability is of importance in this case, often the thermionic source is preferred for these measurements. Alternatively, ultra-high-vacuum SEM-STEM machines with field emission sources have also been used for such measurements.

Typical Auger electron signals are $10^{-4}$–$10^{-6}$ times the incident beam current, with some improvement obtained by specimen tilt. Thus, for $I_b$ of $10^{-7}$ A, the Auger current, $I_{Auger}$, is the order of $10^{-12}$ A, which, as shown later, determines the measurement resolution, or the measurement time for a specific voltage resolution. The work of Janssen *et al.* (1980) has employed an ultra-high-vacuum SEM-STEM machine for voltage measurements, using the shift of the secondary electron distribution curves close to the origin. This experiment used a cylindrical mirror analyzer with a restricted entrance slit, and the feedback loop was also implemented to obtain a voltage resolution of 40 mV. However, the technique used does not differ from the secondary electron schemes discussed earlier.

The more recent work of Pantel *et al.* (1982) claims that earlier observations showing poor voltage resolution was due to contaminated samples. These authors have used the shift in secondary electron distribution curves and have also used the shift of the Auger peak for silicon $KL_{23}L_{23}$ transition at 1628 eV, and also the silicon $LVV$ transition at 92 eV, to measure voltage. The resolution obtained by these authors is about $+20$ mV shift, with 60 s for recording a single line scan, and almost 300 s for the energy spectrum case. Thus, the open-loop measurement would take 5 min to resolve one point, and the closed-loop case may be a little faster. Using this technique, the barrier height of a silicon $p-n$ junction was measured as $450 \pm 20$ mV, which is low. However, the advantage of using Auger electrons is that those emitted at the higher energies are not affected by the local fields. Other work in this area has been performed by Tiwari *et al.* (1980) and Waldrop and Harris (1975).

## VII. Estimate for Minimum Measurable Voltage

The minimum measurable voltage (Gopinath, 1977) is dependent on the noise in the measurement signal, the specimen, and the type of analyzer used. The previous sections have discussed the three types of analyzers used in the voltage measurement schemes: the retarding potential analyzer, the cylindrical sectoral analyzer, and the cylindrical mirror analyzer. The cylindrical mirror analyzer is now widely used in Auger electron analysis instruments since the signal-to-noise ratio is shown to be higher in these applications when compared to other analyzers. The question of the suitability of this in the voltage measurement context in the SEM has not been previously discussed. In the present section we analyze the minimum detectable voltage for restricted aperture analyzers (the cylindrical mirror, the 127°, and the 63.5° sectoral analyzers) and the unrestricted aperture analyzers (retarding potential analyzer), and compare these two classes. For a practical measurement system, a feedback loop of the form shown in Fig. 17 must be used. The floating-specimen system is equivalent to the feedback analyzer–electrode

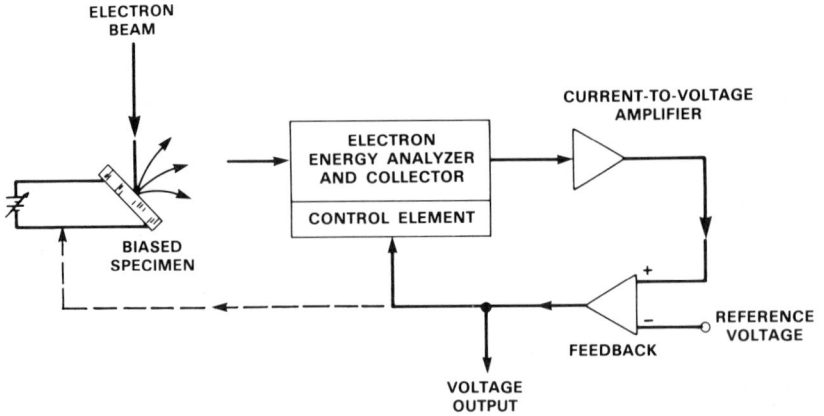

FIG. 17. Schematic diagram of feedback loop system with energy analyzer for voltage measurement.

scheme for the purpose of this analysis as shown in Fig. 17. The Auger electron voltage contrast case is also included.

The analysis follows a simple pattern for both types of analyzers: the slope of the collected current to some control electrode potential $V$, which is equivalent to the electron energy in eV, or the slope of the collected current to the retarding potential in $V$, at the operating point at which the collected current is held constant at $I_h$, is given by

$$g_m = \frac{\delta I}{\delta V}\bigg|_{I = I_h} \tag{5}$$

Thus, if the combined noise sources in the system can be represented by a noise current, $\bar{I}_n$, then the noise output voltage is given by

$$\bar{V}_n = |\bar{I}_n / g_m| \tag{6}$$

For a given signal-to-noise ratio $n$, the minimum detectable voltage change is

$$V_{min} = \frac{n\bar{I}_n}{|g_m|} \tag{7}$$

The results for the two types of analyzers evaluate $g_m$ and $\bar{I}_n$ in terms of the beam current and the analyzer parameters.

## A. Restricted Aperture Analyzers

The energy distribution curves used in these analyzers is shown in Fig. 2b. Let us assume that the feedback loop keeps the current constant at $I_h$, beyond the 3 eV peak. Suppose that $I_s$ is the total secondary electron current due to all electrons emitted with energies less than or equal to 15 eV energy; suppose also that the energy spectrum peaks at 3 eV and falls to almost zero at 15 eV. Assuming straight-line approximations to the curves as in Fig. 18, the total low-energy secondary current given by

$$I_s = \frac{15}{2} I_{max} \tag{8}$$

where $I_{max}$ (A/eV) is the peak value of the electron energy distribution curve. The peak collected current is $RI_{max}$, where $R$(eV) is the resolution of the analyzer, assumed to be constant. Hence the slope of the curve at $I_h$ is given by

$$|g_m| = RI_{max}/12 = RI_s/90 \tag{9}$$

These analyzers have restricted entrance apertures, which are assumed to subtend the solid angle $\Omega$ steradian at the emission point. The secondary electrons are known to be emitted with a cosine polar distribution normal to the emitting surface. Assume that this aperture is placed at the peak of this distribution, then the fraction of electrons entering the analyzer is given by $(\Omega/\pi - \Omega^2/4\pi^2)$. The slope of the energy distribution curve $g_m$ is reduced by this factor. Let the secondary emission coefficient for all electrons in the range of 0 to 15 eV energy be given by $\delta$, and hence $I_s = \delta I_b$, where $I_b$ is the beam current. The expression for $g_m$ becomes

$$|g_m| = \left(\frac{\Omega}{\pi} - \frac{\Omega^2}{4\pi^2}\right)\frac{R\delta I_b}{90} \tag{10}$$

The noise current $\bar{I}_n$ is a combination of the noise on the collected current, the noise due to the measurement and loop amplifiers, and noise due to contamination effects which is essentially a low-frequency drift of the emitted

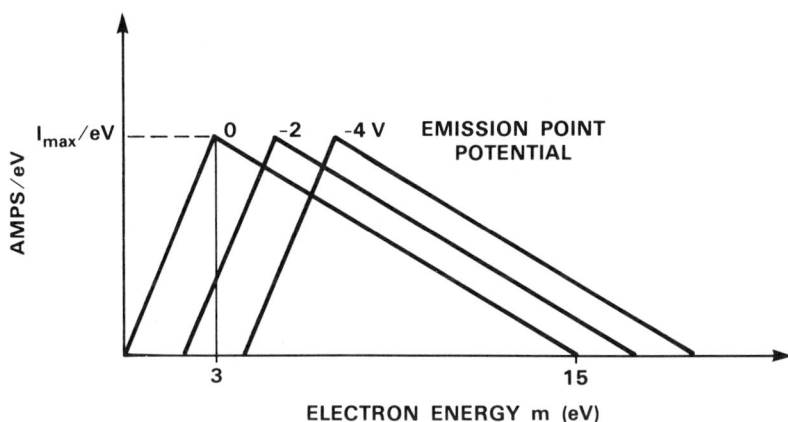

FIG. 18. Straight-line approximation to the energy distribution curves of Fig. 2(b).

current. In the ideal case, the collected current noise is the largest, and the other sources are negligible. The collected current is shot-noise limited, with a small factor to account for partition noise which in the present case may be omitted. Thus

$$\bar{I}_n = \sqrt{2e\,\Delta f\,I_h} \tag{11}$$

where $\Delta f$ is the measurement loop bandwidth and $I_h$ is the collected current held constant. Let us assume that $I_h$ is some fraction $\alpha$ times the peak current $RI_{max}$. Therefore

$$\bar{I}_n = \left[ \tfrac{4}{15} e\,\Delta f \left( \frac{\Omega}{\pi} - \frac{\Omega^2}{4\pi^2} \right) R\delta\alpha I_b \right]^{1/2} \tag{12}$$

Substituting for $g_m$ and $\bar{I}_n$ in Eq. (7):

$$V_{min} = 90 \left( \frac{4e\alpha\,\Delta f}{15(\Omega/\pi - \Omega^2/4\pi^2)\delta RI_b} \right)^{1/2} \tag{13}$$

Thus, for $n = 2$, $\alpha = 0.5$, $\Delta f = 10$ Hz, $R = 0.05$ eV, $\delta = 0.2\Omega = 0.11\pi$, and $I_b = 10^{-8}$,

$$V_{min} \simeq 80 \quad \text{mV}$$

If the collection angle is effectively improved to $2\pi$ by collection fields (which is unrealistic) and $R$ is dropped to 0.1 eV, then $V_{min}$ is about 5.9 mV. Increasing the value of $R$ improves $V_{min}$, but in the limit of large $R$ no electron energy analysis takes place and therefore voltages may not be measured.

## B. Unrestricted Aperture Analyzers

The unrestricted aperture analyzer discussed here is the retarding-potential analyzer. The diagram in Fig. 6 shows the integrated energy distribution curves and its movement with emission-point potential. These curves may be approximated by straight lines.

Thus, the slope $g_m$ is given by $I_s/15$ and in terms of beam current this is $\delta I_b/15$. The analyzer has a finite integrated resolution, which reduces the slope of these curves. This may be accounted for by a nondimensional factor, less than unity, given by $R_{int}$, and thus:

$$|g_m| = R_{int}\delta I_b/15 \tag{14}$$

As before, in the ideal case the noise on the collected current $I_n$ is the only noise component considered. Now the collected current includes the back-scattered electron current given by $\delta I_b$ and a fraction of the emitted low-energy secondaries given by $\eta I_b$. Following the discussion of Bennet (1960) on secondary emission the noise current is given by

$$\bar{I}_n = \{2e \, \Delta f(\eta + \alpha\delta)(1 + \eta + \delta)I_b\}^{1/2} \tag{15}$$

If $\eta + \delta \ll 1$, this becomes

$$\bar{I}_n = [2e \, \Delta f(\eta + \alpha\delta)I_b]^{1/2} \tag{16}$$

Note that $\eta$ is the backscattering coefficient and includes all electrons emitted with energy above 15 eV energy. In the present case the minimum detectable voltage change is given by

$$V_{min} = \frac{15n}{R_{int}} \left( \frac{2e \, \Delta f(\eta + \alpha\delta)(1 + \eta + \alpha\delta)}{\delta^2 I_b} \right)^{1/2} \tag{17}$$

Thus, for $n = 2$, $\alpha = 0.5$, $R_{int} = 0.95$, $\eta = 0.3$, $\delta = 0.2$, $\Delta f = 10$ Hz, and $I_b = 10^{-8}$, $V_{min} \simeq 2$ mV. If the collected current excludes the backscattered electrons, $\eta = 0$, as with the grided collector or with a shielded scintillator, this figure improves to $V_{min} \simeq 0.9$ mV.

## C. Auger Electron Measuring System

The Auger electron measuring system would take the form of a cylindrical mirror analyzer which is scanned to determine the position of the particular Auger peak with and without bias. The change in energy of the peak gives the emission-point potential. An automatic voltage measurement system would take one of two forms: an open-loop and a closed-loop system. The open-loop system would energy analyze the emitted electrons with and without bias in the region of the chosen peak. The resolution of the analyzer approximately determines the minimum detectable voltage.

A closed-loop system would attempt to keep the collected current constant at some slope of the Auger peak. Suppose the collected current at the Auger peak be given by $RI_{peak}$, where $R(eV)$ is the analyzer resolution. Let the width of the Auger line be $W(eV)$, the peak value be $I_{peak}(A/eV)$, and the total current within the Auger line be $I_{Aug}$, and hence

$$RI_{peak} \simeq (2/W)RI_{Aug} \tag{18}$$

Suppose the restricted aperture effect of the analyzer is $F_{ap}$, and the total Auger line current related to the beam current is given by $F_{Aug}$. Then, the slope on either side of the peak is given by

$$\frac{\partial I}{\partial V} = \frac{4}{W^2} R F_{ap} F_{Aug} I_b \tag{19}$$

The collected current is set at some fraction $\alpha$ of $RI_{peak}$, and assuming this is shot-noise limited, then

$$\bar{I}_n \simeq [(4/W)e\,\Delta f\,\alpha RF_{ap}F_{Aug}I_b]^{1/2} \tag{20}$$

Thus, the minimum detectable voltage in this case is given by

$$V_{min} = \frac{nW^2}{2}\left(\frac{e\,\Delta f\,\alpha}{WRF_{ap}F_{Aug}I_b}\right)^{1/2} \tag{21}$$

Now let $n = 2$, $W = 5$ eV, $\alpha = 0.5$, $R = 1$ eV, $F_{ap} = 0.1$, $F_{Aug} = 10^{-6}$, $\Delta f = 10$ Hz, $I_b = 10^{-8}$ A, and hence

$$V_{min} \simeq 0.3 \quad V \tag{22}$$

Suppose $\Delta f = 0.01$, then $V_{min}$ is approximately 9.4 mV, which approaches the claims of Pantel et al. (1982).


## D. Discussion

It would appear on the basis of this analysis that the restricted-aperture analyzers, by their structure, have poor voltage resolution due to the smaller collected currents. The retarding-potential analyzer provides better voltage resolution despite the collection of backscattered current. In practice, the ease of installation of the planar retarding-potential analyzer or its modified versions which do not perturb normal SEM operation settles the choice of analyzer. Auger measurement systems, however, are confined to special cases.

Note that the minimum detectable voltage takes the form

$$V_{min} = \frac{1}{K}\left(\frac{\Delta f}{I_b}\right)^{1/2} \tag{23}$$

where $K$ is the measurement system figure of merit, $\Delta f$ is the measurement bandwidth, and $I_b$ is the beam current. Thus, if results are required to be obtained quickly, $\Delta f$ becomes large, and to maintain voltage resolution $I_b$ also has to become large. Typical values of the figure of merit $K$ are in the $10^7$ range.

In practice, $1/f$ noise from amplifiers in the measuring loop becomes dominant at reduced bandwidths. Thus, to measure the 1 mV sine wave Menzel and Kubalek (1979a) had to modify their system further by additional pulsing and lock-in amplification and detection, so that the low bandwidth was achieved by integration of only the ac components. However, in contrast to this, the 10 mV peak value sine wave measured by Tee and Gopinath (1977) used the simple four-grid analyzer ($\eta \simeq 0$) with a low-bandwidth loop. A detailed analysis of noise including amplifier noise in the retarding-potential measurement system for the grided collector scheme has been given by Thomas *et al.* (1980), for interested readers. Extension of this analysis to other systems has not yet been carried out.

For studies on fast circuits, say waveforms of 1 MHz, with $I_b$ of $10^{-8}$ A, and $K \simeq 10^7$, $V_{min}$ is approximately 1 V. The measurement loop bandwidth in general is not much above several kilohertz, which suggests that for 1 V resolution it would be better to use qualitative contrast, where resolution is of the order of 0.5 V and bandwidth is about 2 MHz or larger, depending on the video amplifier.

The recording of micrographs with quantitative voltage contrast information is an interesting problem. Opinions vary in regard to the minimum signal-to-noise ratio and the number of levels and pixels required to provide adequate information (see Oatley, 1972). The above analysis provides a guide to the bandwidth that may be obtained for a specified voltage resolution and signal-to-noise ratio. To convert this quantitative information into a micrograph is, of course, to throw away a large part of it. Thus, it may be better to obtain qualitative voltage contrast micrographs, and then possibly obtain quantitative information in selective areas through line scans or spot measurements. The spatial resolution is related to the beam current and in turn to the voltage resolution.

Observation of faster waveforms requires sampling techniques, which are discussed in the following sections.

## VIII. Observation of Fast Voltage Waveforms and Dynamic Voltage Distribution

The present and subsequent sections discuss the principles of sampling techniques and their use in scanning electron microscopy. The stroboscopic mode, of course, is one variant of the sampling scheme. In earlier sections, voltage contrast and its linearization have been discussed in some detail, together with the frequency bandwidth limitation. Voltage contrast may be "seen" and recorded within the bandwidth of the typical secondary electron collection and detection system. Thus, with the electron beam held at a spot on the device or integrated circuit, any potential variations on the specimen appear by means of voltage contrast on the secondary video signal which may be displayed on an oscilloscope trace. The rate of this variation that can be displayed or recorded in time is determined by the maximum frequency of operation of the secondary video chain. Typically, this frequency may be a few megahertz but with additional modifications may be raised to a maximum of about 80 MHz. The contrast signal will not be linear, but this may not matter in all cases. Such a system has been called real-time voltage contrast by Ostrow *et al.* (1981), who reported some convincing observations with a video chain frequency of 16 MHz. When the contrast is linearized, the response time of the linearization system drops to about a few kilohertz or less. Thus, real-time observations only record quantitative waveforms to this maximum frequency.

With devices and circuits currently operating from low frequencies to several tens of gigahertz, the alternative to real-time operation is to resort to sampling schemes similar to those used in sampling oscillography. This technique, used for recording periodic waveforms, is illustrated in Fig. 19. In this method, the measurement is performed by taking successive voltage samples on successive (or different) occurrences of the periodic waveform, each sample at a slightly later delay time, along the waveform. The sampled voltage levels are displayed on the time axis corresponding to their sampling instants, determined by their delay times. This may be adapted for the SEM, when the periodic event, which in the present context is a voltage variation at a particular point on a device or IC, is at a higher frequency than the band-width of the voltage contrast detector and loop amplifier frequency response. Thus, with the beam held on a spot on the specimen, the electron beam may be pulsed to provide the sampled output. Alternatively, the output of the voltage contrast detector and amplifier may be sampled using a gate, but this scheme requires that the detector amplifier have a flat frequency response above the maximum frequency of the periodic event up to the gate. Since the bandwidth of the detector amplifier is low, a few MHz with qualitative voltage contrast, a few kHz with linearized contrast, this latter scheme is generally not implemented.

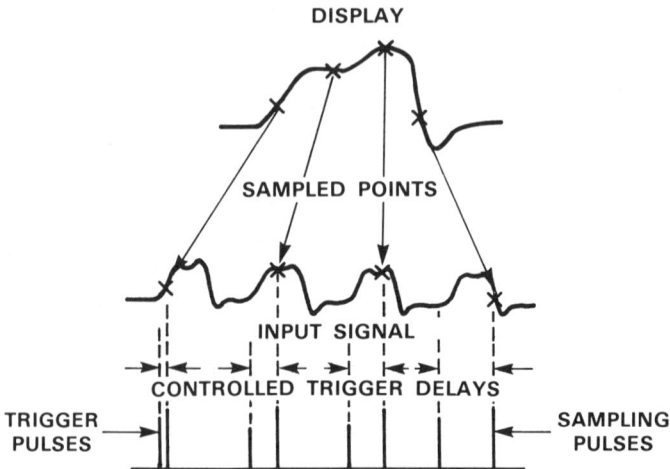

FIG. 19.   Sampling technique for recording periodic waveforms.

The sampling SEM thus requires a method of pulsing the electron beam to sample the event, a method of increasing the delay time of the beam sampling pulse after some chosen number of samples, a method of processing the signal if required, and a method of displaying the output of the sampled signals. Oscilloscope-type sampling circuitry with suitable modifications to interface with the SEM, together with an oscilloscope or an $X-Y$ recorder for display and a means of storage if required, provide the necessary means of performing sampled observations. The dynamic voltage distribution at a particular delay time (phase) may be obtained by holding the sampling beam pulse delay time constant, with the sampled output fed to the usual SEM video display. Thus a stroboscopic micrograph at this delay time (or phase) may be obtained to illustrate the dynamic voltage contrast distribution. Micrographs at different delay times or phases over the entire periodic waveform show the voltage distributions at these delay times or phases.

The circuitry involved in such a system may utilize modified sampling oscilloscope modules or boxcar integrators. We will not discuss these in any detail. The major task in this scheme is the pulsing of the electron beam with negligible spot degradation. In the following sections various beam pulsing schemes are discussed, and some examples of the sampling SEM and the stroboscopic mode of operation are included.

## IX. Electron-Beam Pulsing in the SEM

The electron-beam pulse for the sampling system may be obtained by pulsing the electron gun or by operating on the beam. In this section, the various methods of obtaining beam pulses are discussed.

### A. Electron Gun Pulsing

Electron guns with tungsten filaments and $LaB_6$ sources are usually of the triode design, with a control electrode in the form of a cylinder with a hole for the electron beam to pass through. This control electrode is the Wehnault cylinder, which is usually self-biased by connecting it through a series resistor to the cathode; varying the value of the resistor changes the beam current. All SEMs have the electron gun cathode at the required negative accelerating potential, and the anode and the remainder of the electron column are at ground potential. This ensures that only the cathode and its control electrode are at this large negative potential. The Wehnault cylinder may be biased to be more negative than its usual self-biased condition, and then the gun is turned off. A positive pulse restores the bias to its normal self-biasing condition (about

100 V below the cathode potential) and turns the gun on for the period of the pulse. This technique has been demonstrated by Weinfeld and Bouchoule (1976), based on earlier work of others.

There are two disadvantages to this scheme: first, the pulse has to be produced at the high voltage, and therefore the pulse generator and supplies are required to float. The trigger source to the pulse generator also requires isolation: optoelectronic couplers may be used here. The other disadvantage is that the capacitance to be charged and discharged is significant (several tens of picofarads), so that large current drive for very short pulses is required and the voltage swing required is also large. The advantage is that the beam spot degradation is small. However, the scheme has not found much favor.

## B. Electron-Beam Deflection Pulsing

In this scheme, the electron beam passes through a region of electric or magnetic field which deflects the beam away from its normal path down the column. Apertures defining this path prevent the deflected beam from traveling down the column when off axis. Thus, by providing pulsed voltage or current excitation to this deflection field, the beam may be pulsed on or off as required. In general, magnetic field deflection schemes are slow, and confined to frequencies less than 100 kHz. Electric field deflection has been used up to 9.1 GHz repetition frequencies, with beam pulses of about 1 ps or smaller in width.

Since this method of beam pulsing is most commonly used, various aspects of it are discussed in some detail.

### 1. Deflection Pivot and Apparent Source Movement

Consider an electron beam of voltage $V_b$ traveling centrally between plates $h$ long, distance $d$ apart, and symmetrically excited to $\pm\frac{1}{2}V_p$, respectively, as shown in Fig. 20. The beam emerges from the plates, with transverse displacement, at the edge of the plate given by

$$x = V_p h^2 / 4V_b d \tag{24}$$

and the deflection angle $\alpha_0$ is given by

$$\tan \alpha_0 \simeq \alpha_0 = \frac{v_x}{v_y} = \frac{V_p h}{2V_b d} = \frac{x}{y_0} \tag{25}$$

where $y_0$ determines where the final beam direction vector intersects the original undeflected path as seen in this figure. Substituting for $x$ from above, $y_0 = h/2$. Therefore, the beam may be assumed to pivot at the center of the

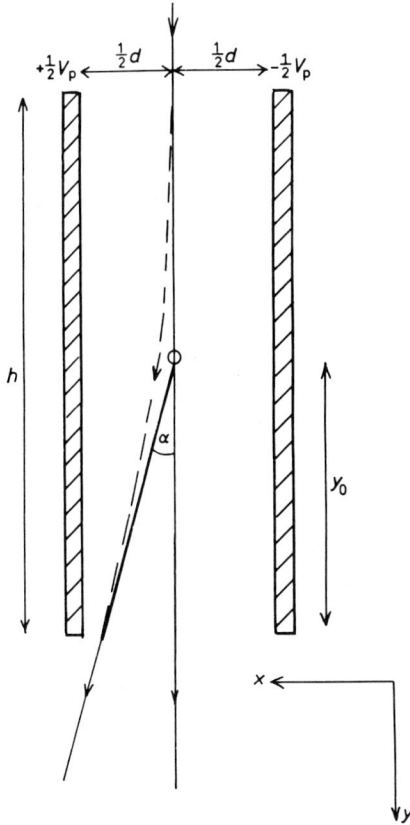

FIG. 20. Deflection of electrons passing through a pair of deflection plates.

deflection plates, and this concept simplifies the electron optical ray diagrams in the following sections.

If the deflected ray is traced back to the plane of the source, it may be assumed that the source has also apparently moved in the deflection to a new apparent position which is the intersection of this ray with the source plane. In practice the plate fields together with the fringe fields, which have been ignored in the discussion, acting on a finite-diameter beam, result in deflection and defocusing, which cause the apparent source to move in a curve. However, since only small-angle deflections are considered here, the apparent movement of the source is assumed to be in the source plane.

Cutoff of the beam occurs when the beam is deflected beyond the edge of a centrally placed chopping aperture. The distance from the pivot point to the chopping aperture along the electron optical axis is called the throw of the chopping system. For a given size of aperture, the larger the throw, the smaller the deflection required and the smaller the plate excitation voltage $V_p$.

## 2. Transit Time

The time taken by the electron to traverse the length of the deflection field ($h$ in the previous section) determines the maximum frequency to which the excited plates deflect the electrons efficiently. Thus, under ac excitation at a frequency $\omega$ radians per second, the deflection angle is given by

$$\alpha = \alpha_0 \frac{\sin(\omega h/2v_y)}{\omega h/2v_y} \tag{26}$$

where $v_y$ is the longitudinal velocity of the beam and $\alpha_0$ is given above in Eq. (25). Note that $\omega h/v_y$ is the electron transit angle $\theta$ (phase delay at frequency $\omega$), and zeros of the above term occur at $\theta = 2n\pi$, $n = 1, 2, \ldots$ It is preferable to work with $\theta$ in the range 0 to $2\pi$, since higher transit angles are inefficient ($\alpha/\alpha_0$ is much less than unity) and also produce lateral displacement of the beam.

The paper by Lischke et al. (1983), which estimates the transmit time through the plates in terms of the deflection angle $\alpha$, is given by

$$\tau = \left(\frac{2V_b m}{e}\right)^{1/2} \alpha_0 \frac{d}{V_p} \tag{27}$$

where $m$ is the mass of the electron and $e$ is its charge. This determines the minimum half-width of the beam pulse that may be generated with these plates. Used in combination with Eq. (25) or separately using Eq. (26), Eq. (27) enables plate design to be performed.

## 3. Deflection Cutoff or Chopping Aperture

The beam normally travels down the electron optical axis of the column (if well aligned!). The beam is turned off by deflecting it across an aperture placed on the electron optical axis, beyond the deflection structure, often called the deflection cutoff or chopping aperture.

In practice, the electron optical column of the SEM has several alignment "splash" apertures of diameters between 0.5 and 2 mm and also a small-diameter (50–200 $\mu$m) final lens aperture, often placed close to or at one of the principal planes of the final lens. This final lens is usually a pinhole lens or some variant thereof, and the final aperture together with the raster double-deflection scheme enables the beam to travel through the center of this lens defined by the final lens aperture, when scanned in the usual raster fashion. The final aperture has demagnified virtual images above the other lenses of the electron optical column, and the final aperture image (FAI) below the deflection structure is often used as the cutoff or chopping aperture. The size and axial position of these final aperture images will depend on the excitation of the lenses, but in general will be close to the lens focal planes. The diameters of these FAIs are likely to be very small depending on the strengths of the lenses. An electron traveling through this image aperture reaches the specimen, but if it is obstructed at this image aperture, it will not reach the specimen, even though it may travel down the column but physically may only be obstructed by the final aperture. The FAI is, therefore, a centrally aligned tiny aperture, and serves well as the chopping aperture, even though its position may not be known very accurately.

In the case of a single-lens instrument, the final aperture may, in turn, also act as the chopping aperture or alternatively, when this is not available, a real chopping aperture will have to be provided.

### 4. Chopping Structure Position

In a multilens electron optical column, the deflection structure may be placed between the gun and top lens or between the first and second or between subsequent lenses. The chopping system is not usually placed between the final lens and the specimen. The region between the final lens and the preceding lens usually houses the stigmator and the beam raster scanning system, and therefore is not generally used for beam chopping. However, the comments and derivations for the latter cases apply if the chopping structure is placed here, except that the FAI is replaced by the real physical final aperture.

### 5. Deflection Structure between Gun and First Lens

The beam emerges from the gun as a diverging cone of electrons, and the extremal ray is defined by a limiting top aperture, shown in Fig. 21. The chopping aperture is assumed to be the final aperture image and for practical purposes to be of negligible diameter when compared to the beam diameter. To be cut off the diverging beam requires to be deflected so that its extremal ray through the structure is just beyond this final aperture image diameter, and if this is negligibly small, this extremal ray crosses the axis at this FAI plane. Using the concept of pivoting at the center of the deflection structure, the minimum angle of deflection to cut off the beam may be estimated, with the symbols and distances as given in Fig. 21.

Let the diameter of the gun crossover be given by $d_G$. The angle of divergence of extremal ray $\gamma_{g1}$ to the vertical, determined by the top limiting

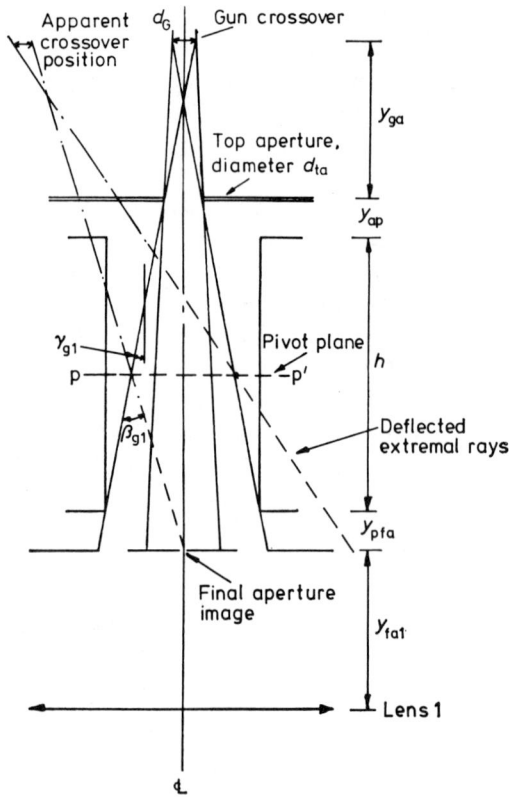

FIG. 21.   Ray diagram of deflection of a finite-diameter beam with the deflection structure between the gun and the first lens.

aperture diameter $d_{ta}$, is given by

$$\gamma_{g1} = (d_{ta} + d_G)/2y_{ga} \tag{28}$$

where $y_{ga}$ is the distance from the crossover to this top aperture. The diameter of the beam halfway through the deflection structure at the pivot plane pp′ is

$$d_{pp'} = d_{ta} + 2\gamma_{g1}(y_{ap} + h/2) \tag{29}$$

where $y_{ap}$ is the distance between the top aperture and the top of the deflection structure and $h$ is the length of the deflection structure. Simplifying:

$$d_{pp'} = d_{ta} + \frac{d_{ta} + d_g}{y_{ga}}\left(y_{ap} + \frac{h}{2}\right)$$

$$\simeq d_{ta}(y_{ga} + y_{ap} + h/2)/y_{ga}, \qquad \text{for} \quad d_G \ll d_{ta} \tag{30}$$

The angle through which this extreme ray has to be deflected to pass through the final aperture image of negligible diameter is

$$\beta_{g1} = \frac{d_{ta}(y_{ga} + y_{ap} + h + y_{pfa})}{2y_{ga}(y_{pfa} + h/2)} \tag{31}$$

where $y_{pfa}$ is the distance from the deflector to the final aperture image. Note that with this deflection the extremal ray passes through the final aperture image, and a slightly larger angle cuts off the beam. From Eq. (25), for a beam accelerated through $V_b$ the required deflection voltage $V_p$ across plates $h$ long, distance $d$ apart, can be estimated. As the beam is deflected it pivots at the center of the deflector. Thus, the rays in the beam appear to emerge from an apparent crossover which is to the side of the true crossover, and the apparent crossover continues to move with increasing deflection. The rays continue to pass through the final aperture image until cutoff occurs. This apparent movement of the source is seen at the specimen as a displacement of the final beam spot, and when the deflection excitation is rapid, appears as an elongated beam spot.

The apparent source size due to deflection is always larger than $d_g$. The ratio of apparent to real source diameter in the direction of the deflection is given by the degradation ratio $\delta$ as

$$\delta = \frac{(\beta_{g1} - \gamma_{g1})(y_{ga} + y_{ap} + h + y_{pfa})}{\gamma_{g1}y_{ga} - d_{ta}/2} + 1 \tag{32}$$

This ratio is always greater than unity, since $(\beta_{g1} - \gamma_{g1})$ is always positive and the denominator is $d_g/2$, and always less than the numerator.

The paper by Lischke et al. (1983), using different notation, derives the movement of the spot $\Delta r$ at the specimen as

$$\Delta r \simeq (y_{ga} + y_{ap} + h/2)\gamma_{g1} M \tag{33}$$

where $M$ is the magnification of the column, and the other terms are as shown in Fig. 21. Note that $M$ is always less than unity in the SEM. The assumption made here is that the distance from the crossover to the pivot plane $(y_{ga} + y_{ap} + h/2)$ is much less than the distance from the crossover to the first principal plane of the lens. Note that $\gamma_{g1}$ is given above in Eq. (28).

While these expressions provide estimates of deflection angle and degradation ratio, the various terms may not be known exactly. Thus, in practice, these equations provide a guide to the magnitude of these effects. Thus, by reducing the diameter $d_{ta}$ of the top aperture, the angle of deflection and degradation are reduced.

## 6. Deflection Structure between First and Second Lens

The deflection structure is placed between the first and second lenses of a three-lens column, and, therefore, a converging beam passes through the deflector. Inevitably, this type of arrangement may result in a magnification of the gun crossover in the first lens, depending on the position of the second crossover. For ideal chopping, the second crossover should be at the plane of the final aperture image, but if it were so, the microscope would never be in focus. In practice, the crossover is above the final aperture image, and may even be at the pivot plane of the deflector for first-order degradation-free operation. This is the crossover position generally adopted in electron-beam lithography machines where chopping degradation has to be negligible. However, the penalty is the larger deflection angle required to cut off the beam, which may not be important for pulsing up to several tens of megahertz. For higher frequencies, fast rise times, and short beam pulses, this may not be acceptable and an alternative reduced degradation setting may have to be used with this chopping structure position.

The second crossover is assumed to be above the final aperture image, and below the pivot plane. The final aperture image as before is assumed to be negligibly small so that the deflected ray has to cross the axis at this plane for cutoff.

From Fig. 22, using similar triangles, the angle of the extremal ray to the vertical is given by

$$\gamma_{12} = \frac{y_{g1}d_{ta} + d_g(y_{1a} + y_{ap} + h + y_{psc})}{2y_{g1}(y_{ap} + h + y_{psc})} \tag{34}$$

and the angular deflection to cut off the beam is given by

$$\beta_{12} = \frac{\gamma_{12}(y_{ap} + h + y_{pfa}) - d_{ta}/2}{h/2 + y_{pfa}} \tag{35}$$

The movement of the crossover of the scheme is determined by the field of view. Thus, all rays that travel past the top aperture and the final aperture image reach the specimen. Therefore, these two apertures define the angular field of view and any rays before or after deflection within this cone can reach the specimen. While the images of the other splash apertures placed along the column appear in this space, the above definition of the field of view may be pessimistic, and therefore adopted for the following discussions. Thus, the semiangle of the field of view is given by

$$\zeta_{fv} = \frac{d_{ta}}{2(y_{ap} + h + y_{pfa})} \tag{36}$$

The semiangle subtended by the second crossover at the final aperture image is given by

$$\xi_i = d_G(y_{1a} + y_{ap} + h + y_{sc})/2y_{g1}y_{scfa} \tag{37}$$

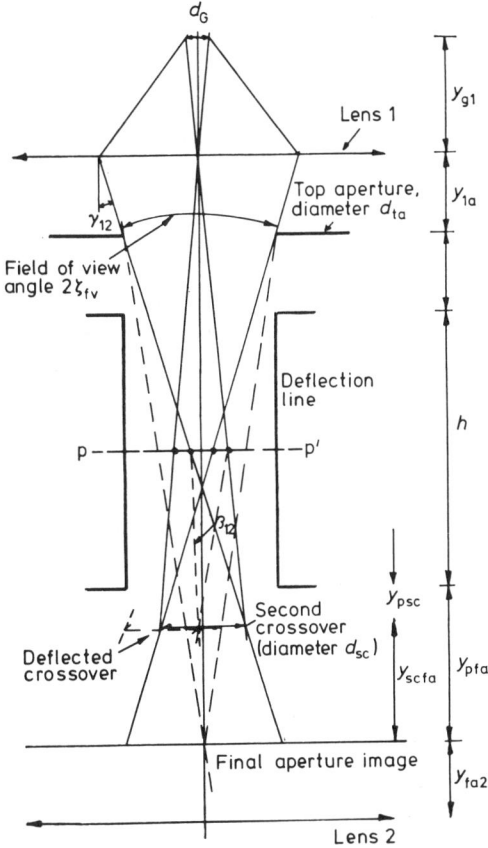

FIG. 22.    Ray diagram of deflection of a finite-diameter beam with the deflection structure between the first and second lenses.

Note that if $\xi_i > \zeta_{fv}$ the system field of view is limited and therefore negligible chopping degradation occurs. The alternative $\xi_i < \zeta_{fv}$ gives rise to spot size degradation, as the image moves across the field of view until cutoff occurs. The worst degradation that can occur in this scheme results in the apparent crossover size diameter increasing from $2\xi_i y_{scfa}$ to $2\zeta_{fv} y_{scfa}$. If cutoff occurs before all or any of this movement is seen, then the degradation is smaller than this ratio. Thus, the spot size degradation may be carefully controlled by varying the position of the second crossover and the size of the top aperture. Further details are discussed by Gopinath and Hill (1976).

## 7. Practical Considerations

In the current SEM columns, the space between the top and second lens may not accessible, and this would also be the case with two-lens instruments.

Therefore the trend has been to place the deflection plates between the gun and the first lens. In order to reduce the degradation to a minimum, the top aperture is chosen to be small, of the order of 50 to 100 $\mu$m diameter, or alternatively a movable pair of knife edges are placed there. With such a small aperture, alignment to the electron-optical axis may be a problem, and may be aided by alignment coils in the gun–lens space.

Another innovation, first discussed by Hieke *et al.* (1975), is to reduce the space between the plates to a distance as small as 100 $\mu$m. Thus, the length of plates may also be correspondingly reduced to circumvent the transit-time limitations at low beam voltages of 1 or 2 kV. The chopping system used by Menzel and Kubalek (1979b) adopted these suggestions. The plates are mounted on micrometer movements, and the design may move either plate in the same plane separately or together. In the latter case, the plate gap is set to a suitably small value, and then the plates moved to intercept the beam by one side, and then the other, and then a central position is used. If the plate distance is 100 $\mu$m or less, then the limiting aperture is not needed any longer, as it may be assumed in Fig. 23 that the distance $y_{ap}$ becomes zero. Thus, chopping voltage drive may be obtained directly from low-voltage drive circuits with buffer sharpening stages. In practice, the contamination of the plates is significant and therefore these require frequent cleaning.

For stroboscopy, it is necessary for the beam pulses to be synchronous with the periodic event being observed, together with a means of varying the phase of these pulses. The choice often adopted is to have the beam deflection plates with dc excitation, so that the beam is normally off, and the excitation, a flat-topped pulse to the plates, turns the beam on for the duration of the flat top. An alternative is to have the beam normally off, and to sweep the beam across the chopping aperture to turn it on during the sweep. This latter case requires a pair of orthogonal blanking plates below the main plates appropriately excited so that the reset of the beam during the falling edge of the excitation pulse does not generate a second beam pulse. The third alternative is to excite the plates sinusoidally or with a symmetric square wave, at half the periodic event frequency, with the beam normally on, and excited off. In this case the pulses become smaller in width as the drive voltage becomes larger. All three methods have been used for stroboscopic studies, but for sampling mode operation the former two choices have been adopted. The third method of sinusoidal excitation for a normally-on beam may possibly be used with the addition of orthogonal plates being excited to turn the beam off during every second zero of the sine wave.

Current usage is the first of the above schemes for most circuits or devices operating up to several tens of megahertz, where pulses of the order of 0.1–1.0 ns are adequate. For higher-frequency operation, the flat-topped turn-on pulse becomes almost sinusoidal, and therefore pulse widths become wide.

The second alternative above, beam off normally and swept on across the chopping aperture with reset blanking, yields smaller pulses. However, in this case the beam is on during the sweep period and thus degradation may be worse than in the previous case. For very high frequencies in the microwave region up to 9 GHz, sinusoidal excitation at half-frequency was the choice of Gopinath and Hill (1976). Fujioka and Ura (1983) have shown that the pulse width with deflection chopping may also degrade due to the velocity modulation of the beam. They also suggest stigmation would reduce degradation effects in all cases considered above.

## 8. Multiple Plate Structures

When the plate structure discussed above needs to be designed to have a small voltage drive, the usual method is to lengthen the plates. However, this is not always feasible, as the transit time determines the smallest pulse that can be generated, which increases as the beam voltage goes down. Thus, traveling-wave structures provide a method of stacking plates in parallel to reduce the excitation voltage. The disadvantage of such a structure is that the axial velocity of the beam must be synchronous with the pulse propagation velocity along the structure and therefore the beam voltage is fixed. Despite this limitation early work by Gopinath (1970) and Gopinath and Hill (1976) used a meander line above a ground plane, a helix within a tube, and a screened meander line to obtain very short beam pulses, of the order of a few picoseconds. However, the limitation of the fixed beam voltage has resulted in this not being generally adopted.

More recently, work on degradation-free high-speed beam blankers has resulted in double-deflection plate structures in the electron-beam microlithographic context. Lin and Beauchamp (1973) suggested that having the plate structure pivot plane at the beam crossover would, to first order, result in negligible movement of the final beam spot, and, thus, negligible degradation. The problem with this scheme is that the plate voltage to cut off the beam for a specific position of cutoff aperture is generally larger than the other schemes. Thus, a method of overcoming this problem is to position two sets of plates on either side of the crossover, and feed the second set with the same voltage, after the correct delay. One such scheme has been discussed by Kuo et al. (1983). A variant of this is to use two pairs of plates in which the gaps taper toward the crossover point on either side, as proposed by Rose and Zach (1984). This paper also derived the dynamic paraxial ray equations for this structure. In this case, as in the Kuo et al. paper, the chopping aperture is at the plane of the crossover. The delay used for the excitation of the second set of plates determines the beam voltage.

## 9. Beam Bunching Scheme

Velocity modulation of an electron beam results in beam bunching, and this effect is used in the generation or amplification of microwave signals in the klystron. The space-charge effects in the typical SEM beam are small, and therefore simple velocity modulation may be used to obtain very narrow and tight pulses, with no loss of current. The velocity of the electrons is perturbed sinusoidally as they pass through the gap of a reentrant cavity, and the faster electrons catch up with the slower electrons to form bunches. In combination with a second cavity placed orthogonally to produce deflection blanking, Ura et al. (1978) were able to obtain 0.2 ps pulses at 1 GHz. This is in contrast to the 1–10 ps pulses obtained by Gopinath and Hill (1976) using deflection chopping with a traveling-wave deflection line. The beam bunching results have been compared with the deflection blanking results recently by Fujioka and Ura (1983), who also provide charts for deflection voltages required for various parameters. Since the beam buncher is usually a single-frequency structure requiring mechanical retuning, this method of beam pulsing has not generally been adopted.

## X. Stroboscopic and Sampling-Mode Operation of the SEM

As discussed in Section VIII, the major requirements for sampling-mode operation of the SEM are the sampling of the relevant periodic voltage contrast signal and the synchronization of the sampling gate and the periodic event being examined. Since the voltage contrast signal is usually limited to the video bandwidth without linearization, the sampling is usually performed by pulsing the beam as discussed in Section IX. The synchronization of the periodic event being examined with the sampling beam pulses is crucial in this mode. In addition to this, it is necessary that the synchronization loop system perform the automatic stepping of the sampling delay after the voltage measurement at each delay time has been completed. With the beam held on spot at some position on the specimen, the voltage waveform may be measured by the sampling system stepping through the waveform. Alternatively, the delay position is held constant and the beam is scanned in a line or in a raster over the specimen. The voltage contrast may then be displayed as y deflection in the case of the line scan and in the raster scan or alternatively as intensity modulation of the display tube to provide a micrograph. The micrograph has usually been confined to qualitative contrast, but occasionally the quantitative results have been thus displayed with rather long exposure times to obtain reasonable contrast. One method of quantizing the display micrograph is by color coding for various magnitudes.

Figure 23 shows the similarity between the sampling oscilloscope and the sampling SEM. The sampling head and sampling gate of the sampling oscilloscope are replaced by the SEM beam pulsing unit, and the output of the sampling gate is replaced by the voltage measurement feedback loop.

The initial work of Thomas *et al.* (1976) on the sampling SEM modified a commercially available sampling oscilloscope circuit. Current systems are

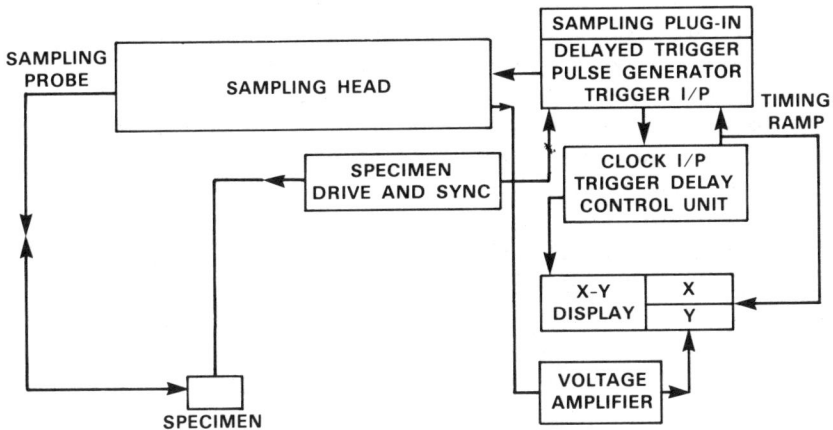

(a)

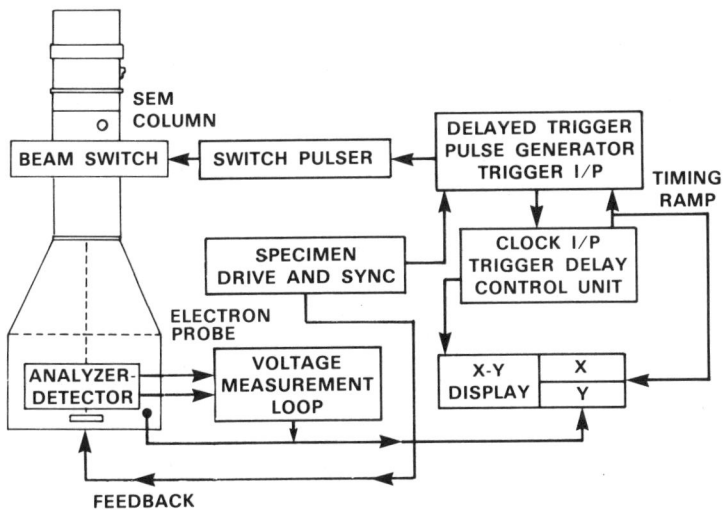

(b)

FIG. 23.  Similarity of the schematic diagrams of (a) the sampling oscilloscope and (b) the sampling scanning electron microscope.

custom built for the SEM or alternatively use commercial boxcar integrators.

Unlike the sampling oscilloscope, it is necessary for the voltage measurement system to settle, and this may require the averaging of a large number of samples at every delay time. Thus, the delay time is held for a preset time or a preset number of samples. This increased delay for integration of the signal is also provided by boxcar integrators, and therefore these are also used in current systems.

Thus the currently available system would employ the Hieke-type chopping plates, modified for easy alignment, with a Feuerbaum-type analyzer with a feedback loop and possibly a lock-in system, a beam pulser driven from a sampling unit delayed pulse generator, or a boxcar integrator which is triggered by the input to the IC or device. Display may be the conventional SEM display tube, oscilloscope, or $x-y$ recorder as required with any signal processing system.

The time resolution of the system is determined by the width of the sampling pulse $\tau_s$, which in the SEM is the beam pulse width. The maximum frequency that can be resolved is of the order of $1/2\pi\tau_s$. Thus the typical beam pulse of 100 ps width will accurately resolve waveforms with frequencies of up to about 3 GHz. Some authors claim that the factor 2 should be omitted from this expression. The shorter the pulse width, the higher the frequency that can be resolved or alternatively the faster the rise time that can be recorded. The

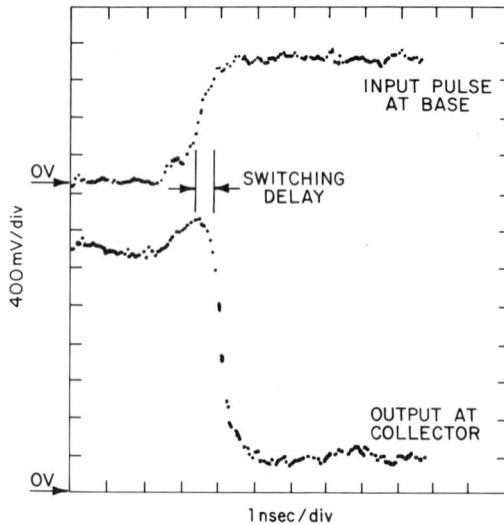

FIG. 24.   Propagation delay measurement on switching transistor 2N708 (Gopinathan and Gopinath, 1978).

pulse width and circuit considerations also determine the time window over which the waveform can be displayed.

The waveform recording mode has been used to record delays in switching times of various devices and circuits. Figure 24 shows an example of the switching delay of a 2N708 (Gopinathan and Gopinath, 1978).

Figure 25 shows a voltage contrast micrograph at various delay times of a NAND gate (Siemans S193, Menzel and Kublalek, 1983b). Linearized and isolated micrographs of this same circuit have also been published by these authors together with others, but the signal-to-noise ratio is poor (Balk *et al.*, 1976).

XI. VOLTAGE CONTRAST WITH SYNCHRONOUS AND
ASYNCHRONOUS PULSED BEAMS

The application of voltage contrast, qualitative and linearized, with and without sampling, in electron-beam probing of ICs and devices has produced several schemes with names such as voltage coding (see Section IV), logic-state mapping, frequency tracing, and mapping. These are briefly outlined in this section.

In voltage coding the operating frequency of the IC is synchronized with a line-scan frequency of the electron probe which may be at a TV line frequency. The voltage contrast micrograph is an image of the logic states and state changes within the IC in the form of bright (logic state 0 or less) and dark (logic state 1 or higher) bands on the interconnects (Lukianoff and Touw 1975). In logic-state mapping, the beam is pulsed in synchrony with the logic IC input pulse, but after each line scan, for example, in the $x$ direction, the delay time or phase of the beam pulses is increased by a specific amount. Thus, the normal micrographs show black and white bars along the $y$ direction and conductors carry the input pulses as shown in Fig. 26c. The micrograph shows the propagation of pulses along the $y$ direction, in which time delay along the $y$ direction can be measured. Lines with subharmonics should have wider bars, and those at harmonic frequencies, narrower bars. When the line scan is along the $x$ direction, and thus the frame scan is in the $y$ direction, the bars appear in the $x$ direction, as can be seen in Fig. 26d. This technique, advocated by Crichten *et al.* (1980), is extremely useful in the testing of digital ICs.

Another technique proposed by Collin (1983), which is applicable both to linear and digital integrated circuits, is to use a continuous beam and pass the output video signal with voltage contrast through a lock-in amplifier tuned to the frequency of interest. Thus, the output of the lock-in amplifier modulates the brightness of the display CRT to image only those lines, interconnections,

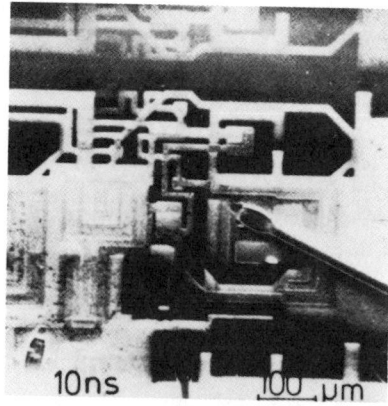

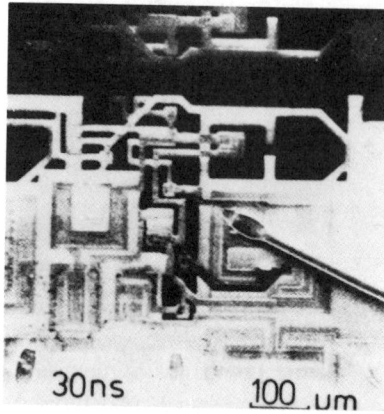

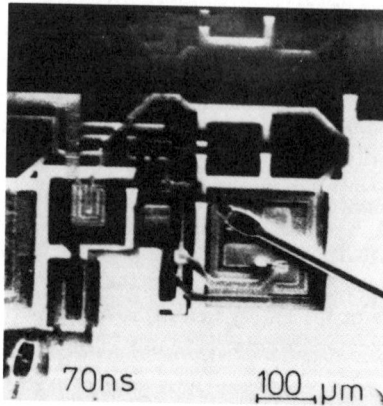

Fig. 25. Stroboscopic micrographs of transition of logic states on a passivated bipolar NAND gate at three delay times (reproduced with permission from Menzel and Kubalek, 1983b, and FACM Publishers Inc., Mahwah, NJ).

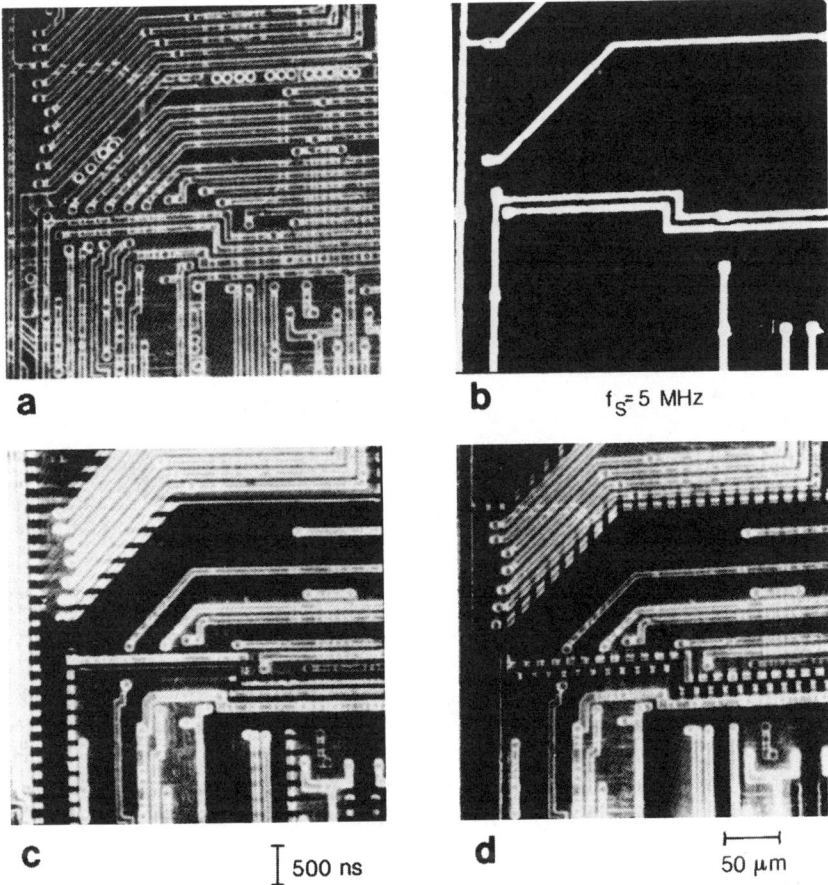

FIG. 26. Details of a 8086 microprocessor: (a) secondary electron micrograph, (b) frequency tracing micrograph at 5 MHz, (c) logic state micrograph, linescan horizontal, and (d) logic state micrograph, linescan vertical (reproduced with permission from Brust *et al.*, 1984).

and devices operating at the desired frequency. The limitation of this method of frequency monitoring is that the maximum detectable frequency is that of the video chain, typically 2–3 MHz.

An alternative, suggested by Brust *et al.* (1984), is to use an asynchronously pulsed beam and extract the lower sideband frequency, designed to be within the video chain bandwidth, for display. Thus, the secondary current is $\delta$ times the product of the primary beam current and $\xi$ times the voltage on the specimen, $\xi$ having dimensions of $V^{-1}$. Thus

$$I_s = \delta I_b \xi V_p \tag{38}$$

Suppose the primary beam is pulsed at some angular frequency given by $\omega_b$ radians per second, and also suppose that the specimen voltage varies as $\omega_s$; then

$$I_s = \delta\left(I_{b0} + \sum_n I_{bn}\cos(n\omega_b t + \theta_{bn})\right)$$

$$\times \xi\left(V_{s0} + \sum_n V_{sn}\cos(n\omega_s t + \phi_{sn})\right) \tag{39}$$

Considering only the fundamental terms gives

$$I_s = \delta\xi(I_{b0} + I_{b1}\cos\omega_{b1}t)[V_{s0} + V_{s1}\cos(\omega_s t + \theta_{s1})] \tag{40}$$

or

$$I_s = \delta\xi[I_{b0}V_{s0} + I_{b1}V_{s0}\cos\omega_b t + I_{b0}V_{s1}\cos(\omega_s t + \theta_s)$$

$$+ \tfrac{1}{2}I_b V_{s1}\{\cos[(\omega_b + \omega_s)t + \theta_s] + \cos[(\omega_b - \omega_s)t - \theta_s]\} \tag{41}$$

Note that this results in a constant term together with frequency components of $\omega_b$ and $\omega_s$, and also the sum and difference frequencies of $\omega_b + \omega_s$ and $\omega_b - \omega_s$, respectively. If $\omega_b$ and $\omega_s$ are much higher than the video bandwidth, then the only terms that will be amplified by the video amplifier are the constant term and the difference frequency term $\omega_b - \omega_s$, provided this is within the video band. Now, the signal passes through a band-pass filter after the video amplifier, which has a passband centered at $\omega_b - \omega_s$, and then through an envelope detector which feeds into the normal video chain to obtain micrographs. Figure 26a shows a normal secondary micrograph of the 8086 microprocessor and Fig. 26b, the frequency tracing case.

Frequency mapping uses the same difference signal except the beam blanking frequency is varied after each line scan. The blanking frequency $\omega_p$ varies from some minimum to a maximum value depending on the frequencies of interest. Thus, the usual form of micrograph is obtained except that voltage contrast is seen only on lines within the scanned frequency range. However, the frequencies of interest show up in bright streaks along the line-scan direction and the micrograph is dark where the frequency does not fall within the scanned range. The scheme is very useful in failure analysis if modes of fault detection are not suitable.

## XII. Conclusions

Voltage contrast has found use in the semiconductor device and integrated circuit area for scientific investigation, testing, and fault detection. It is the basis of currently available electron-beam probe testing systems. Linearization of voltage contrast has resulted in voltage changes of the order of

0.5 mV being measured. In practice this resolution is never reached, typical figures being in the several tens of millivolts.

The bandwidth of these measuring systems is small, of the order of several kilohertz; the bandwidth of qualitative contrast is that of the video-amplifier chain, which is only a few megahertz. Examination of dynamic periodic events and voltage waveforms at frequencies above these requires the use of sampling techniques, and this has resulted in the sampling and stroboscopic modes of operation. To facilitate locating areas of specific interest in integrated circuits, there have been developments in the form of voltage coding, logic-state mapping, frequency mapping, and tracing. All these techniques are available in commercial systems. The basic measuring techniques, the analyzers, and beam blankers have evolved over the past ten years to their present state, and they will continue to improve with time. The area of current interest is the measurement of voltage through dielectric layers, which has been dealt with only briefly here. Electron-beam test systems, custom built for the semiconductor industry, with increased automation, computer control, and digital signal processing, will become more popular in time.

## REFERENCES

Balk, L. J., Feuerbaum, H. P., Kubalek, E., and Menzel, E. (1976). *Scanning Electron Microsc.* 615–624.

Beaulieu, R. P., Cox, C. D., and Black, T. M. (1972). *Proc Reliability Phys. Symp., 10th* 32–35.

Bennet, W. R. (1960). "Electrical Noise." MacGraw-Hill, New York.

Brust, H.-D., Fox, F., and Wolfgang, E. (1984). *In* "Microcircuit Engineering 84" (A. Heuberger and H. Beneking, eds.) pp. 411–425. Academic Press, London.

Collin, J. P. (1983). *C. R. Journee Electron. Ecole Polytech. Fed. Lausanne* 283.

Crichton, G., Fazekas, P., and Wolfgang, E. (1980). *IEEE Test Conf. Dig. Pap., Philadelphia* 444–449.

Driver, M. C. (1969). *Scanning Electron Microsc.* 403–13

Dyukov, V., Kolemeytsev, M., and Nepijko, S. (1978). *Microc. Acta* **80**, 367–374.

Everhart, T. E. (1968). *Scanning Electron Microsc.* 3–13.

Everhart, T. E., Wells, O. C., and Oatley, C. W. (1959). *J. Electron. Control* **7**, 97–111.

Fentem, P. J. (1974). PhD thesis, University College of North Wales, School of Electronic Engineering Science, Bangor, Wales.

Fentem, P. J., and Gopinath, A. (1974). *J. Phys. E Sci. Instrum.* **7**, 930–933.

Feuerbaum, H. P. (1979), *Scanning Electron Microsc.* 285–96

Fleming, J. P., and Ward, E. W. (1970). *Scanning Electron Microsc.* 465–470.

Fujioka, H., and Ura, K. (1983). *Scanning* **5**, 3–13.

Fujioka, H., Tsijitake, M., and Ura, K. (1982). *Scanning Electron Microsc.* Pt. 3, 1053–1060.

Fujioka, H., Nakamae, K., and Ura, K. (1983). *Scanning Electron Microsc.* Pt. 1, 1157–1162.

Gopinath, A. (1970). *Proc Int. Congr. Electron Microsc., 7th, Aug.-Sept., Grenoble* Pt. 1, 203–204.

Gopinath, A. (1974). *In* "Quantitative Scanning Electron Microscopy" (D. B. Holt, M. D. Muir, P. R. Grant, and I. M. Boswarva, eds.), p. 117. Academic Press, London.

Gopinath, A. (1977). *J. Phys. E Sci. Instrum.* **10**, 911–13

Gopinath, A., and Hill, M. S. (1976). *J. Phys. E Sci. Instrum.* **10**, 229–236.

Gopinath, A., and Sanger, C. C. (1971a). *J. Phys. E Sci. Instrum.* **4**, 334–336.

Gopinath, A., and Sanger, C. C. (1971b). *J. Phys. E Sci. Instrum.* **4**, 610–611.

Gopinath, A., Gopinathan, K. G. and Thomas, P. R. (1978). *Scanning Electron Microsc.* 375–380.

Gopinathan, K. G. (1978). UK Patent application 663/78, Patent Office, London.

Gopinathan, K. G., and Gopinath, A. (1978). *J. Phys. E Sci. Instrum.* **11**, 229–233.

Gopinathan, K. G., Thomas, P. R., Gopinath, A., and Owens, A. R. (1976). *Electron. Lett.* **12**, 501–502.

Gorlich, S., Herrman, K. D., and Kubalek, E. (1984). *In* "Microcircuit Engineering" (A. Heuberger and H. Beneking, eds.), pp. 451–460. Academic Press, London.

Goto, Y., Ito, A., Furukawa, Y., and Inagaki, T. (1981). *J. Vac. Sci. Technol.* **19**, 1030–1032.

Hannah, J. M. (1974). PhD thesis, University of Edinburgh, Department of Electrical Engineering.

Hardy, W., Behara, S., and Cavan, D. (1975). *J. Phys. E Sci. Instrum.* **8**, 789–793.

Hieke, E., Meusburger, G., and Alter, H. (1975). *Scanning Electron Microsc.* 219–225.

Hill, M. S., and Gopinath, A. (1973). *IEEE Trans. Electron Devices* **ED-20**, 610–612.

Huchital, D. A., and Rigden, J. D. (1972). *J. Appl. Phys.* **43**, 2295–2302.

Janssen, A. P., Akhter, P., Harland, C. J., and Venables, J. A. (1980). *Surf. Sci.* **93**, 453–470.

Kuo, H. P., Foster, J., Haase, W., Kelly, J., and Oliver, B. M. (1983). *Proc. Int. Conf. Electron Ion Beam Sci. Technol., 10th* **1**, 78–91.

Kurshed, A., and Dinnis, A. R. (1983). *Scanning* **5**, 25–31.

Lin, L. H., and Beauchamp, H. L. (1973). *J. Vac. Sci. Technol.* **10**, 987–990.

Lischke, B., Plies, E., and Schmitt, R. (1983). *Scanning Electron Microsc.* 1177–1185.

Lukianoff, G. V., and Touw, T. R. (1975). *Scanning Electron Microsc.* 465–471.

MacDonald, N. C. (1970). *Appl. Phys. Lett.* **16**, 76–80.

Menzel, E. (1981). PhD thesis, University of Duisburg.

Menzel, E., and Kubalek, E. (1979a). *Scanning Electron Microsc.* 297–304.

Menzel, E., and Kubalek, E. (1979b). *Scanning Electron Microsc.* 305–317.

Menzel, E., and Kubalek, E. (1983a). *Scanning* **5**, 151–171.

Menzel, E., and Kubalek, E. (1983b). *Scanning* **5**, 103–122.

Nakamae, K., Fujioka, H., and Ura, K. (1981). *J. Phys. D Appl. Phys.* **14**, 1939–1960.

Nye, P., and Dinnis, A. R. (1985a). *Scanning* **7**, 113–116.

Nye, P., and Dinnis, A. R. (1985a). *Scanning* **7**, 117–124.

Oatley, C. W. (1969). *J. Phys. E Sci. Instrum.* **2**, 742–744.

Oatley, C. W. (1972). "The Scanning Electron Microscope." Cambridge Univ. Press, Cambridge

Oatley, C. W., and Everhart, T. E. (1957). *J. Electron.* **2**, 568–570.

Ostrow, M., Menzel, E., and Kubalek, E. (1981). *In* "Microcircuit Engineering" (A. Obsenbrug, ed), pp. 514–518. Academic Press, New York.

Pantel, R., Arnaud D'Avitaya, F., Ged, Ph., and Bois, D. (1982). *Scanning Electron Microsc.* 549–557.

Plows, G. S. (1981). Lintech Instruments, Cambridge (see Menzel and Kubalek, 1983a).

Plows, G. S., and Nixon, W. C. (1968). *J. Phys. E Sci. Instrum.* **1**, 595–600.

Rau, E. I., and Spivak, G. V. (1978). *Scanning Electron Microsc.* 325–332.

Robinson, G. Y., White, R. M., and MacDonald, N. C. (1968). *Appl. Phys. Lett.* **13**, 407–408.

Rose, H., and Zach, J. (1984). *Proc Pfefferkorn Conf, 3rd, Ocean City, MD* **April**, 126–136.

Taylor, D. M. (1978). *J. Phys. D Appl. Phys.* **11**, 2443–254.

Tee, W. J., and Gopinath, A. (1976). *Scanning Electron Microsc.* 595–601.

Tee, W. J., and Gopinath, A. (1977). *Rev. Sci. Instrum.* **48**, 350–355.

Tee, W. J., Farquhar, S. G., and Gopinath, A. (1978). *IEEE Trans. Electron Devices* **ED-25**, 655–659.

Thomas, P. R., Gopinathan, K. G., Gopinath, A., and Owens, A. R. (1976). *Scanning Electron Microsc.* 609–614.

Thomas, P. R., Gopinathan, K. G., and Gopinath, A. (1980). *In* "Microcircuit Engineering" (H. Ahmed and W. C. Nixon, eds), pp. 479–499, Cambridge Unive. Press, Cambridge.

Tiwari, S., Eastman, L., and Rathbun, L. (1980). *IEEE Trans. Electron Devices* **ED-27,** 1045–1054.

Todokoro, H., Fukuhara, S., and Komoda, T. (1983). *Scanning Electron Microsc.* 561–568.

Ura, K., Fujioka, H., and Hosokawa, T. (1978). *Scanning Electron Microsc.* 741–746.

Waldrop, J. R., and Harris, J. S. (1975). *J. Appl. Phys.* **46,** 5214–5217.

Well, O. C., and Bremer, C. G. (1968). *J. Phys. E Sci. Instrum.* **1,** 902–906.

Well, O. C., and Bremer, C. G. (1969). *J. Phys. E Sci. Instrum.* **2,** 1120–1121.

Weinfeld, M., and Bouchoule, A. (1976). *Rev. Sci. Instrum.* **47,** 412–417.

ADVANCES IN ELECTRONICS AND ELECTRON PHYSICS, VOL. 69

# New Experiments and Theoretical Development of the Quantum Modulation of Electrons (Schwarz–Hora Effect)

## HEINRICH HORA

*Department of Theoretical Physics*
*University of New South Wales*
*Kensington 2033, Australia*

## PETER H. HANDEL

*Department of Physics*
*University of Missouri at St. Louis*
*St. Louis, Missouri 63121*

## I. Introduction

Quantum modulation of an electron beam is the inelastic interaction of a laser beam crossing the electron beam where a medium in the crossing area provides the recoil for adding or subtracting multiples of the energy of the laser photons to the electrons. The de Broglie wave field of the electrons then experiences a modulation by the photon energy, where the interference field of the electron waves show a basic difference in the second order between the optical waves and the electron waves (Hora, 1972). The adding and subtracting of the photon energy has been demonstrated directly by measuring the energy spectra of the electrons after the generation of the modulation (Andrick and Langhans, 1976; Weingartshofer *et al.*, 1977, 1979) using molecules in the crossing area.

The reproduction of the photons by emission from interaction of the modulated beams with a nonluminescent target (Schwarz and Hora, 1969) indicated the state of the quantum modulation of the electrons and showed properties (Ghatak *et al.*, 1984) that were completely reproduced by the scattering of molecules (Weingartshofer *et al.*, 1985). Apart from this agreement of the modulation properties, the re-emission process from the nonluminescent target has not yet been reproduced by others (Schwarz, 1971b, 1972b). Since there is a clear experimental reproduction of the modulation process, an indirect reproduction of the re-emission, however, has been found experimentally. It was measured that the transfer of the photon energy via quantum modulation and their reproduction as amplified microwave emission at collisions of the electrons in a nonequilibrium plasma is the mechanism by which an amplification of microwaves appears (Rosenberg *et al.*, 1980; Ben-Aryeh, 1982). Further involvements are with respect to coherence processes and a number of related effects.

Since the reproduction of the quantum modulation (Schwarz–Hora effect) in another way than by emission of the photons in a nonluminescent material was verified experimentally by Andrick and Langhans (1976), the statement by Jaynes (1978), "It does seem to me that definite proof of the nonexistence (of the Schwarz–Hora effect) would be a considerable embarrassment to quantum theorists who found it so easy to account for" was not necessary. Nevertheless, at present only a combined picture of the reproduction and confirmation of the experimental results can be given.

It is the task of this article to consider the experimental facts, endeavoring to achieve the necessary critical objectivity, and further to reflect several theoretical results and aspects as well as to underline the process of quantum scattering as the essential implication. The inclusion of some original results based on the earlier given simplified theories is well justified in order to present a more complete picture of the theoretical knowledge. The implication of several new developments on coherence and self-interaction of electrons, basic noise problems, to underline where similarities with photons [interference between independent photons (Paul, 1986)] and electrons are given or not, will be sufficient to stimulate the consideration of further experiments.

## II. Review of Experiments

The physical processes involved with the quantum modulation are of two kinds: a bremsstrahlung process and a coherence property. The first is a single-particle description (particle picture), while the coherence properties at interference of excited or modified states of the wave field of the electron with its initial state is a basic property of wave picture of the electrons (Favro and Kuo, 1973; Hawkes, 1978).

The bremsstrahlung property is described in Fig. 1. If a free electron (a) with an energy $E_0$ interacts with another particle, e.g., an ion (drawn as $\oplus$), the electron will change its propagation direction, which is causing the emission of a photon of energy $h\nu$. If the recoil to the ion is sufficiently small, the electron then has an energy of $E_0 - h\nu$. This process is the spontaneous emission of bremsstrahlung. Another case, Fig. 1b, is the interaction of a free electron in the neighborhood of an ion at simultaneous incidence of a photon, where the electron can absorb the photon energy. This is free-free absorption and can be considered as inverse bremsstrahlung. It is further possible, as in Fig. 1c, that the interaction of the free electron with the ion takes place in the presence of an incident photon of the energy $h\nu$, which causes the stimulated emission of another photon of the same energy and phase ($2h\nu$), and the electron energy is decreased by the amount of the photon energy. This is the stimulated emission of bremsstrahlung.

Processes (a)–(c) in Fig. 1 were treated quantum mechanically by Gaunt (1930) and by Maue (1932), who arrived at the derivation of the absorption constant of electromagnetic radiation in a fully ionized plasma defined by the collision frequency $\nu_{coll}$ given in the complex refractive index $\tilde{n}$ of a plasma

$$\tilde{n}^2 = 1 - \omega_p^2/\omega^2(1 - i\nu_{coll}/\omega) \qquad (1)$$

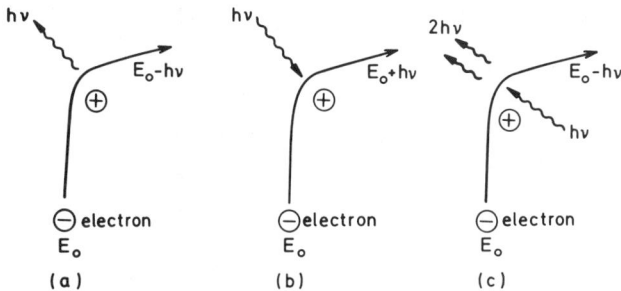

FIG. 1. Scheme for the emission of a photon by the interaction of a free electron $\ominus$ with proton $\oplus$ as (a) bremsstrahlung, or (b) the absorption of a photon as inverse bremsstrahlung, or (c) as stimulated emission of bremsstrahlung.

This optical response function agreed with that derived classically by Kramers (1923), except for some correction factors, the Gaunt factors $g$, changing the collision frequency by a factor of 1–3 versus the Kramers values. The optical constants were also of the same order of magnitude based on the Spitzer and Härm (1953) theory of wide-angle and small-angle scattering of electron by ions and electrons in a plasma.

The question arose, however, whether the classical collision frequency of the electrons derived by Spitzer and Härm (1953), which is in a good agreement with the point-mechanically (hyperbolic) simple 90° scattering of an electron by an ion (Hora, 1981b), has to be modified quantum mechanically at plasma temperatures $T$ above the value $T^*$. The dependence of the collision frequency on the electron density $n_e$ and the plasma temperature $T$ is then given by (Hora, 1981a, 1982)

$$\begin{aligned} v_{coll} &= v_c, & \text{if } T < T^* \\ &= v_c T/T^*, & \text{if } T > T^* \end{aligned} \tag{2}$$

$$kT^* = \tfrac{4}{3}Z^2\alpha^2 mc^2 = \tfrac{4}{3}Z^2 e^2/r_B \tag{3}$$

where $k$ is the Boltzmann constant, $m$ the electron mass, $r_B$ the Bohr radius, $Z$ the number of ion charges, $e$ the electron charge, $\alpha$ the fine-structure constant, and $v_c$ the classical collision frequency. The unique value of the Bohr radius appears then even for free–free collisions. The quantum correction of the collision frequency for temperatures above $T^*$ is the reason for the anomalous resistivity in plasmas (Hora, 1981a).

It is remarkable that the classical collision frequency (apart from a factor of the order of one)

$$v_c = Z\pi e^4 n_e [3^{3/2} m^{1/2}(kT)^{3/2}] \tag{4}$$

using the electron density $n_e$, is exactly the result of the quantum-mechanical treatment (including the second quantization of the electromagnetic field), if the stimulated emission is included in the treatment of the bremsstrahlung and the inverse bremsstrahlung theory only. Treating the inverse bremsstrahlung for the absorption of optical radiation in a plasma without stimulated emission, the quantum-electrodynamical treatment of the collision frequency results in (Hora, 1982)

$$v_{coll}^{QED} = v_c \frac{kT}{\hbar\omega} \frac{2^{3/2}g}{\pi^{1/2}} \sim \frac{1}{T^{1/2}} \tag{5}$$

This is then the quantum-corrected collision frequency in a plasma at high temperatures (anomalous resistivity). The branch of optical absorption in a plasma without stimulated emission is expected, e.g., when an x-ray beam with

1000 eV photon energy irradiates plasma with a temperature much less than 1000 eV (Hora, 1982).

In the usual case in a plasma where the photon energy is much less than the electron temperature, the quantum-mechanical treatment of the absorption constant teaches us that there has to be stimulated emission of bremsstrahlung involved in order to achieve the measured formula that also agrees with the very primitive classical hyperbola trace treatment.

Being aware of these experimentally confirmed theoretical results for inverse bremsstrahlung, some properties of coherence and interference from the wave picture of the electrons should be considered. In Fig. 2a optical interference is shown. If a beam of photons is split as in a Mach–Zehnder interferometer and reunited where a part of the beam has a delay of an optical length L, one will achieve interference patterns of the optical wave field if the initial natural spectral width of the photon beam fulfills the condition of the coherence length. One should take into account that this wave picture is valid as shown experimentally even for the case for which only single photons are used and where the question of whether a photon goes one way or the other is then a question of probability.

The same appears with the interference of waves of electrons, e.g., behind a diffraction screen, as explained by Dirac (1931). Dirac included the condition

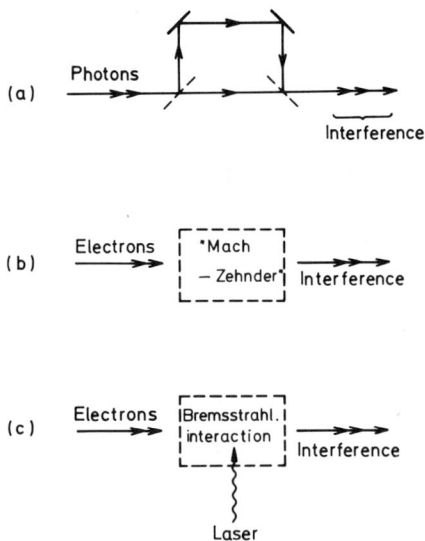

FIG. 2.    Fields of interfering waves: (a) photons passing a Mach–Zehnder interferometer, (b) electrons at an interferometer experiment (Hora, 1970), and (c) electrons where the interferometer experiment is done by a bremsstrahlung process using lasers in the presence of molecules or a crystal.

of coherence lengths as just mentioned in the classical optical sense, where the coherence condition (Hora, 1970) must be fulfilled (Hawkes 1974a,b). Apart from the usual first-order interference, it has been shown that second-order correlation can arrive at interferences even at much larger lengths $L$ than the usual coherence lengths predicted (Hanbury-Brown and Twiss effect (1959; Paul, 1986). It may well be suggested that this second-order correlation is acting also in the interference of electron waves such that the wave properties of the electrons are even stronger and farther reaching than in the modest assumption of the simple first-order correlation coherence conditions—used in the following only—when the wave picture of the electrons have to be added to the their (particle picture) bremsstrahlung processes.

One consequence is then shown in Fig. 2b and c, where an electron beam undergoing a process of interaction as in the Mach–Zehnder interferometer (Fig. 2b) or as a modulation process by the recoil partner in the crossing area of an electron (Hora, 1970) and a laser beam will be able to produce the interference field (Fig. 2c). There the interaction of the modulated electron wave field with ions or molecules can then be expected to result in the emission of a photon of the energy of the modulation.

## A. Molecules in the Interaction Area

In order to extend bremsstrahlung processes from the interaction of electrons with ions to the case of the interaction of electrons with neutral atoms, Geltman (1973) derived the relevant theory. This was the motivation to directly reproduce the mechanisms involved when Andrick and Langhans (1976) prepared an extensive measurement in order to prove that processes Fig. 1b and c, the (induced) absorption and the stimulated emission of bremsstrahlung by incident monochromatic and coherent radiation from a laser into an electron beam in the presence of neutral molecules (instead of the ions in Fig. 1) were interaction partners. The relation to quantum modulation is then given by the fact that the essential properties of the results of the interaction are identical with the result when using a crystal in the interaction area (Schwarz and Hora, 1979; Schwarz, 1971b).

The experiment for the interaction of an electron beam with a crossing laser beam using gaseous argon atoms or other molecules within the crossing area for providing the necessary recoil for the interaction is shown schematically in Fig. 3. In all cases reported by Andrick and Langhans (1976, 1978) and by Weingartshofer et al. (1977, 1979, 1985), the beam was of a cw carbon dioxide laser with an intensity of $3 \times 10^3$ W/cm$^2$ in the first case and with about 1 $\mu$s pulse duration from a TEA laser (Beaulieu, 1970, 1972) with intensities in the interaction area ranging from $10^6$ W/cm$^2$ up to more than 100 times higher in the later cases.

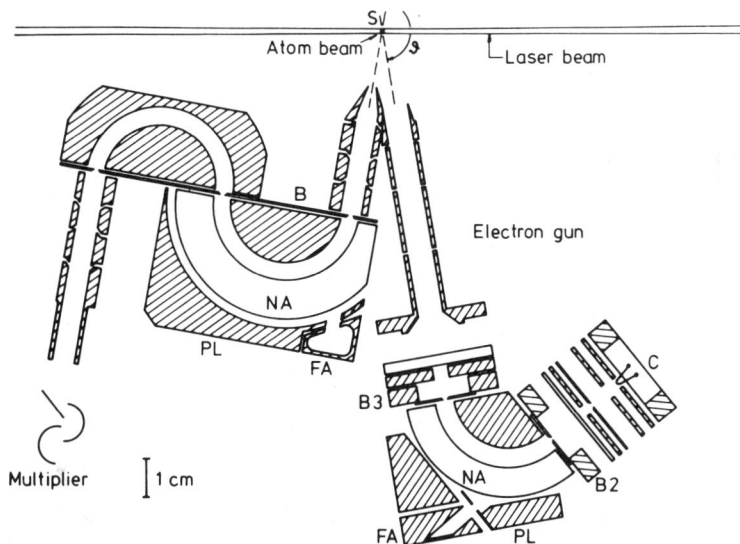

FIG. 3. Experimental apparatus used by Andrick and Langhans (1976) for the electron modulation by a $CO_2$ laser beam in the presence of gas molecules.

The electron beam had an energy of 11 eV, which had an energy spread of only 55 meV. The interaction area between the laser and the electron beams was usually filled with argon gas atoms, but other gases were also treated, e.g., hydrogen molecules in the case of Weingartshofer et al. The electron beam was scattered by the argon (or other) atoms (respectively, molecules) in the usual way, where the detectors for the scattered electrons were placed in such a direction that the scattering by an angle u could be measured. Figure 4 shows the energy spectrum of the electrons as measured by Weingartshofer et al. (1979), where in Fig. 4D the laser was switched off such that the natural spectral width of the electron beam of 55 meV was detected. When the laser pulse was acting, the spectrum had the character shown in Fig. 4a–c. The beam had then been changed by upshifting or downshifting the electron energy by multiples of the photon energy of the laser beam of 0.116 eV.

The scattering angle u was chosen in many of the cases to be 155°, but other angles were also used. The selection of the parameters was suggested by the theory of Kroll and Watson (1973), further evaluated by Krüger and Jung (1978) and Jung and Krüger (1981), since the cross section increases with the fourth power of the laser wavelength, although the smaller photon energy then required a much higher resolution of the electron beam energy, which, however, was a technique available since the beginning of the 1970s. Also the large scattering angle was chosen from the strong effect predicted by theoretical results.

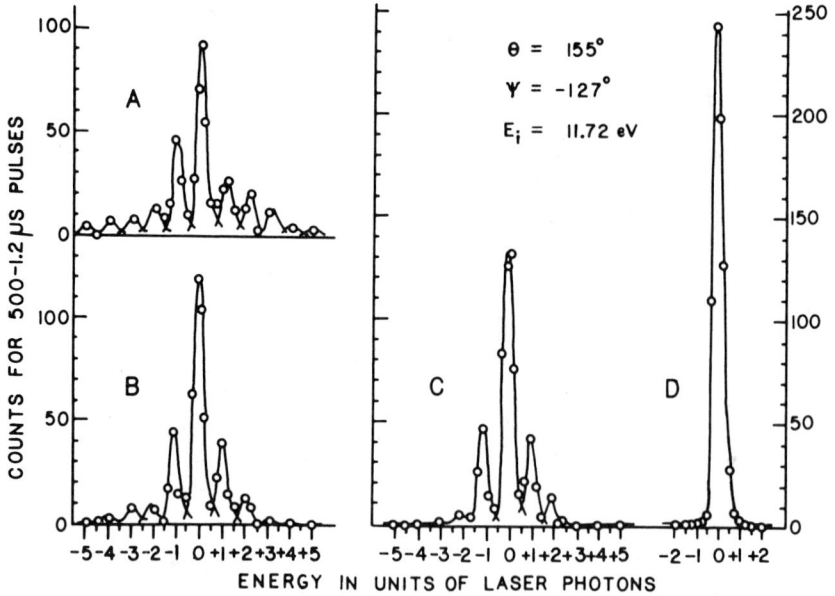

FIG. 4.  Energy spectrum of the electrons after interaction with a $CO_2$ laser photon of energy $\hbar\omega = 0.116$ eV (A)–(C) in the presence of argon atoms (modulation) and (D) before the interaction measured by Weingartshofer et al. (1979).

It should be mentioned that the theory of Kroll and Watson (1973) arrived at a cross section which was proportional to

$$\sigma \sim J_v^2(\mathbf{A} \cdot \mathbf{Q}) \tag{6}$$

where $J_v$ is a Bessel function of the first kind with the index $v$, which is given by the number of photons involved for the upshift or downshift of the electron energy. Its argument is the scalar product of the expression for the dipole moment $\mathbf{A}$ induced by the laser radiation in the atom or molecule; therefore, it is proportional to the electric field vector of the laser light multiplied by the difference $\mathbf{Q}$ of the momentum ($\mathbf{p}$) vectors of the electrons in the final state (f) and in the initial state (i), where momentum conservation requires that

$$\mathbf{Q} = \mathbf{p}_f - \mathbf{p}_i; \qquad |\mathbf{p}_i| = \sqrt{2mE_0}; \qquad |\mathbf{Q}| = \sqrt{2mh\omega} \tag{7}$$

using the electron mass $m$ and the energy of one laser photon $hv = \hbar\omega$.

It should be noted from these results that the interaction of the laser beam with the electrons did not significantly change the scattering direction of the electron by the molecules compared with the nonirradiated case. The relation of the polarization dependence of the effect as given by Eq. (6) was implicitly confirmed by reporting an exponential decay on the polarization.

Since this article contains some critical remarks about experiments and expresses some hesitation when accepting experimental results, it should be noted that one of the well-known experts on the theory in this field at an Austrian university was expressing a great deal of skepticism as to whether one can trust the reported experimental results of Weingartshofer *et al.* When looking at the result reported in Fig. 4 similar to other results mentioned in other reports, it is curious that such a small number of measuring points are given, well noting that the measurement of each point required several hours. It should be mentioned that in the very first measurement of the interaction effect between the electron beam and the laser beam in the presence of the molecules (Andrick and Langhans, 1976) a very large number of experimental points was given (Fig. 5), very precisely showing the effect. It was mentioned that gaining this result required a measuring time in the order of 100 hours. This first measurement, part of a PhD project at a prominent research group in this field (H. Ehrhardt), should not allow any doubt by the persons mentioned before.

Figures 6 and 7 describe the details of the experimental aparatuses used. With respect to the extensive measuring time and the experienced agreement with the quite transparent theory may justify when the theoretical curves in Fig. 4 are used for exploring the effect even if only important points are then measured.

It should be mentioned that the measurements described in this subsection were probably not intended to study the quantum modulation effect and that the motivation arose from the question of studying the inverse bremsstrahlung interaction with un-ionized gaseous atoms or molecules as interaction partners instead of ions in the classical cases of bremsstrahlung studies for

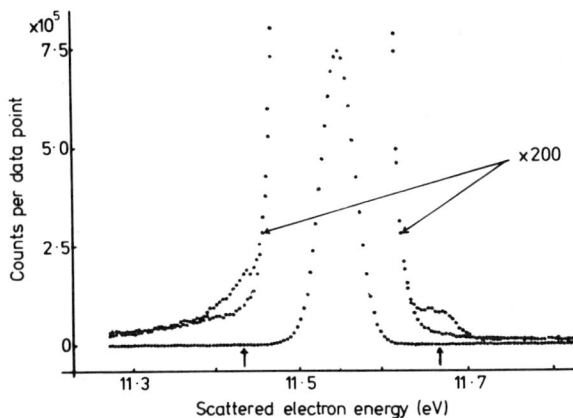

FIG. 5. Detailed measurement of the energy spectrum of the modulated electron beam (Andrick and Langhans, 1976).

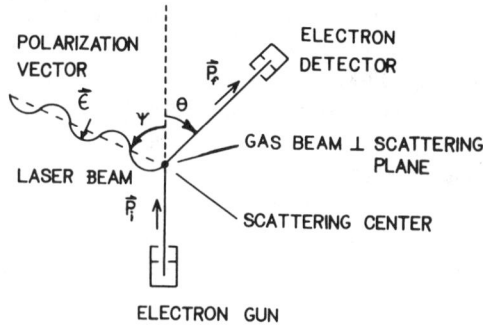

FIG. 6. Scheme of the apparatus used for the modulation experiment by Weingartshofer *et al.* (1979).

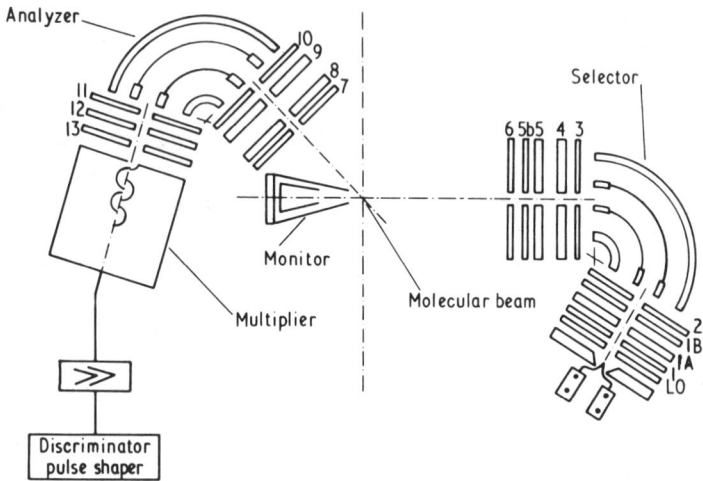

FIG. 7. Diagram of the electron spectrometer used by Weingartshofer *et al.* (1974).

plasma and astrophysics. The design of the experiments was also chosen to be extremely transparent for studying free–free electron transitions well isolated from other effects, which in most other low-energy electron-beam scattering experiments with molecules cannot be isolated and cause many undesirable complications.

## B. Crystals in the Interaction Area

The following similar experiments of the interaction of electron beams with laser beams with a recoil material in the crossing area—in this case crystals—were not primarily motivated by studying bremsstrahlung processes, though their similar results as described in the preceding subsection may suggest this analogy. The motivation was more based on the wave picture of the electrons and not from the particle picture of the bremsstrahlung process.

Helmut Schwarz was motivated by the desire to explore the analogies between optical waves and electron waves by several occasions where the oral reports may trace back to his discussions with Max von Laue in 1943–1944 and by interaction with Dennis Gabor at the CBS Laboratories in Stamford, Connecticut, where Schwarz was consulting. It is not an accident that Gabor (1949) derived the discovery of holography first for the case of electron waves, while success came after the invention of the laser from the optical case in the well-known visible result. How the question of the analogy between optical and electron waves is still essential in the thinking of members of the CBS Laboratory today, long after the death of Gabor, may be seen in Caulfield's article (1984).

As will be discussed later, however, the work on the quantum modulation effect was one of the drastic examples of the difference between optical and electron waves in second order (Hora, 1972).

The motivation of Schwarz was to show how the de Broglie wave of a 50 keV electron with a frequency of $1.2 \times 10^{19}$ Hz can receive a modulation frequency of visible laser light with frequencies about $10^5$ times less. The mechanism is then again the interaction of a laser beam with an electron beam where the upshift or downshift of the energy of the electron wave is in multiples of the photon energy of the laser beam. As a necessary interaction partner for fitting the momentum transfer, a crystal was to be used in the crossing area of the beams. Instead of measuring the upshift or downshift of the electron energy as has been done with argon atoms as interaction partners described in the preceding subsection, Schwarz expected that the superposition of the coherently modified electron waves should arrive at a state of a beat wave. The reproduction of the upshift and of the downshift of the electron wave energy will then be reproduced as optical radiation with the laser frequency when the scattered electron beam hits a nonluminescent target. The entire mechanism as quantum modulation had to be clarified by this consideration (Schwarz and Hora, 1969), as discussed further in Section III.

The scheme of the experiment is described in Fig. 8. An electron beam of 50 KeV energy crossed perpendicular to a 10 W dc argon laser beam focused to about $10^7$ W/cm$^2$ of linear polarization, whose oscillation direction of the electric vector E was parallel to the electron beam.

For the analysis of the scattered electron beam it was then possible, instead of the analyzer of the electron energy spectrum, to use a nonluminescent screen and to observe the emission of the green argon laser light. The crystal used in the crossing area was quartz, but strontium titanate and other crystals

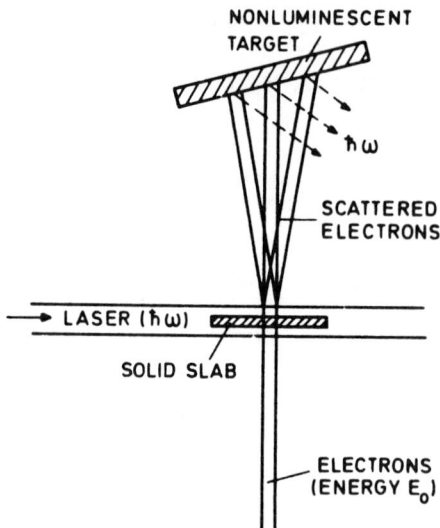

FIG. 8. Scheme of the Schwarz–Hora effect of quantum modulation of an electron beam by a laser beam in the presence of a crystal field (Hora, 1972).

were also used with similar results. Without the laser light, the photography of a luminescent screen showed the electron diffraction pattern of the crystal, while, using a nonluminescent screen with the laser light switched on, the picture of the diffraction pattern was photographed with spots at the same points as the initially detected diffraction pattern. Figure 9 (Schwarz and Hora, 1969) indicates then that the spectrum as in Fig. 4 was not measured by an electron energy analyzer but by the nonluminescent screen from the modulation state of the electrons.

It is remarkable to note that the scattering direction by the crystal was not changed essentially by the interaction with the laser light, similar to the result that the direction was nearly unchanged when scattered by the argon molecules, as described previously in Subsection II,A.

The condition that the interaction has to be described quantum mechanically and not classically, e.g., by a bunching of the waves, was established (Schwarz and Hora, 1969) by considering the length and the momentum of the electrons at oscillation in the laser field. If this product had been much larger than Planck's constant, the classical description would have been appropriate. Since the experimental conditions used laser fields which were not large enough to reach this classical range, the modulation process has to be described quantum mechanically. The limit of the laser field strength (Schwarz and Hora, 1969) is

$$|\mathbf{E}| = 1.02 \times 10^8 \quad V/am \qquad (8)$$

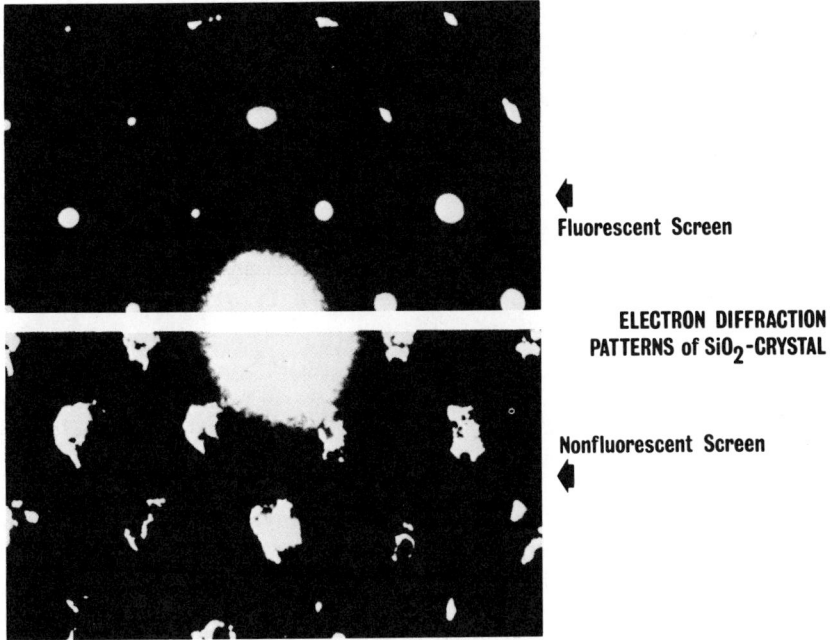

FIG. 9. Photograph of emitted light from diffraction patterns produced by the 50 keV electron beam behind the quartz crystal (Schwarz and Hora, 1969). The upper half is from a luminescent screen, and the lower half from nonluminescent alumina in the color of the laser light. It disappears when the laser is switched off.

below which is the quantum-mechanical case for the 4880 Å argon laser wavelength. This limit results in a laser intensity of $1.43 \times 10^{13}$ W/cm$^2$.

As the limit of the intensity depends on the third power of the frequency for carbon dioxide laser wavelengths, the limit is at $10^{10}$ W/cm$^2$ then. This shows that the measurements with the gas as interaction partner (Weingartshofer et al., 1985) that the used intensities are within the quantum mechanical range too. The decay of their signal from $10^7$ W/cm$^2$ to one-third of the signal at ten times higher intensity may indicate that the signal for their quantum mechanical modulation may go to zero at the limit of $10^{10}$ W/cm$^2$ (see Section III).

The process of modulation in the case of the use of crystals can then be described by the superposition of electron waves

$$\Psi = A_0 \Psi(E_0) + A_1 \Psi(E_0 + \hbar\omega) + A_{-1} \Psi(E_0 - \hbar\omega) + A_2 \Psi(E_0 + 2\hbar\omega) + \cdots \quad (9)$$

of varying amplitude for different values of $n$. $E_0$ is the mean energy of the incident electron beam, and the distance $x$ behind the crystal has to be taken in appropriate geometry from the interaction area.

The difference from optical waves consists in the fact that the factor $x$, the $k$ vector, or wave vector, does not depend linearly on the energy as in the optical case but on the square root of the energy. This then leads to an essential difference between the two waves. While the superposition of two waves leads to beating with a wavelength

$$\lambda_{bn} = \lambda_e E_0/\hbar\omega \tag{10}$$

using the de Broglie wavelength $\lambda_e = 2\pi\hbar/(2mE_0)^{1/2}$, in full analogy to the optical beating process, the mentioned square-root dependence of the energy, however, results in a further long beating wavelength (which does not exist in the optical case) of the value

$$\Lambda_{bn} = 8\lambda_e(E_0/\hbar\omega)^2 \tag{11}$$

This second-order difference between the different wave types has been evaluated (Hora, 1972), and it was found by P. Gräff (personal communication) that it can be used for the nonrelativistic electron wave to derive the Lie generator of the Lorentz transform; it obviously has an essential meaning for the nonrelativistic formulation of quantum mechanics.

The first indication of the long beating wavelength [Eq. (11)] was shown in the first derivation of quantum modulation from scattering theory using the Born approximation (Salat, 1970). Immediately when Schwarz received this manuscript for review he performed a measurement by varying the distance $D$ between the crystal and the nonluminescent screen. The result (Fig. 10) showed a strong variation of the intensity of the radiation from the screen, where a long beating wavelength of

$$\Lambda_{bn} = 1.73(0.01) \quad \text{cm} \tag{12}$$

was measured. The nonrelativistic theory would have concluded a wavelength of 1.6858 cm, while the relativistic treatment (Van Zandt and Meyer, 1970) resulted in 1.646 cm. Based on the Dirac equation and the use of the Varshalovich–Diakonov (1970, 1971b) correction for the dielectric material

$$\Lambda_0 = \frac{8h}{[2mE_0(1 + E_0/2mc^2)]^{1/2}}\left(\frac{E}{\hbar\omega}\right)^2 \frac{1}{[1 - (v^2/c^2)]^{1/2}(1 - c^2/c_M^2)} \tag{13}$$

and further with the use of the Peierls formula for the speed of light $c_M$ in the solid material (Hora, 1977)

$$\left|\frac{c_M}{c}\right| = \frac{\tilde{n}^2 - 1}{\tilde{n}}[1 + \tfrac{1}{5}(\tilde{n}^2 - 1)] \tag{14}$$

with the refractive index $\tilde{n} = 1.58$, a long beating wavelength of 1.740 cm is derived (see Section V).

The first-order coherence theory (Hora, 1970), which did not take into account that the second-order correlation theory may provide much larger coherence lengths [as in the optical case of the Hanbury-Brown–Twiss effect (Paul, 1986)], arrives at a coherence length and a "washing out" of the interference patterns due to the energy width of the initial electron beam of the value $\Delta E$. This results then in the long beating modulation pattern decay on the increasing distance $D$ between crystal and screen, where a half-width for this decay is given by (Hora, 1975) (see Section IV)

$$\Lambda_n^{mod} = \Lambda_{bn} \frac{\hbar\omega}{8\,\Delta E} \tag{15}$$

expressed by the long beating wavelength [Eq. (11)]. Obviously the energetic bandwidth of the electron beam $\Delta E$ has to be very small. Schwarz published the envelopes of the decay of the patterns (Fig. 10) from which an energetic bandwidth in the range of 100 meV or less for the various cases can be indirectly concluded (Favro and Kuo, 1971b) (Fig. 11).

A further result reported by Schwarz (Hora and Schwarz, 1973) is the dependence of the modulation signal from the nonluminescent screen on the polarization direction of the optical radiation, which results in a highly exponential decay of the modulation signal (Fig. 12). This was rather a

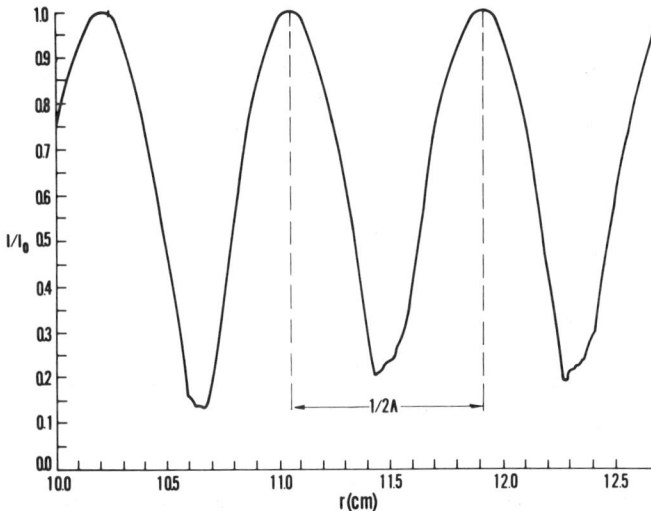

FIG. 10. The optical emission intensity from the nonluminescent screen (as in Fig. 9) for varying distance $D = r$ between crystal and screen. The minima correspond to the long wave beating according to Eq. (11) (Schwarz, 1971b).

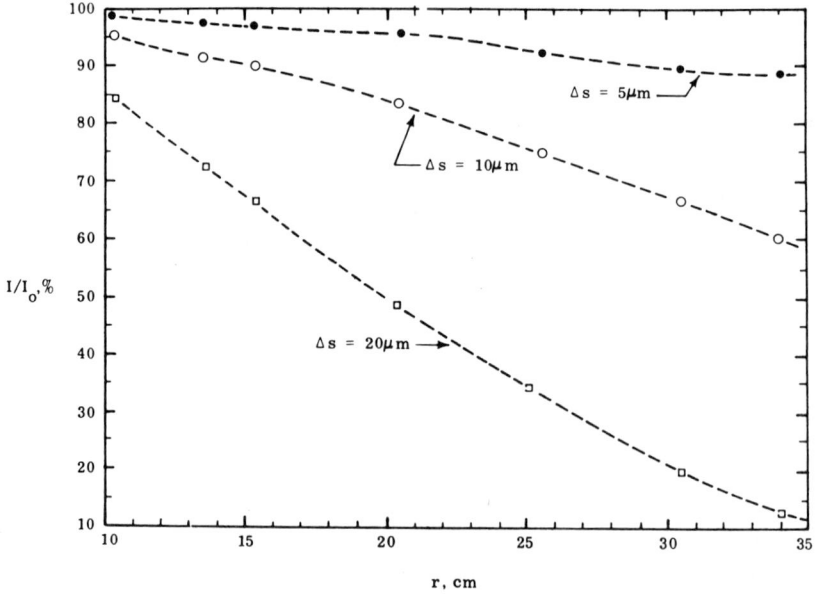

Fig. 11.   Envelopes of the curves for the decay of the modulation signal of increasing $D = x$ [Eq. (15)] for different energy widths $\Delta E$ of the electron beam before interaction with the crystal. Fitting the curves with the measurements allows a determination of $\Delta E$ (Schwarz, 1972a; Kuo, 1971a,b; Favro et al., 1971).

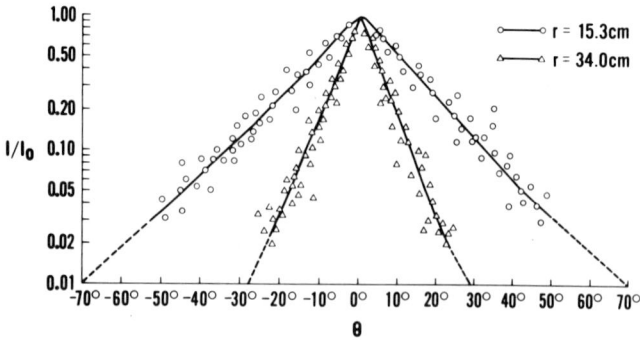

Fig. 12.   Measurement ($\bigcirc$) of the modulation intensity (optical signal from nonluminescent screen) depending on the angle of polarization between the E vector of the laser beam and the direction of the electron beam (Hora and Schwarz, 1973).

surprise, since the curves could in no way be fitted with any $\cos^2$ or higher-power law. The points of measurement, however, correspond very well to the approximately exponential decay given by the Bessel functions of Eq. (6), as the interaction mechanism in the crystal material is of the same kind as that for the dipoles of the gaseous atoms of the cases discussed previously. A more detailed analysis for comparing the exact functions is possible only if more details of the directions of the scattering and the momentum transfer of the experimental arrangement are known. The essential exponential type of decay, however, is a strong proof of the application of the Kroll and Watson (1973) model.

A large amount of scepticism toward the results of Helmut Schwarz and his ideas arose when—after the strong reaction of theoretical exploration of the quantum modulation effect (called then the Schwarz–Hora effect by numerous authors; Physics Abstracts reserved a special entry under this heading)—other authors were not successful in reproducing the experiments. One problem was the necessity for the use of an incident 50 keV electron beam with sufficiently low energy spread in order to prevent the "washing out" of the interference process at too small a distance $D$, as expressed by the modulation length $\Lambda_n^{mod}$ in Eq. (15). The report of negative results (Hadley et al., 1971b; Pfeiffer et al., 1972) led to a verbal reply only (Schwarz, 1971a, 1972b), which did not contribute significantly to a clarification of the facts. In order to achieve an objective measure of the physical processes also in connection with the later experiments on quantum modulation using molecules as described by Andrick and Langhans (1976) and by Weingartshofer et al. (1977) in the preceding subsection, the following facts should be considered:

(1) In both cases, the modulation is a quantum process and may be considered as a bremsstrahlung mechanism, in which the common result is that the electron energy is upshifted or downshifted by one or more laser photons. This is measured in the case of molecules as the interaction partner directly from the electron spectra, and in the case of crystals from the interference patterns of the modulated electrons.

(2) In both cases the scattering angle is not markedly changed by the interaction.

(3) For the polarization dependence there is an exponential variation on the angle between the electric vector of the laser light and the direction of the electron beam (momentum after the scattering) as predicted by Kroll and Watson (1973). This was observed later in the case of the molecule interaction (Weingartshofer et al., 1977) and was observed with crystals prior to 1971 (Hora and Schwarz, 1973).

In defense of Schwarz,[1] one should take into account that, although he used to publish experimental results mainly after he learned theoretical findings (e.g., the long beating wavelength), he could not have known about the

---

[1] One of the authors (H. H.) would like to note that he was unable to observe Schwarz' experiment while it was in progress. The arguments in favor of the results reported by Schwarz are drawn here in an attempt to reach an objective clarification.

polarization dependence according to the Bessel functions. Also with respect to the long beating wavelength, his measurement which was much larger than the classical value despite the fact that relativistic considerations required a shorter value than the classical, it could not have been predicted that Schwarz's unexpectedly high value finally fit the consequent derivation on the basis of the Dirac equation (Hora, 1977), mutually confirming the dielectric response of the crystal according to the model by Varshalovich and Dyakonov (1970, 1971b) if the dispersion relation of Peierls (1976) is used. This may be considered as a mutual confirmation of these models and the reported experiments of Schwarz, although questions may still be raised about the accuracy of these values (Hawkes, 1978).

### III. The Quantum Property of Modulation Derived from a Correspondence Principle for Electromagnetic Interaction

One essential theoretical aspect elaborated in the initial paper on the modulation effect (Schwarz and Hora, 1969) was the clarification that the interaction of the electron beam with the crossing laser beam in the presence of the crystal field was of a quantum-mechanical nature and not a classical bunching (Oliver and Cutler, 1970). Distinguishing between the classical and the quantum range was based on a criterion which can be considered as a correspondence principle for the electrodynamic interaction, which is different from Bohr's (1919) correspondence principle.

The interpretation of the electrodynamic correspondence principle starts from an interpretation of the wave function in the purely statistical sense of Bohr as a distribution function without the unnecessary further extensions of the Copenhagen school (see Appendix A in Hora, 1981b). An electron oscillating in an electromagnetic field in which the maximum momentum $p$ and the maximum elongation $x$ of the quiver motion results in a product $px$ larger than Planck's constant $h$ is considered classical. For this case the classical bunching (Oliver and Cutler, 1970) may be applicable. For the alternative range, the quantum states are defined. The experimental conditions of the modulation effect (Schwarz and Hora, 1969) were in this quantum range.

It turned out from the recent measurements of Weingartshofer et al. (1985) that exactly this distinguishing value of the different ranges was reproduced directly from their experiments. We also found other experiments which can be categorized in the same way. This discussion of the correspondence principle for electrodynamic interaction has the advantage that it immediately can be made visible experimentally (as distinguished from Bohr's correspondence principle) by detecting experiments for varying the parameters of electromagnetic intensities across the limiting value. It also gives a direct understanding of why in the range of radio waves the classical conditions are fulfilled and why in the optical range there are the quantum conditions for the usual intensities, where a change to the classical range occurs only at very high laser intensities.

It should be noted that our discussion of the correspondence principle in the experiments mentioned (Schwarz and Hora, 1969) is a purely theoretical question similar to other considerations (Oliver and Cutler, 1970; Becchi and Morpurgo, 1971; Van Zandt and Meyer, 1970; Varshalovich and Dyakonov, 1970) and is not related to the discussion on the experimental reproduction of the scattering experiments of Schwarz (Hadley et al., 1971b; Pfeiffer et al., 1972; Schwarz, 1971a, 1972b). The ranges of the conditions are independent of this discussion and seem to be reasonably supported by the experiments of Weingartshofer et al. (1977, 1979) on low-energy electron scattering with argon atoms in the presence of a laser field. These experiments have been reproduced, e.g., by the extensive initial results of Andrick and Langhans (1976, 1978).

The criterion of the correspondence principle for the interaction of an electron in a laser field (Schwarz and Hora, 1969) was simply based on the classical motion of an electron without (free) or within an interacting other field within a linearly polarized laser field with an amplitude of the electric field strength $\mathbf{E}$ of radian frequency $\omega$. One can then consider the product of the amplitude of the point-mechanical oscillating motion $x$ and the amplitude of the momentum of the motion $p$ to arrive at a value

$$xp = e^2\mathbf{E}^2/m\omega^3 \tag{16}$$

and conclude that the motion is classical if the value of Eq. (16) is considerably larger than $h/4\pi$. The other case can be expected as a quantum-mechanical behavior of the electron. The threshold field strength $E^*$ of the laser radiation, where $xp$ is equal to $h/4\pi$ (or half of this value), results in the threshold intensity $I^*$ for the borderline of the correspondence principle by

$$I^* = \frac{h}{4\pi}\frac{m\omega^3 c}{8\pi e^2} = 1.403 \times 10^9 \quad \text{W/cm}^2 \quad \text{(for a CO}_2\text{ laser)} \tag{17}$$

which was evaluated for carbon dioxide laser radiation. It should be noted that the choice of $h/4\pi$ was only with respect to the minimum distribution functions for $x$ and $p$ spectra (uncertainty relation).

We have left the question open as to whether the value $h$ only would have had to be used, as in the de Broglie wavelength. Further, the question is not discussed as to why in Eq. (16) the maximum values of the quantities of the quiver motion have been used and not the most probable values. This discussion may cause an inaccuracy of the following within an order of magnitude, which we left open. Nevertheless, the following fit with the experiment seems to be best with the assumptions described previously.

What can be understood from the experiment of Weingartshofer *et al.* (1977, 1985) is that the amplitude of the scattering signal of the energetically unchanged electrons when turned into a scattering direction by the interaction with argon atoms in the presence of a laser field (described as a bremsstrahlung mechanism with the laser-induced dipoles of the atoms as interaction partners, Kruger and Jung, 1978) is a decrease of the relative signal with increasing carbon dioxide laser intensity, as shown in Fig. 13. Though it is not at all defined what kind of law may be behind the connection of the measured points, whether it is some saturation formula, etc., the dashed line drawn would correspond to a preliminarily assumed logarithmic relation. It is remarkable now that this curve points to the abscissa of the diagram close to the intensity of Eq. (17). This well confirms that the measurements of Weingartshofer *et al.* were in the quantum regime as well as for the preceding

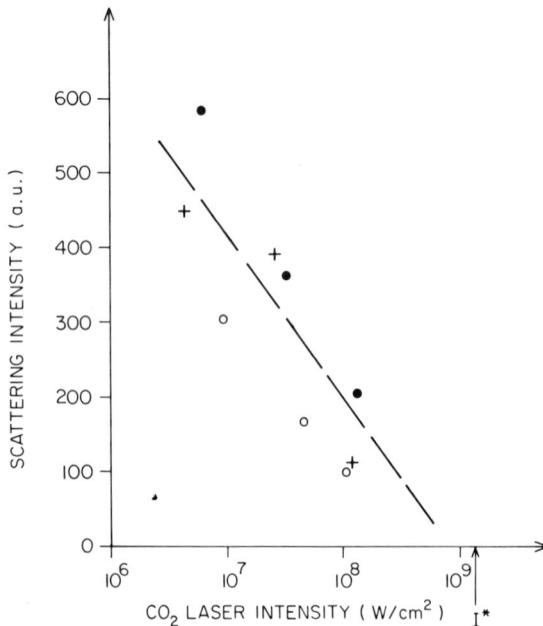

FIG. 13. Measured (Weingartshofer *et al.*, 1985) scattering intensity of electrons after 155° deflection from argon atoms in the presence of carbon dioxide laser radiation of varying intensity. $I^*$ is the intensity of the correspondence principle, Eq. (17), below which the range of quantum-mechanical processes is defined.

experiment of the same process by Andrick and Langhans (1976, 1978), where the laser intensity of about $3 \times 10^3$ W/cm$^2$ is in the quantum regime. Also, the similar scattering experiment using a crystal instead of the atoms for the wavelength of the green argon laser line (Schwarz and Hora, 1969) was in the quantum regime, as we pointed out as a basic relation to confirm the quantum regime.

The question will be, for further experiments of the kind of Weingartshofer *et al.* (1977, 1985) and Andrick and Langhans (1976, 1978), whether there will be a drastic change of the result if the laser intensity is changed from the intensity range below the correspondence principle to the range above.

An experiment is available where possibly both sides from the limit of the correspondence principle are known. For neodymium glass laser radiation, the correspondence limit of the laser intensity is $1.4 \times 10^{12}$ W/cm$^2$. At intensities of $10^{15}$–$10^{16}$ W/cm$^2$—therefore in the classical regime—the ionization process of low-density helium atoms in the laser focus has been measured, and the subsequent ejection of the electrons from the center of the focus in a radial direction by the nonlinear forces (Hora, 1969, 1981b) (of which a special type are the ponderomotive forces) has been detected (Boreham and Hora, 1979; Boreham and Luther-Davies, 1979). The maximum electron energies achieved by the nonlinear force are between 100 and 1000 eV and correspond fully to the transfer of the maximum quiver energy of the electrons in the laser field, of which half is converted into the kinetic energy of radial translation. The electrostatic Debye-length mechanisms are essentially the same as derived later for high-density plasmas (Hora *et al.*, 1984). The ionization process of the atom in the focus can be followed from the temporal sequence of the ionization of the first and then of the second helium electron, fulfilling exactly the Keldysh theory of tunnel-like ionization (Baldwin and Boreham, 1981).

Contrary to this case, for laser intensities of $1.2 \times 10^{12}$ W/cm$^2$ (or probably less because of having no single-mode laser operation and not achieving the diffraction limit), therefore below the threshold of the correspondence principle, the same experiment was performed. Thanks to the extreme resolution of the energy spectrum of the emitted electrons independent of the initial direction using the van der Wiel analyzer, it has been derived from the periodicity of the energy spectrum (Kruit *et al.*, 1983) that the ionization of the electrons was due to a multiphoton effect (Bepp and Gold, 1966), which did not even disturb the nonlinear force acceleration. We assume that this case (Kruit *et al.*, 1983) is basically different with respect to the correspondence principle for laser interaction from the previously mentioned case (Boreham *et al.*, 1979). It is also understandable that the multiphoton process is a typical quantum process for ionization at lower intensities (Kruit *et al.*, 1983), while the Keldysh process (Baldwin and Boreham, 1981) at higher intensities is rather a mechanism closer to the classical range of the correspondence principle.

In order to understand why classical motion of electrons exists in antennas at radio frequencies for generating radiation at typical intensities, whereas in the optical range one has the quantum limit if one does not use the extremely high intensities of lasers, Fig. 14 shows the limit intensity $I^*$ for the correspondence principle for the laser radiation

$$I^* = 1.66/\lambda^3 \quad W/cm^2 \tag{18}$$

where the wavelength $\lambda$ is given in cm. Further, for comparison the laser intensity is given, where the oscillation of the electrons or that of the ions in the laser fields reaches the rest-mass energy $Mc^2$ of these particles. The correspondence principle limit is then equal to the relativistic electron oscillation at a wavelength of $4\pi/3$ of the Compton wavelength $\lambda_e$. For this and shorter wavelengths, any classical oscillation of electrons in the very high-intensity laser field can only be relativistic. For blackbody radiation at such relativistic intensities, the Fermi statistics of electrons breaks down (Hora, 1978; Eliezer et al., 1986).

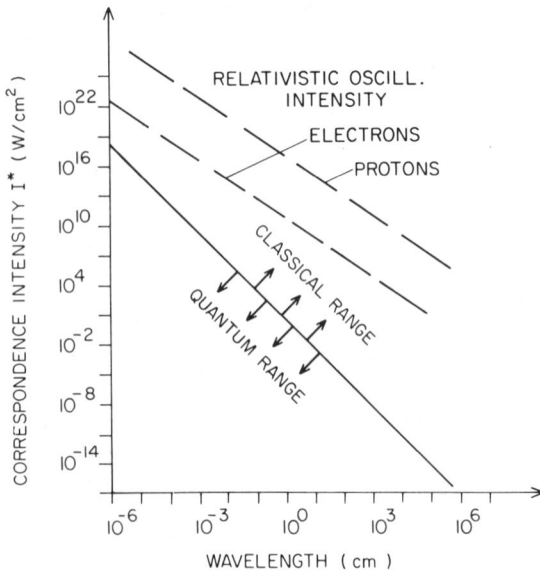

FIG. 14. Intensity $I^*$ of electromagnetic radiation according to the correspondence principle for various wavelengths dividing the classical and the quantum range. $(---)$ give the electromagnetic intensities where the oscillation energy of the electron or proton is equal $Mc^2$ ($M \sim$ mass of the particles).

Following Fig. 14 one may also be able tentatively to comment on the measurement of an amplification of 70 GHz microwave radiation in a waveguide of 0.2 cm width and mW power from a diode as a source when transmitting a highly nonequilibrium gas discharge (Rosenberg *et al.*, 1982). According to Fig. 14, the conditions are in the quantum range. The repetition of this experiment using a Klystron source 150 times more powerful (Palmer, 1983) seems not to show the amplification. The conditions of this experiment are then just on the other side of the limit of the correspondence principle in the classical range and the negative result should not be surprising.

The transition from one range to the other in present-day experiments for the correspondence principle for electromagnetic radiation may be detected in a much more drastic and easier way than was possible by the very complex spectroscopy in the case of Bohr's correspondence principle.

## IV. COHERENCE EFFECTS

The question of the coherence conditions in the quantum modulation of electron beams was treated from the beginning of the theoretical discussion (Hora, 1970; Kondo, 1971). It was emphasized in a review on coherence in electron optics (Hawkes, 1978) that the applications of the theory to electron interference, electron holography, and the Schwarz–Hora effect are not only of interest but also result in "unconventional situations." While the theory of coherence for the case of optics is presented in a large amount of literature and in excellent textbooks, Hawkes (1978) states that "comparatively few papers only have been devoted specifically to the problem of coherence in electron optics." The standard formulation for electron waves with similar notation as for the optical case given by Born and Wolf (1959) for the autocorrelation function and the self-coherence is developed starting from Glaser's (1952) formulations of integrating the components of the wave function in the object plane by a convolution. The main evaluations, however, are based on pointlike sources with ideally monochromatic waves.

Hawkes (1978) emphasizes from the beginning that in electron microscopy the question is very critical that there is an energy spread of the incident electron beam. For achieving the highest possible resolution in the microscope the energy spread and the involved coherence properties were of higher importance than spherical aberration and other errors and obstacles to achieving ideal resolution.

Hawkes (1978) further states that there is "an extremely important difference between the light and electron optical coherence theory," where reference is given to the difference between real functions (in the optical case) and the complex functions in the electron optical case, with Fourier transforms in the real part and Hilbert transforms in the imaginary part. This results in problems of partial coherence and in the fact that the electron wave case has a one-sided transform only with respect to time. The problems of interference and correlation in photon and electron optics (Chang and Stehle, 1972) were discussed in connection with the initial interference experiments of Young (1807) and including the very complex question of the Forrester–Gudmundsen effect (Forrester *et al.*, 1955). The connection of the scattering mechanism of quantum modulation with the Raman effect was discussed by Sinha (1972).

Problems arising from the discussion of the quantum modulation of electrons support this view of the difference between the optical and the electron optical case, where a basic second-order difference exists in the wave functions (Hora, 1972; P. Gräff, personal communication). Furthermore, the problem of the correlation of the second order has not been exhausted in the theory of correlation of electron optical coherence, which in the optical case led to the Hanbury-Brown–Twiss effect (1959) and to the phenomenon of photon bunching and antibunching (Paul, 1986).

Interference effects of quantum modulation (Schmieder, 1972) were also discussed as double modulation and were considered along these lines by Bergmann (1973).

For the coherence in the modulation effect there are different questions to be distinguished. Before we look into the problem of the correlation and the interference patterns and their washing-out behavior, the first question is that the electron in a beam with a Gaussian energy spread $\Delta E$ around an energy $E_0$ does not have [in first-order treatment, if not a higher correlation (Hanbury-Brown and Twiss, 1959) or the $1/f$ noise (Handel, 1975) is considered] a self-interference. The components of the wave function do not permit self-interference from the conditions of the generation of the beam from a filament or from a field-emission process, etc. That is similar to an optical beam deriving from one source. Only if there is a spatial shift of a part of the beam and a reunification with the other part, is an interference pattern produced, as long as the condition of the coherence is fulfilled. The same happens for the interference of parts where in one part the frequency is changed.

The last case exists, e.g., in quantum modulation. This could well be caused by the plasmon-produced shift of the energy of the electrons instead of using laser photons, as in the Schwarz–Hora effect (Hora, 1972). In Fig. 15 we consider the energy spectrum of an electron beam before and after a modulating interaction (in this case by plasmons). Only the fact that the changed state of the electrons results in partial beams with a maximum separated from the main maximum and separated with minor overlap of the

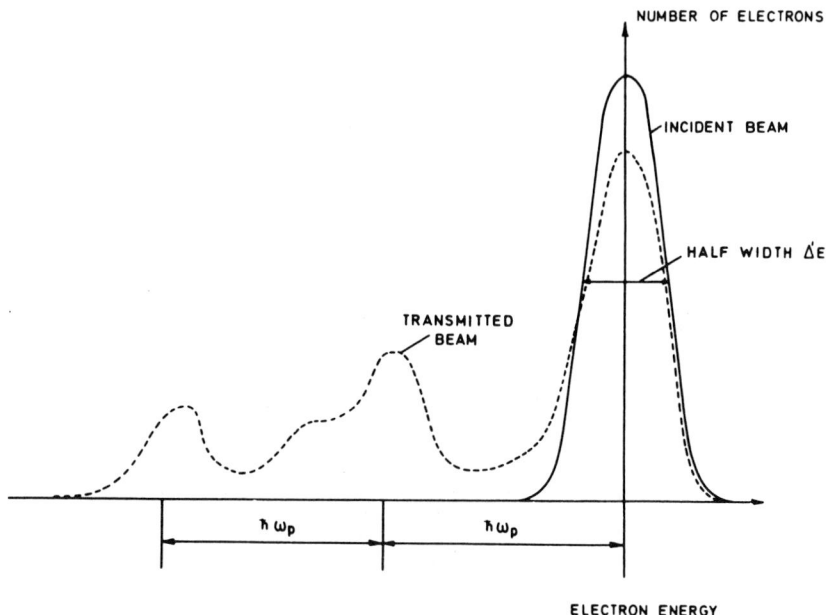

FIG. 15.   Plasmon modulation of the electron beam changes the energy spectrum of the incident electrons into that of transmitted (schematically) (Ruthemann, 1941; Möllenstedt, 1949). The maxima are fairly related to energy losses $\Delta E = \omega_p$ given by the plasmon frequency $\omega_p$ of the slab (Bohm and Pines, 1953). The half-width of the incident bundle is $\Delta E$.

side slopes will result in an interference (beyond the self-interference). This demonstrates that the energy spread of the initial beam is important. The necessary coherence condition is then that the interaction mechanism (by plasmons or by photons in the Schwarz–Hora effect) is temporally or spatially sufficiently concentrated such that the new generated maxima for the energy spectrum are sufficiently separated. This leads to a coherence length $\Delta S$ in order to overcome the self-interference.

The coherence will be a function of the energy width $\Delta E$ of the incident electron beam. The condition of coherence

$$\Delta E \, \Delta t \geqq \hbar/2 \tag{19}$$

is based on a coherence time $\Delta t$ which determines a coherence length $\Delta S$:

$$\Delta S = v \, \Delta t \tag{20}$$

where $v$ is the velocity of the electrons:

$$v = \sqrt{2E_e/m} \tag{21}$$

using the electron mass $m$. Combining Eqs. (19)–(21), the coherence length results in

$$\Delta S \geq \frac{\hbar}{2} \sqrt{\frac{2E_e}{m}} \frac{1}{\Delta E} \tag{22}$$

For the case of the electron bundle of an averaged energy of $E = 50$ keV (Schwarz and Hora, 1969), we find from Eq. (22) (Hora, 1970)

$$\Delta S \geq 1.95 \times 10^{-5} \frac{1}{\Delta E} \quad \text{cm}, \Delta E \text{ in eV} \tag{23}$$

In the experiment the electron beam had an energy width of $\Delta E$ of about, or less than, 0.1 eV, which corresponds fairly well with the temperature of thermionic emission. Therefore, we have for the conditions of the experiment (Schwarz and Hora, 1969) $\Delta S = 2$ $\mu$m. This is large enough compared with the thickness of the slab and with the geometries normal to the electron beam, if the not-too-large angles of diffraction are taken into account.

The case of the modulation of the electron beam by plasmons when penetrating solid slabs (Hora, 1970) touched on the question of the collective interaction of the electron in the solid materials. A further discussion of the collective interaction with Glauber states and many-particle cooperative effects was discussed by Rivlin (1971) and by Makhviladze and Shelepin (1972).

The connection of the problems of Glauber states and photon fluctuations with quantum modulation is further to be seen from the correspondence principle for electromagnetic waves (see the preceding Section III). The fluctuation of photons, bunching and antibunching (Mandel, 1982; Walls, 1983), led to the problem of squeezed states, where a similar distinguishing between quantum and classical states is given as for the electron wave with respect to the correspondence principle (Ben-Aryeh and Mann, 1985). In both cases the distinguishing is due to the quantization, or the atomistic structure of action. Though there is the basic second-order difference between optical waves and electron waves, the case of the electrons with their particle structure is then a useful picture for the understanding of the bunching and antibunching of the photons.

Quantum modulation also extends the knowledge of the properties of the wave state of the electrons. The limited results of the first-order correlations for coherence (Lipkin and Peshkin, 1971) may be considered in a similar way as a first step of the coherence concept as it was in the optical case when the Hanbury-Brown–Twiss effect opened an unexpected extension by the second-order correlation (Paul, 1986). It is indeed very difficult to understand the transfer of a photon by another particle (electron or proton etc.) from one place to another when thinking in the particle picture. In the same way the fact of electron diffraction should not be possible in the particle picture. The quantum-modulation state, however, is a further and a more complex property of the wave picture of the electron than the numerous other known wave properties.

These wave properties of the electron are essentially related to the phase of the electron waves, as demonstrated in the Möllenstedt and Bayh (1961) experiment, which is not necessarily to be interpreted by the Aharonov–Bohm (1959) effect, but simply may be that the vector product of the electron velocity $\mathbf{v}$ and the magnetic flux $\phi$ is quantized:

$$|\mathbf{v} \times \phi|\frac{e}{c} = \hbar$$

which leads to an operator for the magnetic flux (Hora, 1983):

$$\phi = -ice\frac{1}{\alpha}\mathbf{V}_v$$

The interference observed by Möllensteadt and Bayh (1961) is then due only to the fact that $\mathbf{v}$ can never be parallel to $\phi$.

Another phase property of the electrons should be mentioned. The Goos–Haenchen effect for electrons (Hora, 1960) results in a different side shift by a factor of two at grazing incidence at total reflection if the intensity description ($\psi^*\psi$) is used (Renard, 1964), contrary to the use of the phase description (Carter et al., 1971).

We now discuss the further case of coherence and correlation (Hora, 1972) for the superposition of plane waves with separated energy which are permitted to interfere according to a mechanism in which the separation of the partial waves from an initial wave packet fulfills the before-mentioned conditions to overcome self-interference. This is then fulfilled in the cases (1) of the modulation of the electron beam by a laser beam with molecules as the interaction partner (Andrick and Langhans, 1976; Weingartshofer et al., 1977) or (2) with a crystal as an interaction partner (Schwarz and Hora, 1969).

As the result of interaction in both cases (1) and (2), the initial electron energy $E_0$ is up- or downshifted by $\delta E$ (or multiples of this value), and following Hutson (Hutson, 1970) the resulting wave function can be approximately written in the following form:

$$\psi = A\exp\left(\frac{i}{\hbar}(p_0 x - E_0 t)\right) + \exp\left(\frac{i}{\hbar}[(p_0 + p_1)x - (E_0 + \delta E)t]\right)$$

$$+ \exp\left(\frac{i}{\hbar}[(p_0 - p_2)x - (E_0 - \delta E)t]\right) \tag{24}$$

where $E_0 = p_0^2/2m$ represents the energy of the original electron beam,

$\delta E = \hbar\omega$, $\omega$ being the frequency of the laser beam and

$$\delta E = \frac{(p_0 + p_1)^2}{2m} - \frac{p_0^2}{2m} = \frac{2p_0p_1 + p_1^2}{2m} \tag{25a}$$

$$\delta E = \frac{p_0^2}{2m} - \frac{(p_0 - p_2)^2}{2m} = \frac{2p_0p_2 - p_2^2}{2m} \tag{25b}$$

Let $A$ represent the ratio of the amplitude of the resulting wave with the initial energy against that of the up- or downshifted wave. In writing Eq. (24) we have assumed absorption and emission of one photon only. Since, in a practical situation, the electron should be described as a wave packet, we, in this review, assume the electron to be described by a superposition of Gaussian wave packets and from the analysis we explicitly derive the "quantum coherence conditions" that should be satisfied for the modulation effect to be observed.

Before we consider the wave packets we note that if we calculate the probability density using Eq. (24) we obtain

$$\psi^*\psi = A^2 + 4A\cos\left(\frac{(p_2 - p_1)x}{\hbar}\right)\cos\left(\frac{(p_2 + p_1)x}{2\hbar} - \omega t\right)$$
$$+ 4\cos^2\left(\frac{(p_2 + p_1)x}{2\hbar} - \omega t\right) \tag{26}$$

where we have used the relation $\delta E = \hbar\omega$. The second and third terms represent intensity fluctuations at the optical frequency $\omega$ (Van Zandt and Meyer, 1970); further, the periodicity of the amplitude of the second term gives rise to the long beating wavelength $\Lambda_{bn}$ given by

$$\tfrac{1}{2}\Lambda_{bn} = \frac{2\pi\hbar}{(p_2 - p_1)} \cong \frac{8\pi E_0^2}{p_0\hbar\omega^2} = 4\lambda_e\left(\frac{E_0}{\hbar\omega}\right)^2 \tag{27}$$

where

$$\varepsilon = \hbar\omega/E_0 \tag{28}$$

The intensity fluctuations at the frequency $\omega$ would give rise to the excitation of light frequency $\omega$ while interacting with the electrons in the nonluminescent screen (Schwarz and Hora, 1969).

The experiments of Weingartshofer et al. (Weingartshofer et al., 1977) indicate that there is a finite energy width of the electrons and that the width is approximately the same for $E_0$ and $E_0 \pm \hbar\omega$. We thus start with the wave packet

$$\Psi(x, t) = \frac{1}{\sqrt{2\pi\hbar}} \int_{-\infty}^{+\infty} a(p)\exp\left[\frac{i}{\hbar}\left(px - \frac{p^2}{2m}t\right)\right]dp \tag{29}$$

where $a(p)$ represents the wave function in momentum space with $|a(p)|^2\,dp$ representing the probability for the momentum to lie between $p$ and $p + dp$. We may mention that for a simple Gaussian wave packet (with $\langle p \rangle = p_0$) we have

$$\Psi(x,0) = \frac{1}{(\pi\sigma^2)^{1/4}} \exp\left(-\frac{x^2}{2\sigma^2}\right)\exp\left(\frac{i}{\hbar}p_0 x\right) \tag{30}$$

giving

$$a(p) = \frac{1}{\sqrt{2\pi\hbar}}\int_{-\infty}^{+\infty}\Psi(x,0)\exp\left(-\frac{i}{\hbar}px\right)dx$$

$$= \left(\frac{\sigma^2}{\pi\hbar^2}\right)^{1/4}\exp\left(-\frac{\sigma^2}{2\hbar^2}(p - p_0)^2\right) \tag{31}$$

and

$$\Psi(x,t) = \left(\frac{2}{\pi\alpha^2}\right)^{1/4}\exp\left[-\frac{2}{2|\alpha|^2}\left(x - \frac{p_0 t}{m}\right)^2\right]$$

$$\times \exp\left\{\left[\frac{2i}{\hbar}\sigma^4 p_0\left(x - \frac{p_0 t}{2m}\right) - \frac{i\hbar t}{m}x\right]\Big/2|\alpha|^2\right\} \tag{32}$$

where

$$\alpha = \sigma^2 + i\hbar t/m \tag{33}$$

thus

$$\Delta x = \langle x^2 \rangle - \langle x \rangle^2 = \frac{\sigma}{\sqrt{2}}\left[1 + \left(\frac{\hbar^2 t^2}{\alpha^4 m^2}\right)\right]^{1/2} \tag{34}$$

showing the broadening of the packet.

For the electron beam subjected to a modulating electric field we may assume

$$a(p) = C_1\left(\frac{\sigma^2}{\pi\hbar^2}\right)^{1/4}\exp\left(-\frac{\sigma^2}{2\hbar^2}(p - p_0)^2\right)$$

$$+ C_2\left(\frac{\sigma^2}{\hbar^2}\right)^{1/4}\exp\left(-\frac{\sigma^2}{2\hbar^2}[p - (p_0 + p_1)]^2\right)$$

$$+ C_2\left(\frac{\sigma^2}{\pi\hbar^2}\right)^{1/4}\exp\left(-\frac{\sigma^2}{2\hbar^2}[p - (p_0 - p_2)]^2\right) \tag{35}$$

where we have assumed the same $\sigma$ for the three terms—this is indeed what appears to be the case from the experimental data of Weingartshofer *et al.*

(1977); however, the calculations can easily be modified if we assume different $\sigma$'s for the last two terms, only the calculation would become more involved. If we further assume that the Gaussian functions do not overlap greatly, then in calculating $|a(p)|^2$ we may neglect the cross terms; thus the normalization condition would give us

$$|C_1|^2 + 2|C_2|^2 = 1 \tag{36}$$

If we substitute for $a(p)$ from Eq. (35) in Eq. (29) we would get

$$\psi(x,t) = \left(\frac{\sigma^2}{\pi\alpha^2}\right)^{1/4} C_2 \exp\left\{\left[\frac{2i}{\hbar}\sigma^4 p_0\left(x - \frac{p_0 t}{m}\right) - \frac{i\hbar t}{m}x^2\right]\Big/2|\alpha|^2\right\}$$

$$\times (A\Gamma + \Gamma_1 e^{i\beta_1} + \Gamma_2 e^{-i\beta_2}) \tag{37}$$

where

$$A = \frac{C_1}{C_2}; \qquad \beta_{1,2} = \frac{\sigma^4}{\hbar|\alpha|^2}|p_{1,2}x - \delta Et| \tag{38}$$

and $\Gamma$, $\Gamma_1$, and $\Gamma_2$ are Gaussian functions centered around

$$x = \frac{p_0 t}{m}, \qquad \frac{(p_0 + p_1)t}{m}, \qquad \frac{(p_0 - p_2)t}{m}$$

respectively:

$$\Gamma = \exp\left|-\frac{\sigma^2}{2|\alpha|^2}\left(x - \frac{p_0 t}{m}\right)^2\right|$$

$$\Gamma_1 = \exp\left|-\frac{\sigma^2}{2|\alpha|^2}\left(x - \frac{(p_0 + p_1)t}{m}\right)^2\right| \tag{39}$$

$$\Gamma_2 = \exp\left[-\frac{\sigma^2}{2|\alpha|^2}\left(x - \frac{(p_0 - p_2)t}{m}\right)\right]$$

thus

$$\Psi^*\Psi = \frac{\sigma}{\sqrt{\pi}|\alpha|}|C_2|^2|A^2\Gamma^2 + 2A(\Gamma\Gamma_1)\cos\beta_1 + 2A(\Gamma\Gamma_2)\cos\beta_2$$

$$+ \Gamma_1^2 + \Gamma_2^2 + 2\Gamma_1\Gamma_2\cos(\beta_1 + \beta_2)| \tag{40}$$

If we compare the above equation with Eq. (26), we see that we again have the modulation effect due to the terms proportional to $\cos\beta_1$, $\cos\beta_2$, and $\cos(\beta_1 + \beta_2)$; however, the modulation effect will be observed only when $\Gamma$ and $\Gamma_1$ (or $\Gamma$ and $\Gamma_2$) overlap. This will happen when

$$t < \frac{|\alpha|m}{\sigma\,\delta p}\sigma\left|1 + \frac{\hbar^2 t^2}{m^2\sigma^4}\right|^{1/2}\frac{p_0}{\hbar\omega} \tag{41}$$

where we have assumed

$$p_1 \approx p_2 \approx \delta p \approx \frac{m}{p_0} \delta E \approx \frac{mh\omega}{p_0}$$

If $\Delta p_0$ and $\Delta E_0$ represent the momentum and energy spread of the electrons, then

$$\sigma \approx \frac{\hbar}{\Delta p_0} \sim \frac{\hbar}{p_0} \frac{p_0}{\Delta p_0} \frac{\hbar}{p_0} \left(\frac{E_0}{\Delta E_0}\right)$$

Thus Eq. (41) becomes

$$t \lesssim \left(1 + \frac{t^2}{t_0^2}\right)^{1/2} \left(\frac{E_0}{\Delta E_0}\right) \frac{1}{\omega} \tag{42}$$

where $t_0$ is the time in which the wave packet spreads appreciably and is given by (Ghatak et al., 1984)

$$t_0 = \frac{m\sigma^2}{\hbar} \approx \frac{m\hbar}{(\Delta p_0)^2} \approx \frac{\hbar}{2E_0} \left|\frac{E_0}{\Delta E_0}\right|^2 \tag{43}$$

Expressing this by the distance $\Lambda^{mod}$ for the electrons beyond the laser interaction area, this quantum coherence length is given by

$$\Lambda_n^{mod} = t_0 v \approx \tfrac{1}{8} - \Lambda_{bn} \left|\frac{\hbar\omega}{\Delta E_0}\right|^2 \tag{44}$$

where $v = \sqrt{2E_0/m}$ represents the electron velocity.

Case I: $t \ll t_0$. This case describes the condition where the modulation of the electron beam could be detected by its optical frequency density oscillation at a nonluminescent screen. In the case of the experiment (2) with the crystal (Schwarz and Hora, 1969), $\Delta E$ was $\sim 100$ meV and $\hbar\omega = 2.54$ eV. Using $\Lambda_{bn}$ from Eq. (27), in agreement with the theory (Hora, 1970) $\Lambda_n^{mod}$ is then 22 cm. In the case of experiment (1) with molecules, using $\hbar\omega = 0.116$ eV of the $CO_2$ laser and $E_0 = 11$ eV (Weingartshofer et al., 1977), we find $\Lambda_{bn} = 2.65 \times 10^3$ cm and a quantum coherence length $\Lambda_n^{mod}$ of the same order of magnitude, as $\Delta E$ might have been $0.2\hbar\omega$, as given from the experiment. Apart from the fact that the modulation at the $CO_2$ laser frequency would be much more difficult to detect than in the case of the green argon laser line, the other parameters in (1) are chosen at such values that the modulation itself would be washed out too close to the interaction area; only the energy spectrum with the up- and downshift is remaining.

Case II: $t \gg t_0$. For $t \gg t_0$, the condition expressed by Eq. (42) assumes the form $\hbar\omega \lesssim \delta E_0$, which should be satisfied for most experiments; however, for $t \gg t_0$ there would be other effects which ought to be taken into account—for example, the fact that the wave packet is also localized in the $y$ and $z$ directions.

## V. PEIERLS DISPERSION IN THE VARSHALOVICH–DYAKONOV CORRECTION TO EXPLAIN THE RELATIVISTIC LONG BEATING WAVELENGTH

We shall discuss in this section the long beating wavelength of the interference of the modulated electron waves. Initially this long beating was the result of the analysis by Salat (1970) based on the wave-mechanical scattering theory. Following this predicted value, Schwarz (1970b) published measurements where this long-wavelength beating was detected, however at a value which was larger than predicted. The relativistic analysis (Van Zandt and Meyer, 1971) required an even shorter wavelength than the nonrelativistic prediction. It was then shown (Hora, 1977) that the relativistic treatment with inclusion of the dielectric correction of Varshalovich and Dyakonov (1971) arrived at the exact value measured if the Peierls formula (Peierls, 1976) for the dielectric dispersion was used. There are other results for this dispersion, e.g., that of Landau and Lifshitz (1960), and the result of Schwarz (1971b) could then be used both to prove his experiments as well as to prove the correctness of the Peierls formula for the dispersion of dielectric solids.

As explained in Section IV, the superposition of electron waves of energies $E_0$, $E_0 \pm \hbar\omega$, $E_0 \pm 2\hbar\omega$, etc., results in an interference with beating if the electron states are coherent but separated beyond self-coherence. The exact superposition of the nonrelativistic electron waves results in a beating wavelength

$$\lambda_{bn} = \lambda_e E_0 / \hbar\omega \tag{45}$$

($\lambda_e = 2\pi\hbar/\sqrt{2mE_0}$ is the de Broglie wavelength) which is in analogy to optical beating. As a typical difference between electron waves and optical waves, the second-order calculation of the superposition of electron waves results in a long beating wavelength

$$\Lambda_{bn} = 8\lambda_e (E_0/\hbar\omega)^2 \tag{46}$$

The modulation of the electron wave with the laser frequency $\omega$ is emitted with a modulation length $\Lambda_n^{mod}$ describing the distance from the crossing area of electron and laser beams (Schwarz, 1971b):

$$\Lambda_n^{mod} = \Lambda_{bn}\hbar\omega/8\,\Delta E \tag{47}$$

The measurement of the long beating wavelength $\Lambda_{bn}$ of the order of centimeters was possible only (Schwarz, 1971b) by realizing a value of $\Delta E \ll \hbar\omega$, as has been described in detail by Schwarz.

For the experimental conditions (Schwarz and Hora, 1969; Schwarz, 1971b) of $E_0 = 50$ keV, $\hbar\omega$ corresponded to a laser wavelength of 4880 Å, and the value of $\Lambda_{bn}$ from Eq. (46) is

$$\Lambda_{bn} = 1.6858 \quad cm \tag{48}$$

while Schwarz had measured

$$\Lambda_{bn}^{exp} = (1.73 \pm 0.01) \quad cm \tag{49}$$

This experimental value could be taken from the maxima of the curves immediately, as a correction by attenuation was negligible in this case of sufficiently small $\Delta E$, as could be seen from the negligible attenuation on $x$.

The fact that electrons of 50 keV energy have some relativistic properties had been discussed by several authors (Van Zandt and Meyer, 1971; Varshalovich and Dyakonov, 1971b). The electron velocity $v$ can be derived from the expression of the kinetic energy

$$E_0 = m_0 c^2 \left| \frac{1}{\sqrt{1 - v^2/c^2}} - 1 \right| \tag{50}$$

where

$$1 - \frac{v^2}{c^2} = \frac{1}{(1 + E_0/m_0 c^2)^2} = \frac{1}{1.205}$$

and finally

$$v = 1.23 \times 10^{10} \quad cm/s \tag{51}$$

This relativistic behavior has to be taken into account when electron waves of different energies are superimposed. We first have to show that a similar superposition is possible as in the nonrelativistic case.

We use the Dirac equation of the electron in the following formulation:

$$\left| \boldsymbol{\beta} \cdot \left( \frac{\hbar}{i} \Box + e\mu_0 V \right) - i m_0 c \right| \boldsymbol{\Psi} = 0 \tag{52}$$

which without loss of generality can be expressed by two solutions $\Psi_a$ and $\Psi_b$, each being two-dimensional matrices by the use of the spin matrices $\tau$ instead of the four-dimensional Dirac matrices $\boldsymbol{\beta}$ and $\boldsymbol{\Psi}$:

$$\boldsymbol{\tau} \cdot \left( \frac{\hbar}{i} \nabla + e\eta_0 \mathbf{A} \right) \Psi_b + \left( -\frac{\hbar}{c} \frac{\partial}{\partial t} + e\eta_0 i\Phi \right) \Psi_a - i m_0 c \Psi_a = 0 \tag{53}$$

$$\boldsymbol{\tau} \cdot \left( \frac{\hbar}{i} \nabla + e\eta_0 \mathbf{A} \right) \Psi_a - \left( -\frac{\hbar}{c} \frac{\partial}{\partial t} + e\eta_0 i\Phi \right) \Psi_b - i m_0 c \Psi_b = 0 \tag{54}$$

The potential $V$ in Eq. (52), given by the vector potential $\mathbf{A}$ and $\Phi$, has either to be only slowly varying in time or to be periodic for one frequency, to allow a separation.

$$\Psi_b = -\frac{ic}{E + e\mu_0 c\Phi + m_0 c^2}\tau \cdot \left(\frac{\hbar}{i}\mathbf{V} + e\mu_0 \mathbf{A}\right)\Psi_a \tag{55}$$

Expressing

$$E = m_0 c^2 + E_0 \tag{56}$$

we find from Eq. (55)

$$\Psi_b = -\frac{i}{2m_0 c[1 + (E_0 + e\mu_0 c\Phi)/2m_0 c^2]}\tau \cdot \left(\frac{\hbar}{i}\mathbf{V} + e\mu_0 \mathbf{A}\right) \tag{57}$$

As $E_0 < m_0 c^2$ in the case considered, we can approximate

$$\Psi_b \approx -\frac{i}{2m_0 c}\tau \cdot \left(\frac{\hbar}{i}\mathbf{V} + e\mu_0 \mathbf{A}\right)\Psi_a \tag{58}$$

and similarly $\Psi_a$ as a function of $\Psi_b$.

This result of Eq. (58) immediately indicates that the superposition of $\Psi_b(E_0)$ and $\Psi_b(E_0 + h)$ or similar cases can be considered in a linear way, at least as long as $E_0 \ll m_0 c^2$. This can be performed with scalar wave functions, only the energies and momenta have a relativistic correlation. This means that the results of the beating wavelengths $\lambda_{bn}$ and $\Lambda_{bn}$ are conserved and only the relativistic formulation of the de Broglie wavelengths has to be used (in the approximation):

$$\lambda_e = \frac{h}{\sqrt{2mE_0(1 + E_0/2mc^2)}} \tag{59}$$

Using the experimental values of $E_0 = 50$ keV, we arrive at a value of

$$\Lambda_{bn}^{rel} = \frac{8h}{\sqrt{2mE_0(1 + E_0/2mc^2)}}\left(\frac{E}{\hbar\omega}\right)^2 = 1.646 \quad \text{cm} \tag{60}$$

which is less than the nonrelativistic value and therefore more different from the experimental value than the nonrelativistic value.

D. A. Varshalovich (personal communication) expressed his concern on the discrepancies between the experimental and the calculated long beating wavelength equations, though a solution of this problem may just be derived from their result (Varshalovich and Dyakonov, 1971b) that a relativistic dielectric correction is necessary for the long beating wavelength $\Lambda_{bn}$. If the velocity of the laser radiation in the dielectric slab, $c_M$, differs from the vacuum velocity of light, $c$, a correction factor

$$C_{VD} = \frac{1}{1 - (v^2/c^2)(1 - c^2/c_M^2)} \tag{61}$$

is necessary, resulting in the long beating wavelength $\Lambda_0$ by combination with the uncorrected relativistic value:

$$\Lambda_0 = \frac{8h}{\sqrt{2m_0 E(1 + E_0/2mc^2)}} \left(\frac{E}{\hbar\omega}\right)^2 \frac{1}{1 - (v^2/c^2)(1 - c^2/c_M^2)} \tag{62}$$

Usually it is assumed that $c_M^2 = c^2/n$, but we use the expression for the residual photon momentum (Hora, 1974) for solids (Peierls, 1976)

$$\left|\frac{c_M}{c}\right| = \frac{\tilde{n}^2 - 1}{\tilde{n}} |1 + \tfrac{1}{5}(\tilde{n}^2 - 1)| \tag{63}$$

if $n$ is the refractive index of the dielectric slab in the crossing area of the electron and laser beams, where $n$ is to be taken for the laser frequency. The complication due to a possible modification known from the momentum of the photons in plasma (Hora, 1974) and the eventual similarity to the case of polarizable media (Peierls, 1976) will be neglected for the discussion here.

Using $n = 1.58$ for the experimental conditions of the experiment (Schwarz, 1971b), we arrive at

$$\Lambda_0 = 1.740 \quad \text{cm} \tag{64}$$

which agrees exactly with the measurements.

## VI. A Comprehensive Scattering Model for Quantum Modulation

In this section we will use a Green's function method in order to construct the exact form of the scattered Schrödinger field in the presence of an electromagnetic wave representing a laser beam present in the sample which scatters the incoming monoenergetic beam of electrons. Knowing the exact form of the outgoing scattered wave describing the electrons, we will also be able to calculate the intensity of the light emitted later in the Schwarz–Hora effect at any time and at any point in space.

The Green's function used here will be the Green's function of the Schrödinger equation in an external electromagnetic field. This method has been successfully applied by Kroll and Watson (1973) to the study of multiphoton exchange amplitudes in the presence of a laser beam. In our case we are interested in the absorption and emission of photons from a laser beam in the process of electron diffraction or scattering by a crystal. We will call $V(r)$, and sometimes simply $V$, the more or less periodic potential encountered by the beam of electrons in the crystalline sample. Describing the electrons by de Broglie waves of incoming and outgoing wave vectors $k_0$ and $k$, respectively, in the absence of the applied laser field, the crystal's diffraction pattern will be described in the first Born approximation by the matrix elements $V_{k_0,k}$ (Sherif and Handel, 1982).

Let $A(r, t)$ be the vector potential which describes the electromagnetic field of the light coming from the laser. Each laser mode is characterized by a wave vector $k$ of magnitude $k = \omega/c$ and by a polarization $\sigma$. The vector potential can be expanded in a Fourier series over an arbitrarily large box of volume $\Omega$.

It is most convenient to describe the electromagnetic field in terms of plane waves. The vector potential is taken in the radiation gauge as

$$A(r, t) = \sum_{k,\sigma} (\hbar^2 c/\Omega\omega_k)^{1/2} u_{k,\sigma} [a_{k,\sigma}(t) e^{ik \cdot r} + a^*_{k,\sigma}(t) e^{-ik \cdot r}] \tag{65}$$

The polarization vectors $u_{k,1}$ and $u_{k,2}$ are mutually orthogonal unit vectors perpendicular to $k$.

The Schrödinger equation for an electron moving in a vector potential $A$ and scattering potential $V$ is

$$(1/2m)(-i\hbar\nabla - eA/c)^2\Psi + V\Psi = i\hbar\dot{\Psi} \tag{66}$$

An overdot has been used to indicate the time derivative. The electromagnetic field is treated as a classical field at this point. In order to eliminate the $A^2$ term from Eq. (66), we write

$$\Psi \equiv \exp\left(\frac{-i}{\hbar} \int^t \frac{e^2}{2mc^2} A^2 \, dt'\right)\Phi \tag{67}$$

Thus, Eq. (17) is reduced to

$$((-\hbar^2/2m)\nabla^2 + (ieh/mc)A \cdot \nabla + V)\Phi = i\hbar\dot{\Phi} \tag{68}$$

It is convenient to consider first the influence of a single electromagnetic mode, i.e., a single term from Eq. (65). Therefore, we take $A = a\cos(\omega t + \gamma)$, where $\gamma$ is an initial phase constant, and we treat $V\Phi$ as a perturbation source term. The solution for Eq. (66) is an incoming plane wave plus scattered waves given by the integral equation

$$\Phi_{k_0}(r, t) = \phi_{k_0} - \int d^3x' \int_{-\infty}^t dt' \, GV(r')\Phi_{k_0}(r', t') \tag{69}$$

Here $\phi_{k_0}$ is the solution of the homogeneous equation, i.e., with $V = 0$, and can be written in the form

$$\phi_{k_0} = e^{ik_0 \cdot r} \exp\left[-\frac{i\hbar}{2m} \int^t \left(k_0^2 - \frac{2ek_0 \cdot A}{c\hbar}\right)dt\right] \tag{70}$$

$G$ is the Green's function which satisfies the equation

$$[(-\hbar^2/2m)\nabla^2 + (ie\hbar A \cdot \nabla/mc) - i\hbar \, \partial/\partial t]G = \delta(r - r')\delta(t - t') \tag{71}$$

Given $\mathbf{A} = \mathbf{a}\cos(\omega t + \gamma)$, $G$ can be found to be

$$G = i/(2\pi)^3\hbar \int d^3k\, e^{i\mathbf{k}\cdot(\mathbf{r}-\mathbf{r}')} \exp\{-(i\hbar/2m)[k^2t - 2e\mathbf{k}\cdot\mathbf{a}\sin(\omega t + \gamma)/\hbar c\omega]\}$$

$$\times \exp\{(i\hbar/2m)[k^2t' - 2e\mathbf{k}\cdot\mathbf{a}\sin(\omega t + \gamma)/\hbar c\omega]\} \tag{72}$$

In the first Born approximation we set $\Phi_{\mathbf{k}_0}(\mathbf{r}',t') = \phi_{\mathbf{k}_0}(\mathbf{r}',t')$ in the integral present in Eq. (69) and obtain for the scattered wave

$$\frac{i}{(2\pi)^3\hbar} \int d^3x' \int_{-\infty}^{t} dt'\, V(\mathbf{r}') \int d^3k\, e^{i\mathbf{k}\cdot(\mathbf{r}-\mathbf{r}')} \exp\{-(i\hbar/2m)$$

$$\times [k^2t - 2e\mathbf{k}\cdot\mathbf{a}\sin(\omega t + \gamma)/\hbar c\omega]\} \exp[(i\hbar/2m)(k^2 - k_0^2)t']$$

$$\times \{\exp[ie(\mathbf{k}_0 - \mathbf{k})\cdot\mathbf{a}\sin(\omega t' + \gamma)/2mc\omega]e^{i\mathbf{k}_0\cdot\mathbf{r}'}\}$$

Using the relation

$$e^{i\beta\sin(\omega t + \gamma)} = \sum_{n=-\infty}^{\infty} J_n(\beta)e^{in(\omega t + \gamma)} \tag{73}$$

where $J_n(\beta)$ is the $n$th-order Bessel function, we expand the expression contained in the second pair of curly brackets in a Fourier series. Then Eq. (69) takes the form

$$\Phi_{\mathbf{k}_0,n}(\mathbf{r}, t) - \phi_{\mathbf{k}_0,n}(\mathbf{r}, t) = \frac{-i}{(2\pi)^3\hbar} \int d^3x' \int_{-\infty}^{t} dt'\, V(\mathbf{r}') \int d^3k\, e^{i\mathbf{k}\cdot(\mathbf{r}-\mathbf{r}')}$$

$$\times \exp\{-(i\hbar/2m)[k^2t - 2e\mathbf{k}\cdot\mathbf{a}\sin(\omega t + \gamma)/\hbar c\omega]\}$$

$$\times \exp\left(\frac{i\hbar}{2m}(k^2 - k_0^2)t'\right)\left(\sum_{n=-\infty}^{\infty} J_n(\beta)e^{in(\omega t' + \gamma)}\right)e^{i\mathbf{k}_0\cdot\mathbf{r}'} \tag{74}$$

After performing the integration over $t'$ we use a contour integration method for $k$. Then Eq. (74) is reduced to

$$\Phi_{\mathbf{k}_0,n}(\mathbf{r}, t) - \phi_{\mathbf{k}_0,n}(\mathbf{r}, t) = \frac{-m}{(2\pi)\hbar^2} \sum_{n=-\infty}^{\infty} J_n(\beta)e^{in\gamma}(e^{ik(n)r}/r)$$

$$\times \exp\{i\hbar[k^2(n)t - 2e\mathbf{k}(n)\cdot\mathbf{a}\sin(\omega t + \gamma)/\hbar c\omega]/2m\}$$

$$\times \int d^3x' e^{i\mathbf{k}(n)\cdot\mathbf{r}'} V(\mathbf{r}')e^{i\mathbf{k}_0\cdot\mathbf{r}'} \tag{75}$$

where

$$\beta = -e(\hbar\mathbf{k} - \hbar\mathbf{k}_0)\cdot\mathbf{a}/mc\hbar\omega$$

$$= -e\mathbf{Q}\cdot\mathbf{a}/mc\hbar\omega \tag{76}$$

and $\mathbf{Q}$ is the momentum transfer. In Eq. (75) $k(n)$ is defined by

$$[\hbar k(n)]^2/2m = (\hbar k_0)^2/2m - n\hbar\omega \tag{77}$$

The total scattered wave can be written as

$$
\begin{aligned}
\Psi_s = [-m/(2\pi)\hbar^2 r] &\sum_{n=-\infty}^{\infty} e^{ik(n)r} \\
&\times \exp\{-i\hbar[k^2(n)t - 2ek(n)\cdot\mathbf{a}\sin(\omega t + \gamma)/\hbar c\omega]/2m\} \\
&\times V_{k(n),k_0}J_n(\beta)e^{in\gamma}
\end{aligned}
\tag{78}
$$

where

$$V_{k(n),k_0} = \int e^{-i k(n)\cdot\mathbf{r}'}V(r')e^{i k_0\cdot\mathbf{r}'}\,d^3x'$$

is the scattering matrix element calculated without consideration of the interaction with the electromagnetic field oscillators.

Let us focus first only on the terms with $n = 0$, $+1$, and $-1$ in Eq. (78). Taking into account that for Bessel functions of integer order $n$ we have $J_{-n} = (-1)^n J_n$, with the help of Eq. (77) we obtain a beat term of frequency $\omega$ in the density (and in the current density) of the scattered electrons

$$
\begin{aligned}
|\psi_s|^2_\omega &= 2(m/2\pi r\hbar^2)^2|V_{k_0,k(0)}V_{k_0,k(1)}|J_0(\beta)J_1(\beta) \\
&\quad \times [\cos(\varphi - \delta) - \cos(\varphi + \delta)] \\
&= (m/\pi r\hbar^2)|V_{k_0,k(0)}V_{k_0,k(1)}|J_0(\beta)J_1(\beta) \\
&\quad \times 2\sin\varphi\sin\delta
\end{aligned}
\tag{79}
$$

where

$$
\begin{aligned}
\psi = \arg(V_{k(0),k_0}V_{k_0,k(1)}) &+ \gamma - (m\omega/\hbar k_0)r + \hbar\omega t \\
&- (2e/c\omega)[k(0) - k(1)]\cdot\mathbf{a}\sin(\omega t + \gamma)
\end{aligned}
\tag{80}
$$

and

$$\delta = (k_0 r/2)(m\omega/\hbar k_0^2)^2 \tag{81}$$

If we replace $k(1)$ with $k(-1)$ and change the sign of the terms $-(m\omega/\hbar k_0)r$ and $+\hbar\omega t$ in Eq. (80), the result will be to a very good approximation only a change in the sign of $\varphi$. This fact, used in the derivation of Eq. (79), is based on the inequality $|\mathbf{k}(1) - \mathbf{k}(0)| \ll k_0$, which is generally satisfied in Schwarz–Hora experiments. Indeed, from Eq. (77) we obtain

$$k(1) - k_0 = -m\omega/\hbar k_0 - (k_0/2)(m\omega/\hbar k_0^2)^2 \tag{82}$$

and

$$k(-1) - k_0 = m\omega/\hbar k_0 - (k_0/2)(m\omega/\hbar k_0^2)^2 \tag{83}$$

Experimental values of 50 keV for the energy of the electrons correspond to a wave vector $k_0 = 1.2 \times 10^{10}$/cm, and the wavelength of roughly 500 nm of the laser light used corresponds to $\omega = 4 \times 10^{14}$/s, or $\hbar\omega = 4 \times 10^{-13}$ erg $= 0.25$ eV. Therefore, $\omega/ck_0 = 10^{-6}$, and the above-mentioned inequality is satisfied by a factor of the order of $10^6$. Furthermore, $m\omega/\hbar k_0^2 = 0.25$ eV/$100$ keV $= 2.5 \times 10^{-6}$, which leads to a maximal effect ($\delta = \pi/2$) at $r = 40$ cm. For smaller energies this value of $r$ is much smaller, being proportional to $k_0^3$.

If the laser light is concentrated to a flux of $10^7$ W/cm$^2$, the value of $\beta$ will be of the order of 1 for not too small angles of scattering. At energies below 50 keV, or at small scattering angles, the terms with $|n| > 1$, as well as their oscillating (or beat) contributions to the charge and current densities, can be neglected.

The beat current density $j(r)$ is obtained in lowest order from Eq. (79) by multiplication with $e\hbar k_0/m$:

$$j = 2(e\hbar k_0/m)(m/\pi r\hbar^2)|V_{k_0,k(0)}V_{k_0,k(1)}|J_0(\beta)J_1(\beta)$$
$$\times \sin\varphi \sin\delta \qquad (84)$$

This current and the higher-order beat currents, as well as all their harmonics, could be the source of re-emitted light at the level of the screen. The beat current is negligible at $r = 0$, and attains a maximal value at a macroscopic distance. The re-emission is due to correlations in the positions of the electrons emerging from the first interaction, correlations induced by the laser light and reflected in the pair correlation function, or in the charge and current densities which oscillate in time and space.

## VII. RE-EMISSION OF RADIATION FROM A QUANTUM-MODULATED BEAM

For the detection of the modulated beam behind the area of interaction of the electron beam and the laser beam within a medium for fitting the recoil, the re-emission of the laser photons from the electron beam was used. This was to occur when the electrons hit a nonluminescent target. One necessary condition is that the distance from the interaction area is less than the modulation length $\Lambda^{mod}$ [Eq. (15)] (Schwarz and Hora, 1969; Hora 1975). This is different from the detection of the modulation process by the energy spectrum of the electrons after the interaction (Andrick and Langhans 1976; Weingartshofer et al., 1977). Some of the experimental and theoretical results to the question of the re-emission, sometimes called the "demodulation," should be considered in this section.

## A. Experimental Results

The maximum intensity of the re-emission was one or two orders of magnitude less than the intensity of radiation emitted from the luminescent screen when this was used instead of the nonluminescent screen if the screen was at a distance corresponding to the maxima of the long beating wavelength. From this it can be determined that about one photon was emitted per 10–100 incident modulated electrons from the nonluminescent screen.

The fact that there is a re-emission was simply understood as a nonresonant excitation of electrons in the nonluminescent screen since the incident electrons arrive at the target with the rhythm of the optical frequency. A long discussion ensued over the question that the reproduction of the green frequency of the argon laser used for the interaction with the laser beam after modulation has nothing to do with the argon atoms used before during the preparation of the experiment. This had to be performed under extremely high vacuum, and the surfaces including the nonluminescent screen had to be cleaned from contamination by argon discharge.

The optical emission of the argon atoms adsorbed at the surface of the screen definitely does not have the same optical frequency as the emission of free argon atoms in vacuum, since the adsorption energy changes the energy levels of the atoms after absorption.

In a further experiment Schwarz (1972a) reported that using a metal instead of the nonluminescent target, the emitted radiation from the diffraction patterns of the metal were about 100–1000 times less intensive. This could simply be understood from the penetration depth of the 50 keV electrons of about 100 times the absorption length of the green radiation in the metal. The re-emission can be assumed to take place during the whole penetration of the electrons in the nonluminescent material. It can be emitted from the rather transparent alumina, while all the re-emitted photons in the metal are absorbed there if they are not originated from the very thin level at the surface, which is 100 times less than the electron penetration depth.

The number reported on the intensity of the photons (0.01–0.1 per electron) could well be understood since there was a high degree of quantum modulation; i.e., the component of the electrons shifted to the first or higher frequency differences given by the laser frequency may then be estimated to be up to 10% or more of the maximum of the central energy intensity spectrum. This would roughly agree with the observations by Weingartshofer et al. (1977, 1979, 1975) on the amplitude of the various maxima of the electron energy spectrum after modulation.

A further indirect proof of the re-emission process from experiments may be derived from the amplification of a microwave in the highly nonequilibrium part of a gas discharge, as observed by Rosenberg *et al.* (1982). The interpretation (Ben-Aryeh *et al.*, 1983) indicated that there is no mechanism of stimulated emission possible to arrive at the observed high level of amplification. The only quantitative agreement was achieved if there was a transfer of the energy exchanged at the interaction of the electrons with the photons of the microwave such that the electrons transfer these to subsequent collisions in the same way as there is a reproduction of photons in the re-emission of the quantum-modulated electrons.

It is interesting to note that the conditions of the experiment of Rosenberg *et al.* (1980, 1982) were well within the quantum regime of the correspondence principle of electromagnetic interaction (see Section III), while a reproduction of the amplification experiment at higher microwave intensity (Palmer, 1983) was in the classical range and did not show the amplification. With regard to the problems of the electron–hole drops in semiconductors, a phonon-induced Auger process was discussed in relation to quantum modulation by Suguna and Shrivastava (1980). On the implication of interference effects, see Korobochko *et al.* (1973), or the double-resolution holography with respect to the Rayleigh limit (Swinger and Nilsen, 1975).

It should further be mentioned that a similar process—if not the identical one—of re-emission after modulation may be that which was detected at the interaction of a beam of energetic electrons by plasmons. The mechanism was predicted by Ferrell (1958) and measured by Steinmann (1961, 1968). The possibility of achieving quantum modulation by the interaction of an electron beam with the plasmons of a solid (Hora, 1972) was discussed and may be involved in the processes observed by Steinmann, including modulation and re-emission.

### B. Theoretical Interpretation of the Modulation/Re-emission Process

Numerous theories have been developed in order to reproduce the re-emission process from the beginning of the discussions. A review was given by E. P. George (1975), evaluating the different intensities derived from the various models. The initial assumption (Schwarz and Hora, 1969) that there is a simple nonresonant reproduction of the optical frequency from the modulated electrons because of the rhythm of the beats of their arrival at the nonluminescent screen was evaluated by Gibbon and Bullough (1972). It seems that the numbers for the intensities achieved may not always fit the reported experimental observations. The analysis of Becchi and Morpurgo (1971, 1972) and Morpurgo (1976) may also arrive at intensities of the re-emission which are accurate only within orders of magnitude. There is no doubt that the electrons in the modulated state will not emit any radiation without interacting with a material, as has been shown by Peres (1979), although a far-field interaction with a diffraction lattice should result in an emission of photons (Mizuno et al., 1973).

The ability for re-emission of modulated electrons was also concluded from a modulation process following the Ewald construction of crystal diffraction (LaFleur, 1971; Janner and LaFleur, 1971). This is in connection with the analysis of the quantum-modulation process in the interaction region of the laser beam and the electron beams, where Elliott (1972) distinguishes a first-order and a second-order Schwarz–Hora effect, where the transmitted beam in the crystal in the interaction area is modulated by a first-order process involving the absorption and emission of the photons, while the diffracted beams are modulated by a second-order process involving an additional transfer of momentum to the crystal lattice.

Another treatment used the mechanism of the so-called transition radiation, which appears when an energetic electron changes from one medium to another (Ginzburg and Frank, 1946; Frank, 1960). The evaluation of the intensity of this radiation was performed by Rubin (1970) and by Varshalovich and Dyakonov (1970, 1971b). The intensity may be few orders of magnitude lower than in the experiment, as was pointed out by Zeldovich (1972), who concluded a higher intensity of white radiation, a question which has to be resolved at this stage. The model by Varshalovich and Dyakonov (1970) was extended to an interpretation of a diffraction pendulum solution (Fedorov, 1979). Also the case of stimulated transition radiation was discussed in connection with the quantum-modulation effect (Oganesian and Engibargan, 1984).

It should be noted that the absolute intensity of the re-emission was estimated quite inaccurately such that no final decision should be derived about the validity of each model. The connection of the transition radiation process with the much more general state of electrons after quantum modulation was also discussed by Arutyunyan and Oganesian (1978) and by Kalmykova (1982). The use of the transition radiation for the explanation of the re-emission of the modulated electrons was given in papers by Kondo (1971) and Marcuse (1971a,b). This work is an alternative interpretation to that of Van Zandt and Mayer (1970), where the generation of the radiation is derived from the nonoverlapping electrons.

Another treatment by George (1975) used the bremsstrahlung models based on the Williams–von Weitzsacker theory (von Weitzsacker, 1934; Williams, 1935). It used the number of photons $n$ emitted in the frequency range $dv$:

$$n(v)\,dv = \frac{2}{\pi}\frac{e^2}{\hbar c}\ln\left(\frac{E_0}{\hbar c}\right)d\ln v \tag{85}$$

which are then diffused in the nonluminescent screen. Summing up the involved processes leads one to an intensity comparable to the observation of the intensity of the re-emitted radiation. The resulting intensity of the re-emission is of the order of magnitude reported in the experiment (Schwarz and Hora, 1969). In a similar way the interpretation of the results led to a model for the calculation of the intensity of the re-emission (Hadley et al, 1971b, 1972).

It should be noted that the bunching process does not necessarily need to be interpreted as a classical process (Oliver and Cutler, 1970; Favro et al., 1970) but can well be interpreted along the lines of the relativistic interactions observed by Piestrup et al. (1975) and Chen et al. (1978). The question of the classical bunching and how to treat the modulation effect by a plane-wave approximation was discussed by Hutson (1970). In this connection the treatment on the basis of the klystron theory should be mentioned (Farago and Sillitto, 1973). It should also be noted that some advantages were pointed out to treating the Schwarz–Hora effect on the basis of stimulated Čerenkov absorption (Bulshov et al., 1972). The quantum modulation for an electron beam when reflected from (instead of transmitting) a medium containing the laser wave was discussed by Haroutunian and Avetissian (1973) with reference to the Čerenkov effect.

Relative to a very far-reaching application of the re-emission reactions of quantum-modulated radiation, Vysotskii et al. (1980) performed some calculations on how a quantum-modulated beam for randomly distributed relativistic electrons could be used for a coherent excitation of the inversion of nuclei. The aim of producing inverted nuclei for the generation of gamma-ray lasers (grasers) is of considerable interest (Baldwin et al., 1982; Wilson et al., 1978). The proposal of an elementary particle heterodyne was discussed by Brautti (1980).

### C. 1/f Noise and the Self-interaction of Electrons in the Handel Effect

When charged particles are scattered in the absence of any simultaneously applied electromagnetic waves, a certain self-modulation of the electron beam is still possible due to its interaction with the zero-point fluctuations of the electromagnetic field. This interaction causes spontaneous bremsstrahlung associated with the scattering process, and leads to quantum $1/f$ current noise in the beam, or to quantum $1/f$ noise in the associated scattering cross sections (conventional Handel effect—in the terminology of van der Ziel (1987)—or quantum $1/f$ noise), as was shown by Handel in 1975 (Handel, 1975). The effect would be too small to be observable, if it were not for the infrared catastrophe which always causes the effect to become dominant at sufficiently low frequencies. We conclude that, other than the Schwarz–Hora effect, which arises from stimulated radiation processes only, this effect is an infrared divergence phenomenon, and arises exclusively from the spontaneous bremsstrahlung process.

This all refers to the conventional Handel effect, which differs from the coherent state Handel effect, also known as coherent quantum $1/f$ noise (Handel, 1983), which is not caused by bremsstrahlung, but rather by the uncertainty in the energy of a physical charged particle (see Subsection E below).

At this point we provide a brief physical explanation of the conventional Handel effect. Consider, for example, Coulomb scattering of charged particles by a fixed charge. The outgoing (scattered) Schrödinger field monitored by a detector at an angle $\theta$ from the direction of the incoming beam contains a main nonbremsstrahlung part and various contributions which lost small amounts of energy $\varepsilon = hf$ due to the emission of a bremsstrahlung photon of arbitrarily low frequency $f$, and therefore have a de Broglie frequency lowered exactly by $f$. The expression of the outgoing scattered current density is quadratic in the emergent Schrödinger field and will contain a major nonbremsstrahlung part, a small bremsstrahlung part, and two cross terms proportional to both the nonbremsstrahlung and the bremsstrahlung parts of the scattered charged particle wave function. These cross terms oscillate with the beat frequency $f$. Photons are emitted at any frequency, and therefore the cross terms will contain any frequency $f$ with an amplitude proportional to the bremsstrahlung scattering amplitude. The fluctuating cross terms will be registered at the detector as $1/f$ noise in the scattering cross section.

For an elementary derivation of the conventional Handel effect we start with the classical (Larmor) formula for the power $2q^2\dot{v}^2/3c^2$ radiated by an accelerated charge. The sudden acceleration $\dot{\mathbf{v}} = \Delta\mathbf{v}\,\delta(t)$ suffered by the charged particle during scattering has a constant Fourier transform $\dot{\mathbf{v}}_f = \Delta\mathbf{v}$, where $\Delta\mathbf{v}$ is the velocity change during the scattering process and $\delta(t)$ the Dirac delta function. Therefore the spectral density $4q^2|\dot{v}_f|^2/3c^3$ of the radiated energy can be written in the form $4q^2(\Delta v)^2/3c^3$ and does not depend on $f$. Dividing by the energy $hf$ of a photon, we get the number spectrum $4q^2(\Delta v)^2/3c^3hf$ of the radiated photons. Since the amplitude of each of the two cross terms is proportional to the amplitude of the bremsstrahlung part of the scattered particle wave function, the fractional spectral density $S_I(f)$ of the observed $1/f$ noise is twice the number spectrum of the emitted photons per scattered charge carrier:

$$I^{-2}S_I(f) = 8q^2(\Delta v)^2/3c^3hf = (4\alpha/3\pi f)(\Delta v/c)^2$$

Here the definition $\alpha = 2\pi q^2/hc$ of Sommerfeld's fine-structure constant was used. This completes our elementary derivation of the conventional (incoherent) Handel effect.

All scattering cross sections and process rates defined for the current carriers must fluctuate with a fractional spectral density given by this expression. Applied to scattering cross sections, this means that the collision frequency, the mean time between collisions, and the mobility of each carrier independently, must all fluctuate with the same fractional spectrum. If $N$ is the number of current carriers in a conducting sample of any nature, this requires the inclusion of a factor $1/N$ into the above expression. For mobility fluctuations, the empirical Hooge formula is thus derived from first principles as a quantum $1/f$ result with the Hooge constant $\alpha_H$ given by the above expression (Hooge, 1969). All $1/f$ noise formulas derived on the basis of the Hooge formula can therefore be taken as quantum $1/f$ results with the appropriate quantum $1/f$ Hooge parameter, but they will provide only the quantum $1/f$ contributions from the scattering cross sections. Therefore they will not describe the experimental results on $1/f$ noise in semiconductors properly in general, until we add the complementary contributions from quantum $1/f$ fluctuations of the surface and bulk recombination cross sections, from quantum $1/f$ fluctuations in tunneling rates, or possible injection–extraction contributions from the velocity changes $\Delta v$ in transitions of the carriers in junctions and at contacts. Some of these complementary contributions turn out to be similar to results of earlier calculations based on the McWhorter and North–Fonger models, in which the correct quantum $1/f$ expression of recombination rate fluctuations replace the rate fluctuations postulated by McWhorter and North (Van der Ziel and Handel, 1985; Van Fliet et al., 1981). McWhorter (1955) had considered transitions to and from traps in the surface oxide layer and North thermal fluctuations of the surface potential as the final cause of $1/f$ noise. Other quantum $1/f$ contributions, finally, do not even bear a formal resemblance to earlier calculations. We conclude that the quantum $1/f$ approach provides both a foundation and a properly weighted synthesis of earlier calculations, as well as additional contributions. At the same time, the quantum $1/f$ approach eliminates all free parameters that had to be introduced in the previous models, leaving only the fine-structure constant as a common factor of all electromagnetic Handel effect contributions.

In this subsection we will derive the quantum $1/f$ effect with the same Green's function method we used in Section VI, where we had considered only one mode of the electromagnetic field (the strongest laser mode) in interaction with the scattered carriers. In the following subsections we present a second-quantized derivation of the conventional Handel effect (Section D) and a derivation of the coherent state Handel effect (Section E).

We will now consider the actual case of interaction with all electromagnetic field modes. The classical vector potential is given by

$$A = \sum_{k',s} a_{k',s} \cos(\omega_{k'}t + \gamma_{k',s}) \tag{86}$$

where $s$ labels the two polarization states. Using Eq. (86) to substitute into Eq. (72), we can write the Green's function as

$$G = \frac{i}{(2\pi)^3 \hbar} \int d^3k \, e^{ik \cdot (r - r')}$$

$$\times \exp\left(\sum_{k',s} -i\hbar[k^2t - 2ek \cdot a_{k',s}\sin(\omega_{k'}t + \gamma_{k',s})/\hbar c\omega_{k'}]/2m\right)$$

$$\times \exp\left(\sum_{k',s} i\hbar[k^2t' - 2ek \cdot a_{k',s}\sin(\omega_{k'}t' + \gamma_{k',s})/\hbar c\omega_{k'}]/2m\right) \tag{87}$$

The Born approximation has been used in the integral Eq. (69) as before. Using Eq. (73) for each mode and each polarization state, the scattered wave can be put in the final form as

$$\Psi_s = \frac{-m}{(2\pi)\hbar^2} \prod_{k',s} \sum_{n_{k',s} = -\infty}^{\infty} (e^{ik(n_{k',s})r}/r)$$

$$\times \exp\{-i\hbar[k^2(n_{k',s})t - 2ek(n_{k',s}) \cdot a_{k',s}\sin(\omega_{k'}t + \gamma_{k',s})/\hbar c\omega_{k'}]/2m\}$$

$$\times V_{k(nk_{,s}),k_0} J_{n_{k',s}}(\beta_{k',s})e^{in_{k',s}\gamma_{k',s}} \tag{88}$$

Here the definitions given in Eqs. (76) and (77) have been used again. If we limit ourselves again to the terms $n_{k',s} = 0$ and 1, we obtain

$$\Psi_s \cong \frac{-m}{(2\pi)\hbar^2 r} \prod_{k',s} e^{ikr} \exp[-i\hbar(k^2t - 2ek \cdot a_{k',s}\sin(\omega_{k'}t + \gamma_{k',s})/\hbar c\omega_{k'})/2m]$$

$$\times [J_0(\beta_{k',s}) + e^{i(\omega_{k'}t + \gamma_{k',s})}J_1(\beta_{k',s})] V_{k,k_0} \tag{89}$$

In Eq. (89) we have used $k(0) \cong k(1) \cong k$, which is justified at low frequencies.

The probability density in the outgoing Schrödinger field given by Eq. (89) has the form

$$|\Psi_s|^2 \cong [m/(2\pi)\hbar^2]^2(|V_{k,k_0}|^2/r^2) \prod_{k',s} [J_0^2(\beta_{k,s})$$

$$+ 2J_0(\beta_{k,s})J_1(\beta_{k,s})\cos(\omega_k t + \gamma_{k,s}) + J_1^2(\beta_{k,s})] \tag{90}$$

We substitute the approximate values of the Bessel function as before and neglect all powers of $\beta_k$ greater than 1 in Eq. (90). This yields

$$|\Psi_s|^2 = (|a|^2/r^2)\left(1 + \sum_{k,s} \beta_{k,s}\cos(\omega_k t + \gamma_{k,s})\right) \tag{91}$$

The autocorrelation function $A(\tau)$ for the scattered wave can be calculated as

$$|\Psi_s|^2_t \, |\Psi_s|^2_{t+\tau} = \frac{|a|^4}{r^4}\left(1 + \sum_{\mathbf{k},s}\beta_{\mathbf{k},s}\{\cos(\omega_{\mathbf{k}}t + \gamma_{\mathbf{k},s}) + \cos[\omega_{\mathbf{k}}(t+\tau) + \gamma_{\mathbf{k},s}]\}\right.$$

$$\left. + \sum_{\mathbf{k},s}\sum_{\mathbf{k}',s'}\beta_{\mathbf{k},s}\beta_{\mathbf{k}',s'}\cos(\omega_{\mathbf{k}}t + \gamma_{\mathbf{k},s})\cos[\omega_{\mathbf{k}'}(t+\tau) + \gamma_{\mathbf{k}',s'}]\right) \quad (92)$$

The ensemble average over the random phases is now considered for Eq. (71) with $\langle e^{i\gamma_{\mathbf{k},s}}\rangle = 0$ and $\langle e^{i(\gamma_{\mathbf{k},s}-\gamma_{\mathbf{k}',s'})}\rangle = \delta_{\mathbf{k},\mathbf{k}'}\,\delta_{s,s'}$.

We obtain

$$A(\tau) = \langle|\Psi_s|^2_t \, |\Psi_s|^2_{t+\tau}\rangle$$

$$= (|a|/r)^4\left(1 + \tfrac{1}{2}\sum_{\mathbf{k},s}|\beta_{\mathbf{k},s}|^2\cos\omega_{\mathbf{k}}\tau\right) \quad (93)$$

By expressing the summation in terms of the integral $[\Omega/(2\pi)^3]\int d^3k$ in Eq. (93), the autocorrelation function becomes

$$A(\tau) = (|a|/r)^4\left(1 + \tfrac{1}{2}\sum_{s=1}^{2}\frac{\Omega}{(2\pi)^3}\int|\beta_{\mathbf{k},s}|^2\cos\omega_{\mathbf{k}}\tau\,d^3k\right) \quad (94)$$

Note that in Eq. (86) the expansion coefficients also depend on the volume $\Omega$ of the normalization box

$$\mathbf{a}_{\mathbf{k},s} = \varepsilon_s c(\hbar/\omega\Omega)^{1/2} \quad (95)$$

From Eq. (76) one then finds

$$\beta_{\mathbf{k},s} = -(e/m\hbar\omega)(\hbar/\omega\Omega)^{1/2}\mathbf{Q}\cdot\varepsilon_s \quad (96)$$

Substituting into Eq. (94), we obtain

$$A(\tau) = (|a|/r)^4\left(1 + 2\alpha A\int_{\omega_0}^{\omega_1}\cos\omega\tau\,c\omega/\omega\right) \quad (97)$$

where the fine-structure constant $\alpha = e^2/4\pi\hbar c$ has been introduced and

$$\alpha A = \frac{2\alpha}{3\pi}\frac{Q^2}{m^2 c^2} \quad (98)$$

is the infrared exponent present both in the infrared radiative corrections and in the spectral density of the fractional bremsstrahlung cross section $\alpha A/f$. The lower integration limit $\omega_0 \geq 2\pi T^{-1}$, where $T$ is the duration of the experiment, defines the frequency resolution. The upper limit $\omega_1 \leq E/\hbar$ is given by the energy $E$ of the particles in the beam. According to the Wiener–Khintchine theorem the spectral density of the density fluctuations $S_n(f)$ is the Fourier transform of the autocorrelation function $A(\tau)$

$$S_n(f) = (|a|/r)^4 [\delta(f) + 2\alpha A/f] \tag{99}$$

Consequently, the spectral density of the fractional density fluctuations $S_{\delta n/n}$ will be

$$S_{\delta n/n}(f) \equiv \langle n \rangle^{-2} \langle (\delta n)^2 \rangle_f = 2\alpha A/f \tag{100}$$

This final result and Eqs. (97)–(100) coincides with the well-known conventional quantum $1/f$ noise formula (Handel, 1975, 1980, 1982) with the infrared radiative correction factor $(f/f_0)^{\alpha A}$ approximated by unity.

If we consider the current density $j = (e/m)\hbar k |\psi_s|^2$ in Eqs. (90)–(93) rather than $|\psi_s|^2$, we obtain in a similar way

$$S_{\delta j/j} = S_{\delta \sigma/\sigma} = S_{\delta \mu/\mu} = 2\alpha A/f \tag{101}$$

which indicates the presence of fluctuations in the current, the cross section $\sigma$, and (for currents in solids) also in the mobility $\mu$, in the recombination speed, or in the tunneling rate, with the same spectral density of fractional fluctuations.

## D. Second-Quantized Derivation of the Handel Effect

In this section we focus on the particles emerging from an interaction of any kind, e.g., scattering, tunneling, emission processes, etc. Our goal is to calculate the pair-correlation function both for the case that the outgoing particles, which are always considered identical and noninteracting, are fermions, and in the case that they are bosons.

### 1. The Case of Fermions

The quantum state of two fermions, both outgoing from the same interaction, but with statistically independent bremsstrahlung energy losses accompanying the process for them, is

$$\frac{1}{\sqrt{2}} \int d\xi \int d\eta \left( e^{ik\xi} + \sum_\kappa b(\kappa) e^{i(k-\kappa)\xi} \right)$$
$$\times \left( e^{ik\eta} + \sum_{\kappa'} \beta(\kappa') e^{i(k-\kappa')\eta} \right) \psi_s^\dagger(\xi) \psi_{s'}^\dagger(\eta) |0\rangle$$

where $s$ and $s'$ are the spins, while $b(\kappa)$ and $\beta(\kappa')$ are the spontaneous bremsstrahlung energy loss amplitudes of the two particles. For any $\kappa$ they differ only by their independent random phases. Here $\psi$ designates the field operators of argument $x, y$, etc., $= u - vt$, and $u$ is in general different for each of the arguments mentioned, while $t$ is considered the same for all. $u$ is the coordinate along the scattered beam, and $v = du/dt$ is the velocity of the particles in the scattered beam.

The operator of the pair correlation is

$$\theta = \sum_{s,s'} \psi_s^\dagger(x_1)\psi_{s'}^\dagger(x_2)\psi_{s'}(x_2)\psi_s(x_1) \tag{102}$$

This corresponds to a density autocorrelation function. Using the well-known anticommutation relations

$$\psi_{s(x)}^\dagger \psi_{s'}(y) + \psi_{s'}(y)\psi_s^\dagger(x) = \delta(x - y)\delta_{ss'} \tag{103}$$

as well as the corresponding homogeneous relations for operators of the same kind, we obtain:

$$\langle S_{\uparrow\uparrow}^0|O_{\uparrow\uparrow}|S_{\uparrow\uparrow}^0\rangle \equiv \langle 0|\psi_\uparrow(\eta')\psi_\uparrow(\xi')\psi_\uparrow^\dagger(x_1)\psi_\uparrow^\dagger(x_2)\psi_\uparrow(x_2)\psi_\uparrow(x_1)\psi_\uparrow^\dagger(\xi)\psi_\uparrow^\dagger(\eta)$$

$$|0\rangle = [-\delta(\eta' - x_1)\delta(\xi' - x_2) + \delta(\xi' - x_1)\delta(\eta' - x_2)]$$

$$\times [\delta(\eta - x_2)\delta(\xi - x_1) - \delta(\xi - x_2)\delta(\eta - x_1)] \tag{104}$$

$$\langle S_{\uparrow\uparrow}^0|O_{\uparrow\downarrow}|S_{\uparrow\uparrow}^0\rangle \equiv \langle 0|\psi_\downarrow(\eta')\psi_\uparrow(\xi)\psi_\uparrow^\dagger(x_1)\psi_\downarrow^\dagger(x_2)\psi_\downarrow(x_2)\psi_\uparrow(x_1)\psi_\uparrow^\dagger(\xi)$$

$$\psi_\downarrow^\dagger(\eta)|0\rangle = \delta(\eta' - x_2)\delta(\xi' - x_1)\delta(\eta - x_2)\delta(\xi - x_1) \tag{105}$$

$$\langle S_{\uparrow\uparrow}^0|O|S_{\uparrow\downarrow}\rangle = \langle 0|\psi_\uparrow(\eta')\psi_\downarrow(\xi')\psi_\uparrow^\dagger(x_1)\psi_\downarrow^\dagger(x_2)\psi_\downarrow(x_1)\psi_\uparrow^\dagger(\xi_1)\psi_\downarrow^\dagger(\eta)|0\rangle$$

$$= \delta(\xi' - x_2)\delta(\eta' - x_1)\delta(\xi - x_2)\delta(\eta - x_1) \tag{106}$$

We also obtain three similar expectation values with all spins reversed. The spin-averaged pair correlation function is then

$$A = \tfrac{1}{4}\sum_{ss'} \langle S_{ss'}|O_{\uparrow\downarrow} + O_{\uparrow\uparrow} + O_{\uparrow\downarrow} + O_{\uparrow\uparrow}|S_{ss'}\rangle \tag{107}$$

Substituting the calculated expectation values, we obtain

$$A(x_1, x_2) = \tfrac{1}{2} + \sum_\kappa |b(\kappa)|^2[2 - \cos\kappa(x_1 - x_2)]$$

$$+ \sum_{\kappa\kappa'} |b(\kappa)^2||b(\kappa)^2|[1 - \tfrac{1}{2}\cos(\kappa - \kappa')(x_1 - x_2)] \tag{108}$$

which yields the fractional spectral density

$$\frac{S_n}{(n)^2} = \frac{S_j}{(j)^2} = 2|b(\kappa)|^2/1 + 4\sum|b(\kappa)|^2 + 2\sum|b(\kappa)|^2|b(\kappa)|^2$$

$$\cong 2|b(\kappa)|^2 = \frac{2\alpha A}{\kappa^{1-\alpha A}} = \frac{2\alpha A}{f^{1-\alpha A}} \tag{109}$$

which is in agreement with our previous results, and also includes a 180° phase shift due to the exclusion principle, which is important only at short distances between the particles. In the final form we have transformed to the frequency $f$.

## 2. Case of Bosons

In this case we replace all anticommutators with commutators and obtain

$$A(x_1, x_2) = 2 + 2\sum |b(\kappa)|^2 [1 + \cos \kappa(x_1 - x_2)]$$
$$+ \sum \sum |b(\kappa)|^2 |b(\kappa')|^2 [1 - \cos(\kappa - \kappa')(x_1 - x_2)]$$
$$\frac{S_n}{(n)^2} = \frac{S_j}{(j)^2} = |b(\kappa)|^2 = \frac{\alpha A}{\kappa} \tag{110}$$

Here at short distances we notice an increase of $A$. Both results generalize our previous results to the case of short distances.

## E. Derivation of the Coherent-State Handel Effect

An electrically charged physical particle should be described in terms of coherent states of the electromagnetic field, rather than in terms of a (Fock) eigenstate of the Hamiltonian. This is the conclusion obtained from calculations of the infrared radiative corrections to any process, performed both in Fock space (where the energy eigenstates are taken as the basis, and the particle is considered to have a well-defined energy), and in the basis of coherent states. Indeed, all infrared divergences drop out in the calculation of the matrix element of the process considered, as it should be according to the postulates of quantum mechanics, whereas in the Fook space calculation they drop out only *a posteriori*, in the calculation of the corresponding cross section, or process rate. From a more fundamental mathematical point of view, both the description of charged particles in terms of coherent states of the field, and the undetermined energy, are the consequence of the infinite range of the Coulomb potential. Both the amplitude and the phase of the physical particle's electromagnetic field are well defined, but the energy, i.e., the number of photons associated with this field, is not well defined. The indefinite energy is required by Heisenberg's uncertainty relation, because the coherent states are eigenstates of the annihilation operators, and these do not commute with the Hamiltonian.

A state which is not an eigenstate of the Hamiltonian is nonstationary. This means that we should expect fluctuations in addition to the (Poissonian) shot noise to be present. What kind of fluctuations are these? The additional fluctuations will be identified here as $1/f$ noise with a spectral density of $2\alpha/\pi f$ arising from each electron independently, where $\alpha = \frac{1}{137}$ is the fine-structure constant.

The coherent Handel effect will be derived in three steps: first we consider just a single mode of the electromagnetic field in a coherent state and calculate the autocorrelation function of the fluctuations which arise from its non-stationarity. Then we calculate the amplitude with which this mode is represented in the field of an electron. Finally, we take the product of the autocorrelation functions calculated for all modes with the amplitudes found in the previous step.

Let a mode of the electromagnetic field be characterized by the wave vector $q$, the angular frequency $\omega = cq$, and the polarization $\sigma$. Denoting the variables $q$ and $\sigma$ simply by $q$ in the labels of the states, we write the coherent state of amplitude $|z_q|$ and phase $\arg z_q$ in the form

$$|z_q\rangle = \exp(-\tfrac{1}{2}|z_q|^2)\exp(z_q a_q{}^\dagger)|0\rangle$$

$$= \exp(-\tfrac{1}{2}|z_q|^2) \sum_{n=0}^{\infty} \frac{(z_q^n)}{n!}|n\rangle \tag{111}$$

Let us use a representation of the energy eigenstates in terms of Hermite polynomials $H_n(x)$

$$|n\rangle = (2^n n! \sqrt{\pi})^{-1/2}\exp(-x^2/2)H_n(x)e^{in\omega t} \tag{112}$$

This yields for the coherent state $|z_q\rangle$ the representation

$$\psi(x) = \exp(-\tfrac{1}{2}|z_q|^2)\exp(-x^2/2) \sum_{n=0}^{\infty} \{(z_q e^{i\omega t})^n/[n!(2^n\sqrt{\pi})]^{1/2}\} H_n(x)$$

$$= \exp(-\tfrac{1}{2}|z_q|^2)\exp(-x^2/2)\exp(-z^2 e^{-2i\omega t} + 2xz e^{i\omega t}) \tag{113}$$

In the last form the generating function of the Hermite polynomials was used. The corresponding autocorrelation function of the probability density function, obtained by averaging over the time $t$ or the phase of $z_q$, is, for $|z_q| \ll 1$,

$$P_q(\tau, x) = \langle |\psi|_t^2 |\psi|_{t+\tau}^2 \rangle$$

$$= [1 + 8x^2|z_q|^2(1 + \cos \omega\tau) - 2|z_q|^2]\exp(-x^2/2) \tag{114}$$

Integrating over $x$ from $-\infty$ to $\infty$, we find the autocorrelation function

$$A^1(\tau) = (2\sqrt{\pi})^{-1/2}(1 + 2|z_q|^2 \cos \omega\tau) \tag{115}$$

This result shows that the probability contains a constant background with small superposed oscillations of frequency $\omega$. Physically, the small oscillations in the total probability describe a particle which has been emitted, or created, with a slightly oscillating rate, and which is more likely to be found in a measurement at a certain time than at other times in the same place. Note that for $z_q = 0$ the coherent state becomes the ground state of the oscillator, which is also an energy eigenstate, and therefore stationary and free of oscillations.

We now determine the amplitude $z_q$ with which the field mode $q$ is represented in the physical electron. One way to do this is to let a bare particle dress itself through its interaction with the electromagnetic field, i.e., by performing first-order perturbation theory with the interaction Hamiltonian

$$H' = A_\mu j^\mu = -(e/c)\mathbf{v} \cdot \mathbf{A} + e\phi \tag{116}$$

where $\mathbf{A}$ is the vector potential and $\phi$ the scalar electric potential. Another way is to Fourier expand the electric potential $e/4\pi r$ of a charged particle in a box of volume $V$. In both ways we obtain

$$|z_q|^2 = (e/q)^2 (hcqV)^{-1} \tag{117}$$

Considering now all modes of the electromagnetic field, we obtain from the single-mode result of Eq. (115)

$$A(\tau) = C \prod_q (1 + 2|z_q|^2 \cos \omega\tau) = C\left(1 + 2\sum_q |z_q|^2 \cos \omega\tau\right)$$

$$= C\left(1 + 2(V/2^3\pi^3)\int d^3q\, |z_q|^2 \cos \omega\tau\right) \tag{118}$$

Here we have again used the smallness of $z_q$ and have introduced a constant $C$. Using Eq. (117), we obtain

$$A(\tau) = C\left(1 + 2(V/2^3\pi^3)(4\pi/V)(e^2/2hc)\int \frac{dq}{q}\cos \omega\tau\right)$$

$$= C\left(1 + 2\frac{\alpha}{\pi}\int \cos(\omega\tau)\frac{d\omega}{\omega}\right) \tag{119}$$

Here $\alpha = e^2/4\pi hc$ is the fine-structure constant $\frac{1}{137}$. The first term in large parentheses is unity and represents the constant background, or the dc part. The autocorrelation function for the relative, or fractional density fluctuations, or for current density fluctuations in the beam of charged particles, is obtained therefore by dividing the second term in large parentheses by the first term. The constant $C$ drops out when the fractional fluctuations are considered. According to the Wiener–Khintchine theorem, the coefficient of $\cos \omega\tau$ is the spectral density of the fluctuations, $S_{|\psi|^2}$, or $S_j$ for the current density $\mathbf{j} = e(\mathbf{k}/m)|\psi|^2$

$$\frac{S_{|\psi|^2}^{(f)}}{\langle|\psi|^2\rangle} = \frac{S_j^{(f)}}{\langle j\rangle^2} = 2\frac{\alpha}{\pi}\frac{1}{f} \simeq \frac{4.6 \times 10^{-3}}{f} \tag{120}$$

Here we have included the total number $N$ of charged particles which are observed simultaneously in the denominator, because the noise contributions from each particle are independent. This is the coherent Handel effect.

The coherent effect is related to the conventional Handel effect. If a beam of charged particles is scattered, passes from one medium into another medium (e.g., at contacts), is emitted, or is involved in any kind of tranitions, the amplitudes $z_q$ which describe its field will change. Then, even if the initial state was prepared to have a well-determined energy, the final state will have an indefinite energy, with an uncertainty determined by the difference between the new and old $z_q$ amplitudes, $\Delta z_q$. This, however, is just the bremsstrahlung amplitude $\Delta z_q$. We thus regain the familiar quantum $1/f$ effect, according to which the small energy losses from bremsstrahlung of infraquanta yield a final state of indefinite energy, and therefore lead to fluctuations of the process rate, or cross section, of the process in which the electrons have participated, and which has occasioned the bremsstrahlung in the first place. A calculation of the piezoelectric Handel effect $1/f$ noise performed by Handel and Musha in 1983, which deals with phonons as infraquanta, was phrased in terms of the coherent field amplitudes $z_q$ for the first time, although it is concerned only with the usual quantum $1/f$ effect. It has $\alpha$ substituted by the piezoelectric coupling constant $g$.

The assumptions included in the derivation of the above coherent $1/f$ Handel effect are

(1) The "bare particle" does not have compensating energy fluctuations which could cancel the fluctuations present in the field. The latter are due to the interaction with distant charges, and have nothing to do with the bare particle. Therefore, this assumption is quite reasonable.

(2) The experimental conditions do not alter the physical definition of the charged particle as a bare particle dressed by a coherent state field. This second assumption depends on the experimental conditions.

One way to understand this second assumption is based on the spatial extent of the beam of particles or of the physical sample containing charged particles, and is specifically based on the number of particles per unit length of the sample. According to this model, the coherent state in a conductor or semiconductor sample is the result of the experimental efforts directed towards establishing a steady and constant current, and is therefore the state defined by the collective motion, i.e., by the drift of the current carriers. It is expressed in the Hamiltonian by the magnetic energy $E_m$, per unit length, of the current carried by the sample. In very small samples or electronic devices, this magnetic energy

$$E_m = \int (B^2/8\pi)\, d^3x = (nevS/c)^2 \ln(R/r) \qquad (121)$$

is much smaller than the total kinetic energy $E_k$ of the drift motion of the individual carriers

$$E_k = \sum mv^2/2 = nSmv_2/2 = E_m/s \tag{122}$$

Here we have introduced the magnetic field **B**, the carrier concentration $n$, the cross-sectional area $S$ and radius $r$ of the sample, the radius $R$ of the electric circuit and the "coherent ratio"

$$s = E_m/E_k = 2ne^2S/mc^2 \ln(R/r) = 2e^2N'/mc^2 \tag{123}$$

where $N' = nS$ is the number of carriers per unit length of the sample and the natural logarithm $\ln(R/r)$ has been approximated by one in the last form. We expect the observed spectral density of the mobility fluctuations to be given by a relation of the form

$$(1/\mu^2)S_\mu(f) = [1/(1 + s)](2\alpha A/fN) + [s/(1 + s)](2\alpha/\pi fN) \tag{124}$$

which can be interpreted as an expression of the effective Hooge constant if the number $N$ of carriers in the (homogeneous) sample is brought to the numerator of the left-hand side. Equation (124) needs to be tested experimentally. In this equation $A = 2(\Delta\mathbf{v}/c)^2/3\pi$ is the usual nonrelativistic expression of the infrared exponent, present in the familiar form of the conventional Handel effect. This equation does not include the quantum $1/f$ noise in the surface and bulk recombination cross sections, in the surface and bulk trapping centers, in tunneling and injection processes, in emission, or in transitions between two solids.

Note that the coherence ratio $s$ introduced here equals unity for the critical value $N' = N'' = 2 \times 10^{12}$/cm, e.g., for a cross section $S = 2 \times 10^{-4}$ cm$^2$ of the sample when $n = 10^{16}$. For small samples with $N' \ll N''$ only the first term survives, and for $N' \gg N''$ only the second term remains in Eq. (124).

## VIII. Concluding Remarks

For the quantum modulation of electron beams when interacting with crossing laser beams in the presence of an interacting medium (Schwarz–Hora effect), the initial experimental achievements by Schwarz using crystals as a medium were extended by using molecules as a medium by Andrick and Langhans and by Weingartshofter *et al.* The results are identical in both cases. The energy of some electrons receives an upshift or a downshift by multiples of the photon energy, while the angle of the scattered electrons during the interaction is not much affected by the modulation process in both cases. The dependence of the degree of modulation on the angle between the electrons and the E vector of the laser exponentially changes in both cases.

The re-emission of the photons from the quantum-modulated electron beam when entering a nonluminescent screen was reported only when using crystals. In the case of the experiments with molecules, the modulation length —at least as calculated from the first-order theory—was far too short. Indirect experimental verifications of the modulation state of electrons were concluded from the very strong amplification of microwaves in a non-equilibrium gas discharge (Rosenberg *et al.*). There a modulation state transport of microwave photons is concluded between the collisions. Further, a self-interaction process in the $1/f$ noise (Handel effect) can be based on the modulation processes.

Both processes with crystals or molecules are describable as inverse bremsstrahlung. This is more a particle picture of the mechanism. For the coherence processes as an essential wave property, the theory is not nearly as well developed as coherence in optics.

The second-order correlation, which is similar to the Hanbury-Brown–Twiss effect, has not been treated. The present-day coherence is based on correlations of the first order and results in the reported long beating wavelength (as a basic difference between optics and electron waves) and the modulation lengths. The modulation process turns out to be an additional property of the wave nature of electrons which is not understandable in the particle picture. The possibility of a similar property of the optical case may be indicated.

The basic result of the quantum nature of the modulation—as elaborated in the first paper—was based on a consideration of a correspondence principle for electromagnetic interaction where a borderline is defined between the quantum and the classical interactions. In the experiments with crystals and with molecules, the quantum regime was given. A recent intensity dependence of the scattering observed by Weingartshofer *et al.* directly confirms the borderline. Other contradictory-looking experiments, e.g., that by Kruit *et al.* and by Boreham *et al.* for laser ionization of low-density helium gas and subsequent acceleration of the electrons by the nonlinear force, each turn out to be on different sides of the borderline. This is also illustrated since in one case there is multiphoton ionization and in the other case there is field-effect-like Keldysh ionization.

This borderline of correspondence also explains why electrons in radio waves behave classically and in optics quantum mechanically if not extremely high laser intensities are used. For wavelengths below the Compton wave length, subrelativistic interaction with electrons is always in the quantum range, and classical interaction can only be relativistic. Blackbody radiation of these relativistic intensities causes a breakdown of the Fermi statistics in favor of strong coupling (Eliezer et al., 1986).

The results of the quantum modulation are related to the Forrester– Gudmundsen effect, to the transition radiation (Ginzburg and Frank), stimulated Čerenkov absorption, elementary particle heterodyne, and coherent excitation for the inversion of nuclei for gamma-ray lasers. The far-range interaction of photons (Aspect et al., 1982a,b) may be of interest also for the long-range wave properties of electrons. A further relation is to the work of Farkas (1978), Freeman et al. (1986), and Najorskij (1986).

### REFERENCES

Aharoov, Y., and Bohm, D. (1959), *Phys. Rev.* **115**, 485.
Andrick, D., and Langhans, L. (1976). *J. Phys.* **B9**, L459.
Andrick, D., and Langhans, L. (1978). *J. Phys.* **B11**, 2355.
Arutyunyan, V. M., and Oganesyan, S. G. (1977). *Sov. Phys. JETP* **45**, 244.
Aspect, A., Grangier, P., and Roger, G. (1982a). *Phys. Rev. Lett.* **49**, 91.
Aspect, A., Dalibard, J., and Roger, G. (1982b). *Phys. Rev. Lett.* **49**, 1804.
Baldwin, G. C., Solem, J. C., and Goldanski, V. I. (1982). *Rev. Mod. Phys.* **53**, 687.
Baldwin, K. G. H., and Boreham, B. H. (1981). *J. Appl. Phys.* **52**, 2627.
Beaulieu, A. J. (1970). *Appl. Phys. Lett.* **16**, 504
Beaulieu, A. J. (1972). *In* "Laser Interaction and Related Plasma Phenomena," (H. Schwarz and H. Hora, eds.), Vol. 2, p. 1. Plenum, New York.
Becchi, C., and Morpurgo, G. (1971). *Phys. Rev.* **D4**, 288.
Becchi, C., and Morpurgo, G. (1972). *Appl. Phys. Lett.* **21**, 123.
Ben-Aryeh, Y. (1982). *Int. Quantum Electron. Conf., Munich, June.*
Ben-Aryeh, Y., and Mann, A. (1985). *Phys. Rev. Lett.* **54**, 1020.
Ben-Aryeh, Y., Felsteiner, J., Politch, J., and Rosenberg, A. (1983). *In* "Laser Interaction and Related Plasma Phenomena (H. Hora and G. H. Miley, eds.), Vol. 6, p. 165. Plenum, New York.
Bepp, H. B., and Gold, A. (1966). *Phys. Rev.* **143**, 1.
Bergmann, E. E. (1973). *Nuovo Cim.* **14B**, 243.
Bohm, D., and Pines, D. (1953). *Phys. Rev.* **92**, 609.
Bohr, N. (1919). *Kopenhagen Acad.* **4**, 1.
Boreham, B. W., and Hora, H. (1979). *Phys. Rev. Lett.* **42**, 776.
Boreham, B. W., and Luther-Davies, B. (1979). *J. Appl. Phys.* **50**, 2533.
Born, M., and Wolf, E. (1959). "Principles of Optics." Pergamon, Oxford.
Brautti, G. (1980). *Nuovo Cim. Lett.* **27**, 38.
Bulshov, L. A., Dyhhne, A. M., and Roslyakov, V. A. (1972). *Phys. Lett.* **42A**, 259.
Carter, J. L., and Hora, H. (1971). *J. Opt. Soc. Amer.* **61**, 1640.
Caulfield, H. J. (1984). *Natl. Geogra.* **165**, 366.
Chang, C. S., and Stehle, P. (1972). *Phys. Rev.* **A5**, 1928.
Chen, C. K., Shepperd, J. C., Piestrup, M. A., and Pantel, R. H. (1978). *J. Appl. Phys.* **49**, 41.
Dirac, P. A. M. (1931). "The Principles of Quantum Mechanics." Clarendon, Oxford.
Eliezer, S., Ghatak, A., and Hora, H. (1986). "Equations of State." Cambridge Univ. Press, Cambridge.

Elliott, J. A. (1972). *J. Phys.* **C5**, 1976.

Farago, P. S., and Sillitto, R. M. (1973). *Proc. R. Soc. Edinburgh Sect.* **A71**, 305.

Farkas, Gy. (1987). *In* "Multiphoton Processes" (J. H. Eberly and P. Lambropoulos, eds.), p. 81. Wiley, New York.

Favro, L. D., and Kuo, P. K. (1971a). *Phys. Rev.* **D3**, 2934.

Favro, L. D., and Kuo, P. K. (1971b). *Phys. Rev.* **D3**, 2931.

Favro, L. D., and Kuo, P. K. (1973). *Phys. Rev.* **A7**, 866.

Favro, L. D., Fradkin, D. M., and Kuo, P. K. (1970). *Phys. Rev. Lett.* **25**, 202.

Favro, L. D., Fradkin, D. M., Kuo, P. K., and Rollnik, W. B. (1971). *Appl. Phys. Lett.* **19**, 378.

Fedorov, V. V. (1979). *Phys. Lett.* **75A**, 137.

Ferrell, R. A. (1958). *Phys. Rev.* **111**, 1214.

Forrester, A., Gudmundsen, R., and Johnson, P. (1955). *Phys. Rev.* **99**, 1961.

Frank, I. M. (1960). *Sov. Phys. Usp.* **131**, 702.

Freeman, R. R., McIlrath, T. J., Bucksbaum, P. H., and Bashkansky, M. (1986). *Phys. Rev. Lett.* **57**, 3156.

Gabor, D. (1949). *Proc. R. Soc. Edinburgh Sect.* **A197**, 454.

Gaunt, J. A. (1930). *Proc. R. Soc. Edinburgh Sect.* **A126**, 654.

Geltman, (1973). *J. Quant. Spectrosc.* **13**, 601.

George, E. P. (1975). *Rep. Univ.* New South Wales, July 24.

Ghatak, A. K., Hora, H., Wang Run-Wen, and Viera, G. (1984). *Proc. Int. Conf. High Power Particle Beams, 5th, Sept. 1983.*

Gibbon, J. D., and Bullough, R. K. (1972). *J. Phys.* **C5**, L80.

Ginzburg, V. L., and Frank, I. M. (1946). *Zh. Eksp. Teor. Fiz.* **16**, 15.

Glaser, W. (1952). "Grundlagen der Elektronenoptik." Springer, Beolin.

Hadley, R., Lynch, D. W., Stanek, E., and Rossaner, E. A. (1971a). *Appl. Phys. Lett.* **19**, 145.

Hadley, R., Lynch, D. W., and Stanek, E. (1971b). *Appl. Phys. Lett.* **19**, 145.

Hadley, G. R., Stanek, E. J., and Good, Jr., R. H. (1972). *J. Appl. Phys.* **43**, 144.

Hanbury-Brown, R., and Twiss, R. Q. (1959). *Nature (London)* **177**, 27.

Handel, P. (1975). *Phys. Rev. Lett.* **34**, 1492.

Handel, P. H. (1980). *Phys. Rev.* **A22**, 745.

Handel, P. H. (1983). *In* "Noise in Physical Systems" (M. Savelli *et al.*, eds.), p. 97. Elesevier, Amsterdam.

Handel, P. H., and Musha, T. (1983). *In* "Noise in Physical Systems" (M. Savelli *et al.*, eds.), p. 76. Elesevier, Amsterdam.

Haroutunian, V. M., and Avetission, H. K. (1973). *Phys. Lett.* **44A**, 281.

Hawkes, P. W. (1974a). *Optik* **40**, 539.

Hawkes, P. W. (1974b). *Optik* **41**, 64.

Hawkes, P. W. (1978). *Adva. Opt. Electron. Microsc.* **7**, 101.

Hooge, F. N. (1969). *Phys. Lett.* **A29**, 139.

Hora, H. (1960). *Optik* **17**, 409.

Hora, H. (1969). *Phys. Fluids* **12**, 182.

Hora, H. (1970). *Phys. Stat. Solidi* **42**, 131

Hora, H. (1972). *Proc. Int. Conf. Light Scattering Solids, 2nd* p. 128.

Hora, H. (1974). *Phys. Fluids* **17**, 1042.

Hora, H. (1975). *Nuovo Cim.* **26B**, 295.

Hora, H. (1977). *Phys. Stat. Solidi (b)* **80**, 143.

Hora, H. (1978). *Phys. Stat. Solidi (b)* **86**, 685.

Hora, H. (1981a). *Nuovo Cim.* **64B**, 1.

Hora, H. (1981b). "Physics of Laser Driven Plasmas." Wiley, New York.

Hora, H. (1982). *Opt. Commun.* **41**, 268.

Hora, H. (1983). *Optik* **66**, 57.

Hora, H., and Schwarz, H. (1973). German Pat. No. 2160656; US Pat. No. 3730979.

Hora, H., Lalousis, P., and Eliezer, S. (1984). *Phys. Rev. Lett.* **53**, 1650.

Hutson, A. R. (1970). *Appl. Phys. Lett.* **17**, 343.

Jayes, E. T. (1978). *In* "Novel Sources of Coherent Radiation" (S. F. Jacobs, M. Sargent, III, and M. O. Scully, eds.) p. 1. Addison-Wesley, Reading, Mass.

Janner, A. G. M., and La Fleur, P. L. (1971). *Phys. Lett.* **36A**, 109.

Jung, C., and Krüger, H. (1978). *Z. Phys.* **A287**, 7.

Kalmykova, S. S. (1982). *Sov. Phys. Uzp.* **25**, 620.

Kondo, J. (1971). *J. Appl. Phys.* **42**, 4458.

Korobochko, Y. N. S., Grachev, B. D., and Mineev, V. I. (1973). *Sov. Phys. Tech. Phys.* **17**, 1880.

Kramers, A. (1923). *Philos. Mag.* **46**, 836.

Kroll, N. M., and Watson, M. (1973). *Phys. Rev.* **A8**, 804.

Krüger, H., and Jung, C. (1978). *Phys. Rev.* **A17**, 1706.

Kruit, P. K., Kimman, J. K., Müller, H. G., and Van der Wiel, M. J. (1983). *Phys. Rev.* **A28**, 248.

La Fleur, P. L. (1971). *Lett. Nuovo Cim.* **2**, 571.

Landau, L. D., and Lifshitz, E. M. (1960). "Electrodynamics of Continuous Media." Pergamon, Oxford.

Lipkin, H., and Peshkin, M. (1971). *Appl. Phys. Lett.* **19**, 313.

McWhorter, A. L. (1955). Rep. *MIT Lincoln Lab.* No. 80.

Mandel, L. (1982). *Phys. Rev. Lett.* **47**, 136.

Marcuse, D. (1971a). *J. Appl. Phys.* **42**, 2259.

Marcuse, D. (1971b). *J. Appl. Phys.* **42**, 2255.

Maue, A. W. (1932). *Ann. Phys.* **13**, 161.

Makhviladze, T. M., and Shelepin, L. A. (1972). *Sov. J. Nucl. Phys.* **15**, 335.

Mizuno, K., Ono, S., and Suzuki, N. (1973). *Nature (London)* **244**, 13.

Möllenstedt, G. (1949). *Optik* **5**, 449.

Möllenstedt, G., and Bayh, (1961). *Naturwissenschuften* **48**, 400.

Morpurgo, G. (1976). *Ann. Phys.* **97**, 519.

Najorskij, G. A. (1986). *Nucl. Inst. Meth. A* **248**, 31.

Oganesian, S. G., and Engibargan, V. A. (1984). *Sov. Phys. Tech. Phys.* **29**, 859.

Oliver, B. M., and Cutler, L. S. (1970). *Phys. Rev. Lett.* **25**, 273.

Palmer, A. J. (1983). *Appl. Phys. Lett.* **42**, 1011.

Paul, H. (1986). *Rev. Mod. Phys.* **58**, 209.

Peierls, R. (1976). *Proc. R. Soc. Edinburgh Sect.* **A347**, 475.

Peres, A. (1979). *Phys. Rev.* **A20**, 2627.

Pfeiffer, L., Rousseau, D. L., and Hutson, A. R. (1972). *Appl. Phys. Lett.* **20**, 147.

Piestrup, M. A., Rothbart, G. H., Fleming, R. N., and Pantell, P. H. (1975). *J. Appl. Phys.* **46**, 132.

Renard, R. (1964). *J. Opt. Soc. Am.* **61**, 1640.

Rivlin, L. A. (1971). *JETP Lett.* **13**, 257.

Rosenberg, A., Felsteiner, J., Ben-Aryeh, Y., and Politch, J. (1980). *Phys. Rev. Lett.* **45**, 1787.

Rosenberg, A., Ben-Aryeh, Y., Politch, J., and Felsteiner, J. (1982). *Phys. Rev.* **A25**, 1160.

Rubin, P. L. (1970). *JETP Lett.* **11**, 239.

Ruthemann, G. (1941). *Naturwissenschaften* **29**, 648.

Salat, A. (1970). *J. Phys.* **C3**, 2509.

Schmieder, R. W. (1972). *Appl. Phys. Lett.* **20**, 516.

Schwarz, H. (1971a). *Appl. Phys. Lett.* **19**, 148.

Schwarz, H. (1971b). *Trans. N. Y. Acad. Sci.* **33**, 150.

Schwarz, H. (1972a). *Proc. Int. Conf. Light Scattering Solids, 2nd* p. 125.

Schwarz, H. (1972b). *Appl. Phys. Lett.* **20**, 148.

Schwarz, H. and Hora, H. (1969). *Appl. Phys. Lett.* **15**.

Sherif, T. S., and Handel, P. H. (1982). *Phys. Rev.* **A26**, 596.

Sinha, K. P. (1972). *Curr. Sci.* **41**, 124.

Spitzer, L., Jr., and Härm, (1953). *Phys. Rev,* **89**, 977.

Steinmann, W. (1961). *Z. Phys.* **163**, 92.

Steinmann, W. (1968). *Phys. Stat. Soc.* **28**, 437.

Suguna, A., and Shrivastava, K. N. (1980). *Phys. Rev.* **B22**, 2343.

Swinger, D. N., and Nilsen, C. S. (1975). *IEEE Proc.* **63**, 1074.

Van der Ziel, A. (1986). "Noise in Solid State Devices and Circuits." Wiley, New York.

Van der Ziel, A., and Handel, P. (1985). *IEEE Trans.* **ED-32**, 1802.

Van Fliet, K. M., Handel, P. H., and Van der Ziel, A. (1981) *Physica* **108A**, 511.

Van Zandt, L. L., and Meyer, J. W. (1970). *J. Appl. Phys.* **41**, 4470.

Varshalovich, D. A., and Dyakonov, M. A. (1970). *JETP Lett.* **11**, 411.

Varshalovich, D. A., and Dyakonov, M. A. (1971a). *Phys. Lett.* **35A**, 277.

Varshalovich, D. A., and Dyakonov, M. A. (1971b). *Sov. Phys. JETP* **33**, 51.

von Weitzsäcker, C. F. (1934). *Z. Phys.* **88**, 612.

Vysotskii, V. I., Vorontsov, V. I., and Kuzmin, R. N. (1980). *Sov. Phys. JETP* **51**, 49.

Walls, D. F. (1983). *Nature (London)* **306**, 141.

Weingartshofer, A., Willman, K. W., and Clarke, E. M. (1974). *J. Phys.* **B7**, 79.

Weingartshofer, A., Holmes, J. K., Claude, G., Clarke, E. M., and Kruger, H. (1977). *Phys. Rev. Lett.* **49**, 268.

Weingartshofer, A., Clarke, E. M., Holmes, J. K., and Jung, C. (1979). *Phys. Rev.* **A19**, 2371.

Weingartshofer, A., Holmes, J. K., Sabbagh, J., and Chin, S. L. (1985). *J. Phys.* **B16**, 1805.

Williams, A. (1935). *Kgl. Dansk Vid. Selsk.* **13**, 4.

Wilson, G. V. H., George, E. P., and Hora, H. (1978). *Aust. J. Phys.* **31**, 55.

Young, T. (1807). *In* "Course of Lectures on Natural Philosophy" (J. Johnson, ed.) Vol. 1.

Zeldovich, B. Ya. (1972). *Sov. Phys. JETP* **34**, 70.

# Device Developments for Optical Information Processing

## JOHN N. LEE AND ARTHUR D. FISHER

*Naval Research Laboratory*
*Washington, DC 20375*

## I. INTRODUCTION

Optical techniques have long held an attraction for the processing of massive amounts of information, for example, that encountered in imagery and in data streams from large sensor arrays. Optics intrinsically offers a very high temporal–spatial-throughput bandwidth. Furthermore, the ability to create high-resolution, two-dimensional (2D) data fields and to provide parallel interconnect paths between these 2D information planes with optics could offer an effective avenue for the implementation of highly parallel computations. Fulfillment of this promise is dependent on both the formulation of algorithms and architectures that exploit the unique advantages of optics, and the development of appropriate optical devices. This article is concerned with specific device developments that have led to demonstrations of unique capabilities, and with the developments required for fuller exploitation of optics.

It is important to recognize that algorithmic and device advancements intimately impact each other, so that one cannot occur in isolation from the other. This interplay can be illustrated by the historical developments in optical processing. The Fourier-transforming property of a simple lens has long been known, and the configuring of passive optical systems to perform transformations culminated in the mid-1960s with what many consider the first major success for optical processing—the implementation of the synthetic aperture radar (SAR) algorithm for image formation from side-looking radar (Cutrona et al., 1970). This success, plus the invention of the laser and its application to the making of holograms, sparked the generation of numerous optical processing concepts (Tippet et al., 1965). Most notable among these was the matched-filter correlator (VanderLugt, 1963). A period followed where it was recognized that significantly better optical devices were required for effective concept demonstrations. Not all applications could tolerate off-line processing with photographic film as in SAR. Intensive activity ensued in the development of spatial light modulators (SLMs), particularly two-dimensional, real-time devices that could substitute for photographic film. Progress in the development of the 2D devices turned out much slower than anticipated; however, significant progress was achieved by the mid-1970s with one-dimensional (1D) spatial light modulators, specifically the acousto-optic (AO) type (Berg and Lee, 1983). Impressive demonstrations of capability were achieved in 1D signal processing applications such as spectral analysis and correlation. For example, a 1000-point spectrum analysis performed in a few microseconds is now considered feasible. Additionally, the availability of good AO devices led to the investigation of 2D algorithms and architectures that allow the use of orthogonal 1D input devices to provide 2D parallelism. Such algorithms include folded spectrum analysis, matrix processing, and generation of ambiguity functions and other time–frequency representations of temporal signals (Turpin, 1981).

As improved devices became available for the implementation of optical processors, a second impact arose, apart from the interplay with algorithmic development. Namely, the size, weight, and power-consumption advantages to be obtained with the optical devices often in themselves became the rationale for employing optical processing. The emphasis on these physical parameters is a trend that will continue to drive device developments in the future.

At present, research work is aimed toward applying optical techniques to a broader range of applications, particularly those where conventional electronic approaches are encountering increasing difficulty. For example, there have been exciting recent proposals for exploiting the inherent parallel-processing and interconnect capabilities of optics to implement such intrinsi-cally parallel computational models as associative networks (Psaltis and Farhat, 1984, 1985; Fisher et al., 1984, 1985), cellular automata (Sawchuk and Jenkins, 1985; Yatagai, 1986), or generalized finite-state machines (Huang, 1985). There is hope that these new concepts will be able to address the demanding processing requirements of such sophisticated problem domains as multisensor processing, robot vision, and symbolic processing for artificial intelligence applications. Most of these new optical directions identify shortfalls in specific optical devices such as 2D SLMs.

The types of active devices required for an optical processor are relatively few: light sources, modulators or control devices for the light beams, and photodetectors. The device developments that have impacted optical processing have clearly not all been motivated by optical processing needs. Much of the development has been due to the impetus of other application areas such as optical fiber communications, flat screen and large area displays, and remote sensing. Consequently, the device characteristics stressed in development are not always those that are required for optical signal processing purposes. It is the aim of this article to emphasize those developments in the three classes of devices that do address optical processing needs, and to point out specific promising directions for development.

## II. Light Sources

### A. Coherent Sources

The invention of the laser provided an impetus for increased interest in optical signal processing; it made available high-brightness beams, with the possibility for both amplitude and phase control. Only recently has it been possible to obtain most of the laser parameters required in a practical optical processor; these include: small size, high optical output power, high electrical-to-optical conversion efficiency, single-wavelength operation, mode stability, and long lifetime. Each of these has a readily apparent effect on processor performance; e.g., higher optical power results in a larger potential dynamic range. However, there are interrelationships between these parameters which require that most, if not all, the parameters be achieved simultaneously, e.g., poor mode stability could negate the dynamic range improvement obtainable with higher power. Thus, optical processing often imposes more stringent requirements than many other laser applications.

Gas lasers such as helium–neon (He–Ne) and argon ion, and solid-state lasers such as Nd: YAG, have superior performance in one or two parameters such as power or mode shape. For laboratory experiments and demonstrations of principle, He–Ne lasers are almost invariably used unless specific deficiencies are identified. Where optical power requirements are modest ($< 10$ mW), it is possible to construct compact optical signal processing systems using He–Ne lasers that can be mounted onto standard electronic racks (Lindley, 1983). However, to obtain the best performance over all parameters, it is becoming clear that the best choice, if not already for most applications, then almost certainly in the future, is the semiconductor laser diode, using III–V compounds like GaAlAs. The following discussion considers the implications for optical processors of the various unique properties of laser diodes.

The most prominent aspect of laser diodes is their small size. The lateral dimensions of the diode junction, or stripe where lasing occurs, is a fraction of a micrometer thick (in the direction of current flow), and the in-plane width can range from a few micrometers to more than a millimeter, depending on the type of laser. Figure 1 illustrates the general size and construction for one common type. (Many good reviews of the various types of laser diodes exist in the literature.) The small lasing area results in a high degree of beam

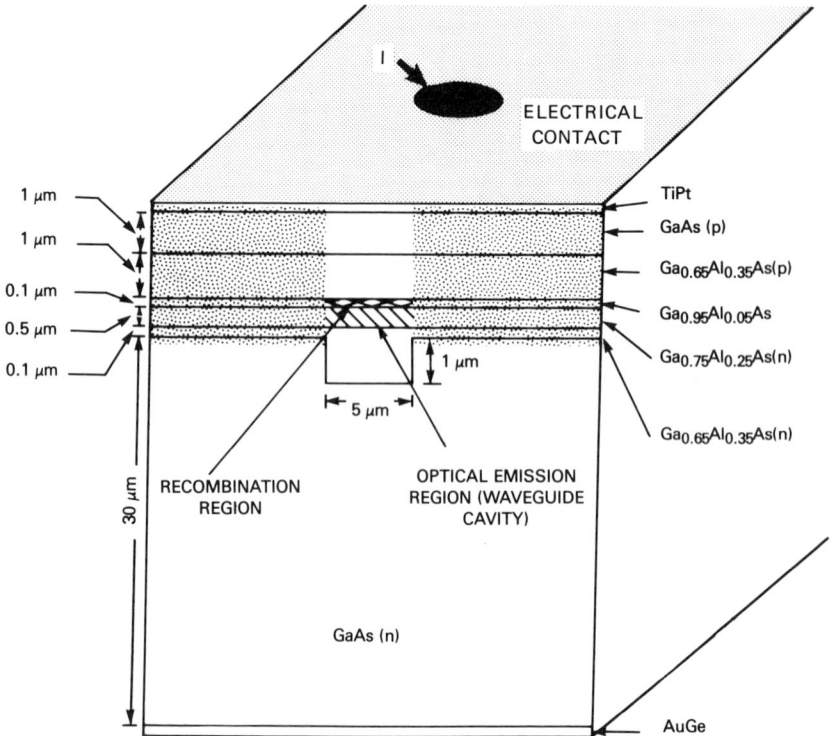

FIG. 1. Construction of a typical large-optical-cavity laser diode, where the optical waveguide cross section ($Ga_{0.75}Al_{0.25}As$ layer) is much larger than the cross section of the active gain region ($Ga_{0.95}Al_{0.05}As$ layer).

divergence due to optical diffraction, with half-intensity beam widths as high as $60°$. Another property is two different beam divergences, arising from disparity between the two lateral dimensions of the laser junction; this makes it difficult to produce a uniform circular beam when using common spherical optics. Obtaining high-quality, low-aberration optical wavefronts requires low $f$-number beam-shaping optics; microscope objectives are commonly used. However, it is important to recognize that microscope objectives are usually designed to operate over a broad wavelength spectrum and are not optimized for particular laser wavelengths nor designed for highly coherent light. Further, most microscope objectives are not intended to produce collimated light, but rather to image at some specific distance (usually 160 mm). Recently, attention has been given to the design of beams-shaping optics for lasers at 800–850 nm and to designing anamorphic optical systems to produce circular beams from laser diodes (Forkner and Kuntz, 1983). Optical output power is another primary parameter of interest in any laser. Laser diode outputs are now comparable to those of the much bulkier gas lasers. Single-element lasers have been developed with both spatial and temporal single-mode operation and with powers of greater than 50 mW at 830 nm (Goldstein et al., 1985) and at 1300 nm (Kobayashi and Mito, 1985; Kitamura et al., 1985). Maximum output power is limited primarily by thermal damage at the laser facets that serve as the laser-cavity mirrors. The threshold for such damage has been raised by a number of techniques. One scheme is to employ a large optical cavity structure (Kressel and Butler, 1975), as illustrated in Fig. 1. In this structure, the optical resonator cross section (the $Ga_{0.75}Al_{0.25}As$ layer in the example of Fig. 1) is much larger than the cross section of the region where optical gain occurs (the $Ga_{0.95}Al_{0.05}As$ layer); hence, the optical power density is reduced by being spread over a larger area at the output facets. A second approach is to selectively dope the surface region of the facet to increase the bandgap relative to that of the remainder of the cavity. The near-surface region thereby becomes relatively transparent to the laser wavelength and absorbs less power (Blauvelt et al., 1982). Finally, there have been beneficial improvements in fabrication techniques.

A monolithic array of laser diodes on a single substrate can provide much higher powers than a single laser. Outputs of up to 2.6 W cw have been reported using 40-element arrays (Scifres et al., 1982). Even higher powers should be available with the addition of more elements to the array. However, the use of arrays introduces new complications. The individual elements produce beams which are coherent with respect to each other, due to the spreading of optical fields between neighboring elements. Unfortunately, the wave function of the adjacent elements has a strong tendency to be in an antiphase relationship (Paoli et al., 1984), which results in a double-lobed far-field beam pattern. Potential techniques for overcoming this problem are to inject into the array the beam of a very-low-power, well-behaved master laser to lock the output wavelength and phase of the entire array (Goldberg *et al.*, 1985), or to couple the lasers to parallel single-mode waveguides that are linked by "Y" junctions (Welch *et al.*, 1986).

Intimately connected with power output is the efficiency of electrical-to-optical power conversion. In laser diodes this conversion efficiency can be as high as 35% (Kressel and Butler, 1975). This may be compared to efficiencies of $< 1\%$ for He–Ne, argon, and Nd: YAG lasers. Thus, another attraction of optical implementations is often the low electrical power required for the light source.

The wavelength of emission of a laser diode depends on the bandgap of the material from which it is fabricated. Lasers have been most commonly constructed for the 800–850 nm region using GaAlAs layers of varying composition. More recently, good-quality lasers have also been produced at around 1300 and 1500nm, using quaternary compounds such as InGaAsP. The motivation for these wavelengths is low-loss and low-dispersion trans-mission in optical fibers. However, for optical processing applications, the shorter wavelength is generally more advantageous. In many processors, diffraction of light by grating-like structures is used to perform operations such as data multiplication and beam steering. Since the diffraction efficiency varies approximately as the square of the inverse of the optical wavelength (Gaylord and Moharam, 1985), lasers at wavelengths shorter than 800 nm are desirable. Lasers diodes at visible wavelengths as low as 653 nm have recently been developed using AlGaInP (Ishikawa *et al.*, 1986; Ikeda *et al.*, 1985). Good mode quality, modest cw power ( $\sim 7$ mW), and room-temperature operation have been achieved at 679 nm. Further reductions in wavelength will be limited by the bandgap and transition efficiencies of various materials. AlGaInP might produce wavelengths as short as 580 nm. While the bandgap of GaP corresponds to an optical wavelength near 500 nm, the bandgap transition is indirect, resulting in poor efficiency (Dean and Thomas, 1966).

In many optical processing schemes, both phase and amplitude of the light must be modulated. Beam coherency and linearity of modulation of laser diodes are therefore of concern. It is often desirable to directly modulate the laser diode beam intensity by control of the electrical drive current. Not only is this method straightforward, but it is possible to modulate at extremely high rates and to obtain linear modulation over large dynamic range, in contrast to many electro-optic/acousto-optic techniques. Figure 2 shows the general characteristic of optical power versus electrical drive current for laser diodes; most notably, optical power increases slowly below the lasing threshold and increases rapidly above threshold. With present lasers it is possible for the region above threshold to be made "kink-free" and linear, i.e., linear optical

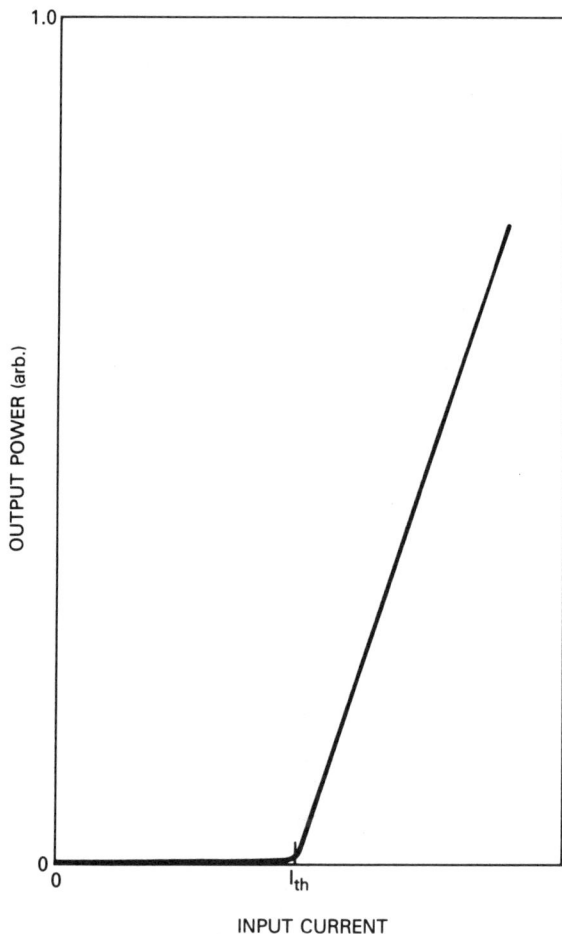

FIG. 2. Typical dependence of optical output power with electrical drive current for laser diodes.

intensity output as a function of drive current. (More detailed discussion of light modulation follows in Section III.) Further, where phase information is important, such as in interferometric optical processors, the coherency of modulated laser diode beams must be maintained. Several studies have been performed to date on the coherency of pulsed laser diode beams (Haney and Psaltis, 1985; Lau et al., 1985). With full depth square-wave modulation, transient thermal effects at the lasing junction due to the modulated current cause shifts of the longitudinal mode or multimode oscillation. It has been concluded that laser diodes that are normally single mode under cw operation can be used as pulsed light sources in interferometric optical processors in which the pulse width of the laser is long compared to its characteristic coherence time constant (usually a few nanoseconds) and short compared to its characteristic thermal time constant (usually several hundred nanoseconds) (Haney and Psaltis, 1985; Lau et al., 1985). For modulation on extremely high-frequency carriers (up to 10 GHz), single longitudinal and transverse mode operation can be maintained at modulation depths of up to about 70% (Haney and Psaltis, 1985; Lau et al., 1985) (since thermal time constants are much longer than the modulation period); above 80% modulation depth, the laser is not maintained above lasing thresholds at all times and multimode oscillation results. Under cw operating conditions the temporal coherence of a laser diode beam is usually good. However, in optical processing applications requiring large dynamic range, the below-threshold, noncoherent, emission shown in Fig. 2 can cause spurious responses.

Spatial mode purity is also important in optical processing applications; spatial nonuniformities introduce extraneous information that can consume much of the dynamic range of the processor. Optical disk applications have driven the development of laser-optics assemblies to provide cw laser beams with very little spatial amplitude and phase spatial distortion. However, relatively little work has been done with respect to the spatial properties of pulsed laser beams.

In noncoherent optical processor applications, the phase of the light is not important to the intrinsic operation, which allows laser diodes to be modulated without regard to coherency or mode stability. Stacked lasers with very large stripe width offer maximum power in pulsed operation and have long been commercially available. The stripe width can be as large as 1.5 mm, producing peak powers of > 100 W in pulses of 50–200 ns duration (Kressel and Butler, 1975; Lockwood and Kressel, 1975). However, due to the large stripe width, the near-field beam pattern will usually exhibit many strong filaments. One method for making the spatial intensity distribution more uniform is to pass the laser beam through an integrating sphere (E. C. Malarkey, personal communication).

## B. Incoherent Sources

Incoherent light sources do not generally have comparable brightness to lasers, and phase is not available as an information-carrying parameter, but their use in noncoherent optical processors does offer advantages such as absence of the phase noise present in high-coherency light beams, fewer critical mechanical alignments, and a variety of well-developed technologies for the light sources. With wideband optical sources, wavelength multiplexing becomes an option for increasing processor throughput.

Minimal size is offered by light emitting diodes (LEDs) based on III–V compounds such as GaAlAs. LEDs are available over the wavelength range of 500–1500 nm. At 800 nm, LED power outputs of > 20 mW are possible with electrical-to-optical conversion efficiencies of about 10%. However, the emission of these high-power LEDs is Lambertian, and most commercial LEDs are encapsulated in a spherical-shaped enclosure which increases outcoupling of the light but results in large-angle emission. Typically, the spectral bandwidth is 30–40 nm. Mode purity and stability are not a concern with LEDs as they are in laser diodes. Another advantage is that the emitted power is less sensitive to temperature, e.g., only a factor of about 2 between room temperature and 100°C (Kressel and Butler, 1975). Hence simplified drive circuitry, with reduced need for temperature stabilization, is possible, especially in pulsed operation. Modulation of LEDs is possible up to 1 GHz, but full depth is obtained only below several hundred megahertz (Kressel and Butler, 1975).

Where wavelengths in the blue or near-ultraviolet region are required, laser sources are usually bulky or nonexistent. High-intensity gas-discharge lamps can be used, such as the mercury and xenon types. However, efficiency is low, and direct pulse modulation generally does not exceed hundreds of hertz.

Another means of data input into noncoherent processors is self-luminous displays. Among the types available are high-resolution, high-brightness cathode ray tubes (CRTs), arrays of LEDs, electroluminescent displays, and plasma displays.

## III. Light Control Devices

In Section II it was noted that laser diode output could be directly modulated by control of the drive current. This modulation, even though potentially high speed, constitutes only a single channel of information. Further, a primary attribute of optical processors is the capability of processing many channels of information in parallel; with a single-channel input only, many applications would require additional memory devices to perform the formatting into an array. Therefore, it is desirable to employ devices that directly input data as a spatial array. The lack of effective means for transducing information spatially onto light beams was a major reason for slow progress in optical processing in the past. Although a variety of devices exists for single-channel modulation of a cw laser beam—ranging from integrated-optic devices optimized for high speed to mechanical devices employed in printing applications—only those devices controlling spatial arrays of data will be treated here.

## A. Modulation—General Considerations

This section will consider some of the light modulation requirements of optical processing and some of the general characteristics of commonly employed electro-optic and acousto-optic modulation techniques.

A very general representation for information is as a spatio-temporal amplitude and phase function which may be expressed as

$$A(x, y, t)\exp\{-j[\Omega t + k_1 x + k_2 y + \phi(x, y, t)]\} \tag{1}$$

Equation (1) is, in general, bipolar and complex. The objective of optical modulation devices is to produce an optical field of the form of Eq. (1) from electrical signals; the electrical signals are most often temporal, and there may be a multiplexity of channels (e.g., from sensors). Since light intensity is a non-negative quantity, biasing schemes must be used in order to represent negative quantities. Temporal phase information may be carried by the phase of a coherent laser beam; however, incoherent light beams require a phase encoding scheme. Spatial phase information can be encoded in a number of ways, including relative location of a spatial grating pattern and multichannel representation of the components of a phasor by light intensities.

It would be desirable to have light intensity directly related to the strength of the modulating signal. However, conventional electro-optic and acousto-optic techniques do not allow this, instead generally relying on alteration of the index of refraction of a transparent medium for phase modulation of the light beam. First consider the classic electro-optic effects. In the most common cases, a change in one of the refractive indices of a birefringent material is induced that is either linear (Pockel effect) or quadratic (Kerr effect) with applied electric field. The phase change produced on a traversing light beam must then be converted into a change in the light intensity. However, the usual manner of employing the change in index to alter light-beam intensity, placing the electro-optic material between crossed polarizers, as illustrated in Fig. 3, results in a $\sin^2$ dependence of the intensity on the refractive-index change. An alternative method is to introduce a periodic spatial variation of the index change. The resultant phase grating diffracts an incident light beam, and either the undiffracted beam or the diffracted beams may be considered as intensity modulated. The intensities of these beams can be obtained through a coupled-mode analysis (Klein and Cook, 1967). (The intensities of the diffracted orders

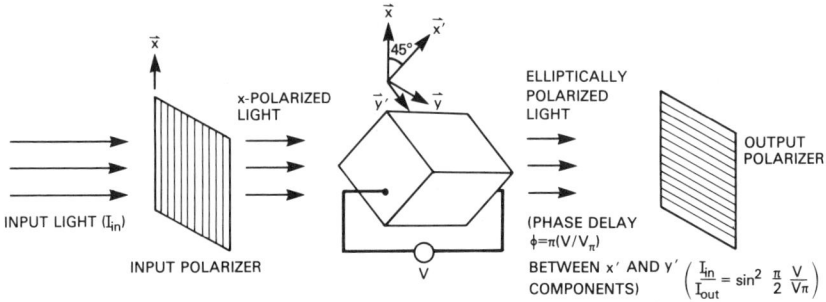

FIG. 3.  Basic electro-optic technique for intensity modulation of a coherent light beam.

are found to be proportional to Bessel functions of corresponding order whose arguments contain a modulation index directly related to the refractive-index change.) A special case of great interest is the thick phase grating operating in the Bragg regime (Kogelnik, 1969); only one diffraction order results, and its intensity variation with index change $\Delta n$ is given by

$$I = \sin^2(k\,\Delta n\, L/2 \cos \theta_0) \qquad (2)$$

where $k$ is the optical wave number ($2\pi/\text{wavelength}$), $L$ is the length of the grating lines, and $\theta_0$ is the angle between the light and the grating lines. An advantage of this diffraction technique is the ease of encoding input phase information onto the carrier.

The spatial-carrier technique is also the basis for 1D acousto-optic modulation techniques, illustrated in Fig. 4. In the example of acousto-optic

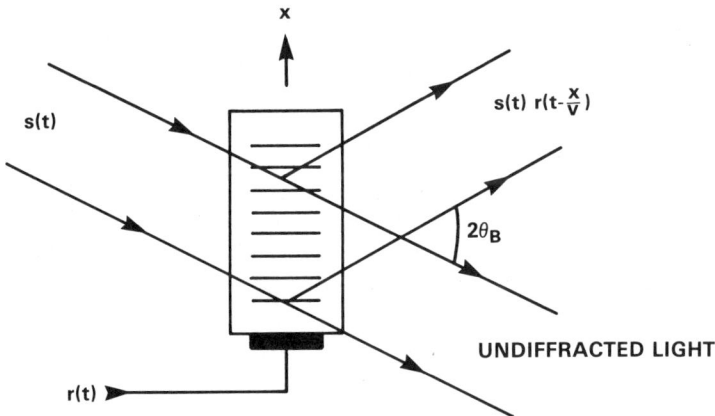

FIG. 4.  Acousto-optic technique for amplitude modulation of light beam by electrical signal $r(t)$ and for complex multiplication of the signals $r(t)$ and $s(t)$.

modulation, a sound (or pressure) wave produces periodic rarefractions and compressions in a photoelastic medium. Refractive-index changes proportional to pressure are produced, and diffraction efficiency in the Bragg regime as a function of pressure is given by a modified version of Eq. (2):

$$\frac{I_1}{I_0} = C_0 \sin^2 \left[ \frac{\pi^2}{2\lambda^2} \left( \frac{n^6 p^2}{\rho v^3} \right) \frac{L}{H} P_a \right]^{1/2} \tag{3}$$

where $C_0$ is a proportionality constant due to overlap between optical and acoustic fields, $\lambda$ is the light wavelength, $n$ is the index of refraction of the medium, $p$ is the photoelastic constant for the particular acoustic wave direction in the medium, $\rho$ is the density, $v$ is the acoustic velocity, $L$ and $H$ are the length and height of cross section of the traveling acoustic column, respectively, and $P_a$ is the acoustic power density (more detailed device descriptions will be given in Section III,B).

The crossed-polarizer and diffraction methods treated so far both result in the same $\sin^2$ functional relationship. However, the linear modulation often desired can be obtained in electro-optic and acousto-optic techniques by either (1) operating at low modulation index, so that Eqs. (2) and (3) are approximately linear (for nonlinearity not to exceed 1%, diffraction efficiency $I_1/I_0$ cannot exceed 2%), or (2) operating about the $\sin^2(\pi/4)$ point in the relationship (a bias intensity level of one-half the maximum intensity results, and modulation depths of $< 20\%$ of the bias level ensure $< 1\%$ nonlinearity). The latter technique has the disadvantage that the high optical bias level can introduce substantial quantum noise into the processor.

Another important consideration in light modulation techniques is dynamic range. Clearly this should be as large as possible, but a number of factors enter into the determination of dynamic range. For the case of a single signal channel, quantum considerations generally set the ultimate limits. Usually photon quantum limits are the most important. In acousto-optic modulation the quantum noise from the acoustic wave is less important, since there are $\sim 10^5$ times more phonons than photons in equal amounts of energy. Otherwise, the noise on the drive signal itself sets the limit; thermal amplifier noise is usually the most important factor not intrinsic to the signal (or to the sensors that may have produced the signal). For multiple channels of information, nonuniformity among the channels is the major factor. The noisiest channel will determine the ultimate accuracy with which all the information may be represented.

The frequent occurrence of a nonlinear transfer function between the input and output of a light-control device has already been noted. This nonlinearity can result in loss of dynamic range when multiple signals are input simultaneously to the light modulator. Spurious signals will occur in the presence of high-level signals. In the electro-optic modulator shown in Fig. 3, intermodulation, or harmonic distortion, terms arise. The level of such spurious signals can generally be calculated. For example, they are related to Bessel functions of corresponding order in the electro-optic modulator of Fig. 3.

A third consideration for modulation is the information throughput. The total input information bandwidth for a modulator is given by the product of the number of resolvable spots (or channels) in the input optical field and the bandwidth of each channel. This is a general result for all forms of channel encoding. As a specific example, consider the case of signal channels encoded according to spatial frequency, but which extend over the entire input device aperture. Such encoding can arise with the use of acousto-optic spectrum analysis devices. For Bragg diffraction in an isotropic medium, light is angularly diffracted as a function of wavelength (or signal frequency) as

$$\theta_B = \sin^{-1}(\lambda/2n\Lambda) \tag{4}$$

where $\Lambda$ is the grating period, $\lambda$ is the light wavelength, and $n$ is the index of refraction. The maximum number of independent channels in such a case is determined by the total range of angular deflection divided by the minimum angular spot size. For small angles of deflection (which is generally the case, at optical wavelengths), the angular spread of an optical beam due to a finite 1D aperture of length $D$ is

$$\delta\theta \simeq \lambda/D \tag{5}$$

The angular deflection of the light beam is directly proportional to the grating spatial frequency $K = 2\pi/\Lambda$; hence $\theta_{max} \propto \lambda K_{max}$ and $\theta_{min} \propto \lambda K_{min}$. Since the spatial bandwidth $B$ can de defined as $B = K_{max} - K_{min}$, one obtains as a figure of merit the space–bandwidth product

$$SBW = DB = (\theta_{max} - \theta_{min})/\delta\theta \tag{6}$$

In the acousto-optic case, where the optical aperture of length $D$ is related to an equivalent time aperture $T$ by the wave velocity $(D = vT)$, one obtains the time–bandwidth product

$$TBW = T(\Delta f) \tag{7}$$

where $\Delta f$ is the temporal bandwidth $(f = vK/2\pi)$. The results in Eqs. (6) and (7) are common measures of capacity in information-processing systems. While one may increase either the time aperture or the bandwidth of a processing system, the time–bandwidth product is generally difficult to increase dramatically. Different classes of devices will be discussed below that have various implementations, but all result in relatively the same time–bandwidth, generally due to a tradeoff between speed and resolution.

## B. One-Dimensional Devices

1D spatial light modulator devices have filled particular roles in optical processing, necessitated by the slower development of 2D light modulators. The 1D devices are obviously appropriate for the processing of a single channel of data. Thus, many early developments in optical processing centered around applications such as high-speed spectrum analysis and correlation of temporal data streams. Some 2D processing problems can be handled using 1D input devices if the problems can be broken down into the product of two or three 1D factors. Such problems can be handled by the use of a modulated light source and two 1D modulators in a mutually orthogonal configuration as shown in Fig. 5. Review articles (Turpin, 1981; Lohmann and Wirnitzer, 1984) discuss this approach in applications such as triple correlation, multidimensional spectrum analysis, and speckle interferometry. Additionally, 1D devices have been employed to perform matrix multiplications, without the need to employ a 2D light modulator for all matrix inputs (Athale and Collins, 1982; Athale and Lee, 1984).

The 1D modulators that have enjoyed the most successful development are the acousto-optic devices. Some of the salient features in the operation of an acousto-optic modulator are illustrated in Fig. 4. The net result of the acousto-optic interaction is that there is a multiplication of the data on the light beam with that on the acoustic wave. More specifically, the light beam amplitude and phase are multiplied by those of the acoustic wave as a result of diffrac-

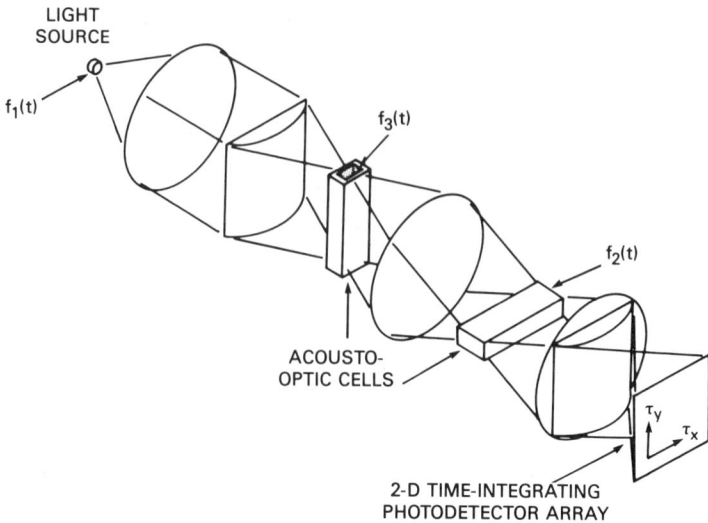

FIG. 5.  Acousto-optical processor for performing triple-product processing.

tion by the acoustically induced refractive-index changes. Additionally, the frequency of the light is shifted by the frequency of the acoustic wave (this can be viewed as either conservation of momentum in a phonon–photon interaction, or as a Doppler frequency shift). In Fig. 4 the acoustic frequency adds to the light frequency. The angle of deflection of the light beam from the incident direction is given by Eq. (4). Figure 4 shows only one diffracted light beam. In general there are many diffraction orders (Raman–Nath regime). The intensities of the diffracted orders can be obtained, as mentioned earlier, by solving a set of coupled-mode equations, to give

$$I_m = J_m^2 \left[ \left( \frac{k \, \Delta n \, L}{2 \cos \theta} \right) \frac{\sin \left[ L(\tan \theta_0)/2\Lambda \right]}{\left[ L(\tan \theta_0)/2\Lambda \right]} \right] \tag{8}$$

where $m$ denotes the diffraction order and $J_m[\cdot]$ is an $m$th-order Bessel function. In Bragg diffraction at incident angle $\theta_B$, only the zero and first orders have significant intensity. The criterion for obtaining Bragg diffraction is that the figure of merit, $Q$, defined as

$$Q = (2\pi\lambda/\Lambda)L \tag{9}$$

be much greater than unity. Equation (9) shows that Bragg diffraction occurs when the thickness of the acoustic column traversed by the light is large. More quantitatively, $Q > 7$ results in $> 90\%$ of the diffracted light being in the first order. For optical processing, operation in the Bragg regime is greatly preferred for maximum efficiency and to minimize extraneous light beams.

The three parameters of most importance in acousto-optic devices are diffraction efficiency, time bandwidth, and absence of spurious signals. The factors affecting these parameters include choice of material, device geometry, transducer design, and device fabrication.

Some of the more popular acousto-optic materials include lithium niobate, tellurium dioxide, lead molybdate, and gallium phosphide. Table I lists a number of the important acousto-optic parameters for these materials. The material parameters that affect diffraction efficiency are grouped in parentheses in Eq. (3), and this grouping is known as the figure of merit:

$$M_2 = n^6 p^2 / \rho v^3$$

[Other figures of merit have been devised as a measure of diffraction efficiency in conjunction with other parameters such as time aperture (Gordon, 1966).] Note the very strong dependences of $M_2$ on index of refraction $n$ and acoustic velocity $v$. Most good acousto-optic materials have been identified through measurement of $n$ and $v$, and tabulations of the material parameters are available (Uchida and Niizeki, 1973). However, to maximize $M_2$ for a given material requires detailed computations. The stiffness tensor of a material is used in solving the Christoffel matrix (Dieulesaint and Royer, 1980), giving

TABLE I

Parameters for Several Important Acousto-optic Crystal Cuts at 633 nm

| Material | | | | | | |
| --- | --- | --- | --- | --- | --- | --- |
| | Acoustic mode | Light polarization[a] | $M_2$ $(10^{-15} \text{ s}^3/\text{kg})$ | Acoustic velocity, $v$ (m/s) | Acoustic attenuation, $\alpha$ (dB/$\mu$s GHz$^2$) | Demonstrated diffraction efficiency (peak) (%/rf W) |
| TeO$_2$ | (110) shear | Circular | 793 | 616 | 18 | 200 (TBW = 2000, BW = 150 MHz) |
| TeO$_2$ | (001) long | $\perp$ | 34.5 | 4200 | 6.3 | |
| TeO$_2$ | (001) long | $\parallel$ | 25.6 | 4200 | | |
| GaP | (111) long | Arb. | 29.0 | 6640 | 3.8 | 48 (TBW = 150, BW = 500 MHz) |
| GaP | (110) long | Arb. | 44.6 | 6320 | | 20 (TBW = 300, BW = 1 GHz) |
| PbMoO$_4$ | (100) long | $\parallel$ | 36.3 | 3630 | 5.4 | |
| LiNbO$_3$ | (100) long | Extraordinary ray 35° to y axis | 7.0 | 6570 | 0.098 | 2 (TBW = 1000, BW = 500 MHz) |
| LiNbO$_3$ | (100) shear | Anisotropic interaction | 13.0 | 3600 | | $\begin{cases} 44 \text{ (TBW = 300, 2–3 GHz band)} \\ 12 \text{ (TBW = 600, 2–4 GHz band)} \end{cases}$ |

[a] With respect to acoustic wave vector.

acoustic-mode velocities and polarizations. The strain tensor associated with these modes is used to calculate a strain-perturbed index ellipsoid which in turn is related to an optical polarization direction to give an effective photoelastic constant, and hence the resultant $M_2$. Since both $n$ and $p$ vary with optical wavelength, $M_2$ must be specified for a particular wavelength ($\lambda = 633$ nm in Table I).

Another material parameter of importance is the attenuation of the acoustic wave as it propagates down the Bragg cell. Acoustic attenuation is usually given as an exponential-decay parameter measured at a particular frequency (e.g., 1 GHz) and is best expressed in units of dB per unit time, since it is the time aperture of a Bragg cell that is affected. Extrapolation to other acoustic frequencies is made empirically as some power of the frequency $f$, with an $f^2$ dependence most commonly noted (Uchida and Niizeki, 1973). In addition to reducing the time aperture of an acoustic cell, the attenuation also reduces the effective diffraction efficiency; for a weekly truncated Gaussian laser beam, it can be shown that the diffraction efficiency is reduced by a factor of $10^{-\alpha t/20}$, where $\alpha$ is the attenuation parameter, and $t$ the total optical aperture. Finally, high attenuation results in heating of the acoustic material, which can lead to effects such as change of acoustic velocity and defocusing of the optical beam.

Material quality must eventually be considered in the construction of an acousto-optic device. The material must have good, homogeneous, optical transmission, and defect levels must be kept low to minimize optical loss and scatter. It is desirable to employ the maximum-size sample through which the acoustic wave can propagate without significant attenuation. Relatively few materials have been sufficiently developed to where high quality is routinely available. Lithium niobate and tellurium dioxide are popular in large measure due to their availability, in spite of their being nonoptimum in $M_2$ and $\alpha$, respectively.

The diffraction efficiency of an acousto-optic cell can be improved not only by choosing materials with high $M_2$, but by increasing the interaction thickness of the acoustic region that the optical beam traverses. The optimum $L$ is that which maximizes $I_1/I_0$ in Eq. (3). However, in most applications, it is necessary to drive the acoustic cell with large-bandwidth rf signals. This bandwidth $\Delta f$ causes a deviation $\Delta \bar{K} = \bar{K} - (\bar{k}_d - \bar{k}_i)$ in the acoustic wave vector $\bar{K}$ from the ideal Bragg condition, i.e., $\bar{K}$-vector conservation condition $\bar{k}_d = \bar{k}_i \pm \bar{K}$ for a fixed pair of incident and diffracted optical wave-vector directions, $\bar{k}_i$ and $\bar{k}_d$. The dependence of diffraction efficiency on interaction width $L$ and bandwidth $\Delta \bar{K}$ can be written as (Dixon, 1967)

$$\frac{I_1}{I_0} \propto \text{sinc}^2\left(\frac{|\Delta \bar{K}|L}{2\pi}\right) \tag{10}$$

In general, $L$ must be limited to $2\pi/|\Delta\bar{K}|$ which is much shorter than the optimum $L$ suggested by Eq. (3). Nevertheless, it is possible to significantly increase the allowable interaction width $L$ for a given bandwidth by employing the natural birefringence of some acousto-optic materials such as lithium niobate. In birefringent materials it is possible to change the polarization of the light beam upon diffraction. Unlike the previously discussed case of "isotropic" Bragg diffraction, the incident and diffracted optical wave fronts in this "anisotropic" case have different magnitudes, i.e., $|\bar{k}_d| \neq |\bar{k}_i|$. Satisfaction of the $\bar{K}$-vector phase-matching condition, $\bar{k}_d = \bar{k}_i \pm \bar{K}$, in both the isotropic and anisotropic cases is compared in Fig. 6a and b by employing geometric $\bar{K}$-space constructions. Notice that in the isotropic case the incident and diffracted angles are equal, i.e., $\theta_i = \theta_d = \theta_B$; whereas in the anisotropic case $\theta_i$ and $\theta_d$ are neither symmetric nor equal. (See Dixon, 1967, for explicit expressions for $\theta_i$ and $\theta_d$ as a function of the acoustic wave-vector direction and magnitude and the orientation of the crystal axes.) In Fig. 6a, with a fixed incident $\bar{k}_i$ vector, changes in $\bar{K}$ with the rf acoustic drive frequency will cause $\bar{k}_d$ to significantly deviate from the optical $\bar{k}$ surface—this deviation is a geometric measure of the phase error $|\Delta\bar{K}| = |\bar{K} \pm (\bar{k}_i - \bar{k}_d)|$ which enters into Eq. (10). In the anisotropic case, it turns out that for a particular value of $|\bar{K}|$ (or acoustic frequency), $\bar{K}$ and $\bar{k}_d$ are orthogonal to each other [more specifically (Dixon, 1967) at $f_0 = v\sqrt{n_i^2 - n_d^2}/\lambda$]. This situation is illustrated in Fig. 6b; now $\bar{K}$ is tangent to the $\bar{k}_d$ surface, and changes in $\bar{K}$ with acoustic frequency will cause only very small phase error deviations $|\Delta\bar{K}|$ from the $\bar{k}_d$ surface. Thus, the phase error can be maintained within a given maximum constraint over

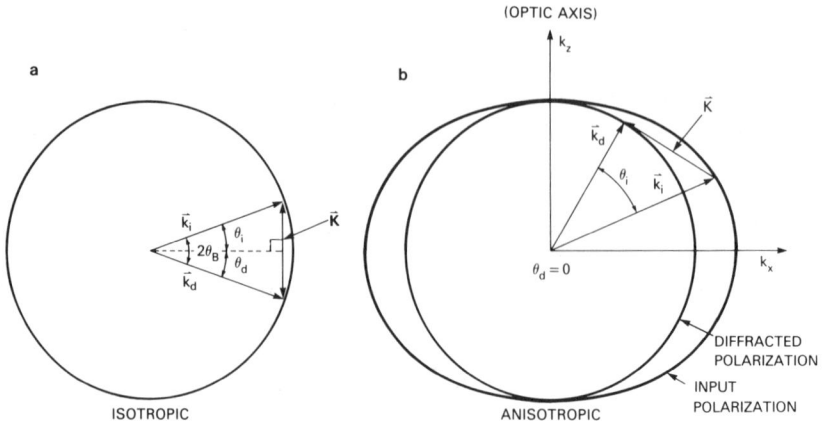

FIG. 6.   $\mathbf{K}$ conservation in acousto-optic interaction for (a) isotropic case and (b) anisotropic case (birefringent material).

a significantly broader bandwidth. Conversely, with a given acoustic bandwidth, this anisotropic phase error $\Delta \bar{K}$ is much smaller than in the corresponding isotropic case, and the interaction width $L$ in Eqs. (3) and (10) can be made much larger with a resulting increase in diffraction efficiency. The achievable increase in diffraction efficiency using anisotropic interaction (Chang, 1983) is illustrated for $LiNbO_3$ in Table I.

1D spatial light modulators utilizing the electro-optic effect have also been constructed. These generally do not employ a traveling-wave phenomenon as in the acousto-optic devices. However, significant space–bandwidth comes at the expense of complex electrical addressing networks. Also, sufficiently high voltages must be applied to obtain the desired depth of modulation.

One successful approach to electro-optic 1D modulators has been to utilize VLSI techniques to produce the addressing circuitry (Johnson et al., 1983). More than 5000 channels have been produced on a silicon chip which was overlaid onto an electro-optic plate made of $LiNbO_3$, as illustrated in Fig. 7. Addressing was serial–parallel, with 32 parallel channels per block. A light beam incident at an angle for total internal reflection interacts with the refractive-index changes induced in the $LiNbO_3$ by the fringing electric fields between electrode pairs. Total internal reflection occurs within the $LiNbO_3$ plate; hence this device is called a TIR electro-optic modulator.

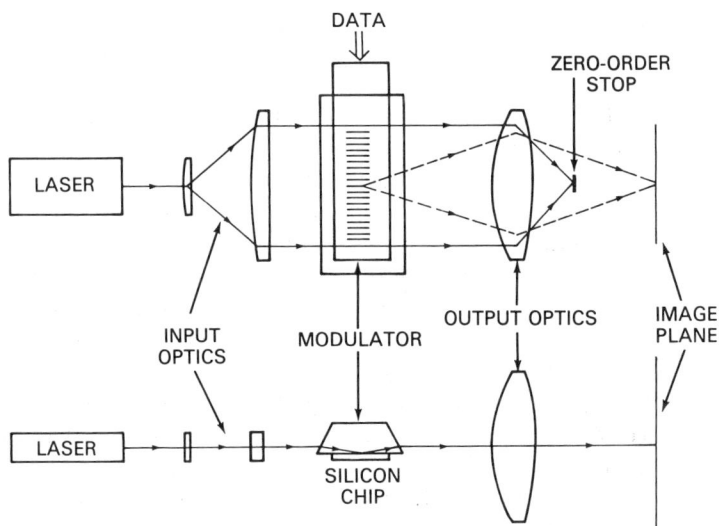

FIG. 7. Multichannel 1D electro-optic $LiNbO_3$ modulator, using VLSI silicon chip for addressing (Johnson et al., 1983).

The refractive-index change in most high-speed modulators varies linearly with electric field, i.e., utilizes the linear electro-optic effect, where $\Delta n$ may be derived from

$$\Delta\left(\frac{1}{n_i^2}\right) = \sum_{j=1}^{3} r_{ij} E_j \qquad (11)$$

where $r_{ij}$ are the electro-optic coefficients for a particular material, and $E_j$ are the components of applied electric field, with $j = 1, 2, 3$ corresponding to the $x$, $y$, and $z$ directions, respectively, and $i = 1, \ldots, 6$ corresponding to $n_x^2, n_y^2, n_z^2$, $n_{yz}^2, n_{xz}^2$, and $n_{xy}^2$, respectively. Many tens to hundreds of volts may be required for good depth of modulation of the light beam by a bulk sample of such electro-optic materials; these voltages may be difficult to provide at high speed (e.g., > gigahertz) without excessively high drive power requirements. One remedy is to employ integrated-optic waveguide structures (to be discussed further shortly) where the light beam can be confined in at least one direction to small dimensions (e.g., several micrometers); then the required modulation can be obtained without unduly high input powers. An alternative is to employ materials that exhibit a strong quadratic electro-optic (or Kerr) effect.

The material that has been most successfully used to produce light modulator arrays employing the quadratic electro-optic effect is the lead–lanthanum–zirconium–titanate (PLZT) system. Thin films of PLZT are commercially available with deposited metal electrodes. By varying the percentages of Pb and La, the Curie temperature of the material and the strength of the index change per unit field at a given temperature can be controlled. Modulator voltages of no more than a few tens of volts are required. However, the speed of the modulation is limited at present to $\sim 100$ kHz by the polycrystalline fine-grained ceramic nature of these materials. Even with fine electrode spacings, modulations much beyond 100 kHz are not possible. There has been recent success in deposition of single-crystal films of PLZT (Kawaguchi et al., 1984), but it is too early to determine if this will provide the hoped-for solution to the speed versus drive-voltage trade-off.

As mentioned earlier, 1D modulators have been used to fill the void caused by the lack of ideal 2D modulators through the use of innovative signal-processing architectures. 1D modulators have another important aspect— they are intrinsically compatible with integrated-optic structures. In integrated optics, light is confined to waveguides that can be either thin and planar, or channels of small cross section. Many reviews and journals for this technology exist (Marcuse, 1973; *J. Lightwave Technol.*, 1983+). Integrated optics has the potential to provide rugged optics in small package sizes at low cost, and is a natural complement to fiber-optic technology for all-optical processing schemes.

Integrated-optic signal-processing devices have been developed using 1D modulators to perform operations such as spectrum analysis (Mergerian *et al.*, 1980), correlation (Verber *et al.*, 1981), and matrix algebra (Verber, 1984). Spectrum analysis and correlation have been performed using acousto-optic devices employing surface acoustic waves (instead of bulk waves) to concentrate acoustic energy in the planar optical waveguides. Electro-optic gratings overlaid on planar waveguides have been used for signal input in the matrix algebra devices.

## C. Two-Dimensional Spatial Light Modulators

Two-dimensional spatial light modulators (2D SLMs) play an essential role in configuring optical information processing systems to efficiently exploit the inherent parallel-processing and interconnection capabilities of optics. An attempt is made here to provide a perspective of where SLM development is today and where it can be expected to be in the near future.

Some of the earliest proposed applications which demanded 2D SLMs included (Pollock *et al.*, 1963; Tippett *et al.*, 1965; Goodman, 1968; Heynick *et al.*, 1973): the utilization of the innate matched filtering capability of optics for real-time pattern recognition on imagery, real-time holography for nondestructive testing, synthetic aperture radar processing, and flat-screen and projection display. The need for high-quality 2D SLMs is becoming even more critical as optics moves into more sophisticated directions such as multisensor processing, robot vision, and symbolic processing for artificial intelligence applications. For example, as mentioned previously, there have been promising recent proposals for highly parallel optical computing architectures which exploit 2D SLMs to implement associative networks (Psaltis and Farhat, 1984, 1985; Fisher *et al.*, 1984, 1985), cellular automata (Sawchuk and Jenkins, 1985; Yatagai, 1986), or generalized parallel finite-state machines (Huang, 1985). A large variety of two-dimensional spatial light modulator designs has been advanced in the past two decades (Flannery, 1973; Thompson, 1977; Casasent, 1977, 1978a,d; Bartolini, 1977; Lipson, 1979; Knight, 1981). Attaining the required performance levels from these SLM devices proved to be more difficult than initially expected; however, recent advances in fabrication techniques and materials are beginning to produce a new generation of high-performance 2D spatial light modulators (Tanguay, 1983, 1985; Effron, 1984; Tanguay and Warde, 1985; Fisher, 1985; Warde and Effron, 1986).

The state of the art of many of the more promising 2D SLM technologies is reviewed in the following sections, with a large number of both electronically and optically addressed devices being discussed. The characteristics of promising SLM technologies are summarized in Tables II and III. (The

TABLE II

Two-Dimensional Spatial Light Modulators

| No. | Name/type | Modulating material | Addressing Optical sensor | Addressing Electronic | Developed at |
|---|---|---|---|---|---|
| O.1 | Phototitus | KD*P (KD$_2$PO$_4$) | Amorph. Se | n | LEP (France) |
| O.2 | Titus | | Si photodiode | n | Lockheed |
| E.3 | | | n[b] | e beam | LEP, CMU, Soderm[a] |
| O.4 | LCLV | Twisted nematic liquid crystal | CdS | n | Hughes[a] |
| O.5 | | | BSO | n | Thom. CSF |
| O.6 | | | BSO | n | Lockheed |
| O.7 | | | Si photodiode | | Hughes |
| E.8 | | | n | CCD | Lockheed, NEC, U. of Co. |
| O.9 | | Ferroelectric liquid crystal | BSO | n | Displaytech |
| E.10 | | | n | Matrix | U. of Edinburgh |
| O.11 | | Guest–host LC | n | Si circuits | Hughes, USC, Xerox |
| O.12 | VGM | Nematic liquid crystal | ZnS | n | UCSD |
| O.13 | PLZT | PLZT | Si phototrans. | n | UCSD |
| E.14 | | | n | Si transist. | Bell |
| O.15 | FERPIC/ | | ZnCdS | n | Sandia |
| O.16 | FERICON/CERAMPIC | | PVK | n | |
| E.17 | | | n | Matrix | |
| O.18 | RUTICON | Deformable elastomer | Amorph. Se | n | Xerox |
| O.19 | | | PVK:TNF | n | Xerox, Harris |
| E.20 | | | n | e beam | IBM |
| O.21 | MLM | Deform. mem. | Si photodiodes | n | Perkin-Elmer |
| E.22 | | | n | Electrodes | Perkin-Elmer |

| | | | | | |
|---|---|---|---|---|---|
| O.23 | DMD | | Si phototrans. | n | TI |
| E.24 | Micro-mechanical | Cantilevered beams ($SiO_2$ or metal film) | n | Si CCD, trans. | TI, RCA |
| E.25 | | | n | Si circuits | TRW, IBM, Telesensory, TI Westinghse, RCA |
| E.26 | TP | Thermoplastic | n | e beam | Harris, NRC,[a] Xerox, Honeywell Kalle-Hoechst,[a] Fuginon[a] |
| O.27 | | | PVK:TNF | n | |
| E.28 | Lumatron | | n | e beam | CBS, ERIM |
| O.29 | PROM | BSO or BGO | BSO or BGO | n | ITEK, USC, Sumitomo[a] |
| O.30 | PRIZ | | | n | USSR |
| O.31 | Volume-holographic | Photorefractive crystals | Photoferroelec. (holo, TWM, FWM) | n | Hughes, others, Thom. CSF, USC, Cal. Tech. |
| O.32 | PICOC | BSO | BSO | n | Lincoln |
| E.33 | Electro-absorption | GaAs (Franz–Keldysh) | n | GaAs CCD | |
| O.34 | | | Photogeneration | n | |
| O.35 | MSLM | EO crystal ($LiNbO_3$, KDP, $LiTaO_3$) | Photocath, MCP | n | MIT, Optron,[a] Hamamatsu[a] |
| E.36 | PEMLM | Deformable membrane | n | e beam | MIT |
| O.37 | | | Photocath, MCP | n | England, NRL |
| E.38 | Liquid-film | Oil-film (Polysiloxan) | n | e beam | |
| O.39 | | | Heat absorption in plastic substr. | n | Switzerland |
| E.40 | | Ethanol | n | | |
| E.41 | Eidophor | Oil-film | n | Resistive heat matrix | Canon |
| | Talaria | | n | e beam | Greytag,[a] GE[a] |
| O.42 | | Deformable gel | Photoconductor | n | Switzerland |

*(continued)*

TABLE II (continued)

| No. | Name/type | Modulating material | Addressing | | Developed at |
|---|---|---|---|---|---|
| | | | Optical sensor | Electronic | |
| O.43 | VO$_2$ | VO$_2$ | VO$_2$ (Heat abs.) | n | Vaught, USC |
| E.44 | | | n | e beam | HP, Singer[a] |
| O.45 | Librascope | Smectic liquid crystal | Liquid crystal (heat absorp.) | n | |
| E.46 | LIGHT-MOD SIGHT-MOD | YIG (Y$_3$Fe$_5$O$_{12}$) (Magneto-optic) | n | Matrix | Litton[a] Semetex[a] |
| E.47 | LISA | | n | | Philips |
| E.48 | Particle suspension opt. tun. arr. | Anisotropic particle suspension | n | Matrix | Bell, Holotronics, VARAD |
| E.49 | Bragg | Multichannel 2D Bragg cell | n | Acoustic beams | Harris |
| E.50 | TIR | LiNbO$_3$ | n | Si circuits | Xerox[a] |
| O.51 | Platelet laser | GaAs | GaAs | n | ETL (Japan) |

[a] SLM is commercially available from this company.

[b] n: not applicable (e.g., optical sensor for E-SLM).

## TABLE III
### TWO-DIMENSIONAL SLM PERFORMANCE

| No. | Name/type | Resolution (lp/mm) (no. pixels) | Sensitivity[g] ($\mu J/cm^2$) | Time response[a] | | |
|---|---|---|---|---|---|---|
| | | | | Write (ms) | Erase (ms) | Store |
| O.1 | Phototitus | $15^a$, $30^b$ | 10 | 0.01 | 0.03 | 1 h |
| O.2 | | $10^a$ | 2 | 2 | <0.5 | 5 s |
| E.3 | Titus | $20^a$, $30^b$ | $n^l$ | $30^i$ | $5^i$ | 1 h |
| O.4 | LCLV | $30^a$, $40^b$ | 6 | 10 | 15 | 15 ms |
| O.5 | | $10^a$, $40^b$ | 20 | 15 | 15 | 20 ms |
| O.6 | | $10^a$ | 1 | 1 | 1 | 20 ms |
| O.7 | | $15^a$ | 1 | 5 | 10 | 10 ms |
| E.8 | | $\{256 \times 256\}^d$ | n | $5^{hi}$ | $10^{hi}$ | 10 ms |
| O.9 | | $10^l$ | 1 | 1 | 1 | seconds |
| E.10 | | $\{32 \times 32\}^d$ | n | 1 | 1 | seconds |
| O.11 | VGM | $\{16 \times 16\}^d$ | n | 200 | 400 | hours[f] |
| O.12 | | $5^a$ | 15 | $10^3$ | $<10^3$ | seconds |
| O.13 | PLZT | $10^d$ | $2^k$, $10^{-3k}$ | $<0.01^e$ | $<0.01^e$ | seconds[f] |
| E.14 | | $10^d$ | n | $<0.01^{h,e}$ | $<0.01^{h,e}$ | |
| O.15 | FERPIC/ | $20^a$, $40^b$ | 5 | <1 | 10 | hours |
| O.16 | FERICON/CERAMPIC | $20^a$, $40^b$ | 5 | 10 | 1000 | hours |
| E.17 | RUTICON | $\{128 \times 128\}^d$ | n | $0.01^h$ | $0.01^h$ | 10 ms |
| O.18 | | $40-120^{a,c,k}$ | 30 | 5 | 4 | 15 min |
| | | $10-45^{a,c,k}$ | | | | |
| O.19 | | | 5 | 30 | 70 | 15 min |
| E.20 | MLM | $15^{a,c}$ | n | $20^i$ | $20^i$ | minutes |
| O.21 | | $10^d$ | 2 | <0.001 | 10 | 10 ms |
| E.22 | | $\{100 \times 100\}^d$ | n | $<0.001^h$ | $<0.001^h$ | $\mu$seconds |

*(continued)*

TABLE III (continued)

| No. | Name/type | Resolution (lp/mm) (no. pixels) | Sensitivity[g] (µJ/cm²) | Time response[a] | | |
|---|---|---|---|---|---|---|
| | | | | Write (ms) | Erase (ms) | Store |
| O.23 | DMD | $\{128 \times 128\}^d$ | 2 | 0.025 | 0.04 | 200 ms |
| E.24 | Micro-mechanical | $\{128 \times 128\}^d$ | n | $0.025^h$ | $0.025^h$ | 200 ms$^f$ |
| E.25 | | $\{18 \times 2\}^d$ | n | $<0.01^h$ | $<0.01^h$ | 0.01 ms$^f$ |
| E.26 | | $40^d$ | n | $30^i$ | $30^i$ | months |
| O.27 | TP | $200{-}1600^{a,c}$ | 5 | 10 | 100 | years |
| E.28 | Lumatron | $70^{a,c}$ | n | $<1^h$ | $1000^i$ | years |
| O.29 | PROM | $6^a, 12^b$ | 5 | $<0.1$ | $<0.1$ | <2 h |
| O.30 | PRIZ | $10^a, 20^b$ | 5 | $<0.1$ | $<0.1$ | <2 h |
| O.31 | Volume-holographic | $>1500^{a,c}$ | $1{-}10^4$ | $<30$ | $<30$ | mseconds |
| O.32 | PICOC | $50^a$ | 1 | 30 | 10 | –hours |
| E.33 | Electro-absorption | $\{16 \times 1\}^d$ | n | $10^{-6\,h}$ | $10^{-6\,h}$ | mseconds |
| O.34 | | | | | | |
| O.35 | MSLM | $10^a, 20^b$ | $3 \times 10^{-6\,e,j}$ | 10 | 20 | days to months |
| | | | | $1^e$ | $1^e$ | |
| E.36 | | $3^a$ | n | $200^i$ | $50^i$ | |
| O.37 | PEMLM | $80^d, 40^a$ | $3 \times 10^{-4\,e,j}$ | $0.5^e$ | $0.5^e$ | days to months |
| E.38 | | | n | $0.5^{e,h}$ | $0.5^{e,h}$ | |
| O.39 | Liquid-film | $4^a$ | 100 | 100 | $<100$ | <0.1 s |
| E.40 | | $\{256 \times 1\}^d$ | n | 1 | 1 | 1 ms |

| | | | | | | |
|---|---|---|---|---|---|---|
| E.41 | Eidophor | 50[a,c] {1023 × 1023} | n | 15[i] | 15[i] | 0.3 s |
| | Talaria | | n | 10 | 10 | 10 ms |
| O.42 | | 10[a] | | | | |
| O.43 | VO$_2$ | 150[a] | 3 | $3 \times 10^{-5}$[h] | <1 | years |
| E.44 | | 20[a] | $2 \times 10^4$ | $10^{-3}$[h] | $10^{-4}$[i] | years |
| O.45 | Librascope | 40[a] | n | 0.005[h] | 0.001[h] | months |
| E.46 | LIGHT-MOD | {128 × 128}[d] | $10^4$ | $10^{-3}$[h] | $10^{-3}$[h] | years |
| | SIGHT-MOD | {512 × 512}[d,e] | | | | |
| E.47 | LISA | {256 × 128}[d,e] | n | 0.015[h] | 0.015[h] | years |
| E.48 | Particle suspension | {16 × 16}[d] | n | 1 | 1 | 15 ms |
| | opt. tun. arr. | | | | | |
| E.49 | Bragg | {32[d] × 120} | n | 0.002[i] | 0.002[i] | 10 ns |
| E.50 | TIR | {5000 × 1}[d] | n | <0.001[h] | <0.001[h] | <1 μs |
| O.51 | Platelet laser | 50[a] | $10^3$ | $10^{-6}$ | $10^{-6}$ | nseconds |

a 50% MTF.

b 10% MTF.

c Spatial bandpass (limits or bandwidth).

d Discrete pixels: pix/mm or { } = No. of pix.

e Expected performance in near future.

f Potential exists for Si memory circuits.

g 90% of half-wave or full contrast, unless noted otherwise.

h Time for one pixel (or line).

i Time for full frame.

j Quantum-limited ($S/N = 10$, at given resolution).

k For different implementations of device.

l n: not applicable.

optically or electronically addressed devices are referred to as O-SLMs or E-SLMs, respectively, and denoted by the either the letter "O" or "E" in the first column of Tables II and III.) Not all variants or developers of each modulator type have been included, and some are mentioned only to provide a historical perspective. A description of the basic functions that are performed by 2D SLMs in optical processing systems also follows and provides an additional perspective for a comparative discussion of the functional and performance capabilities of the various SLM devices. Some comments are also made on the current status of and future directions for 2D SLM technology.

*1. Optically Addressed Devices*

Most optically addressed spatial light modulators can be loosely grouped into one of a few classes based on their overall structure, with some subclassification by modulating material. Many of the earliest optically addressed spatial light modulators adopted the basic sandwich structure illustrated in Fig. 8. In operation, a bias voltage applied to the sandwich is shunted within the illuminated regions of the photoconductor to a voltage-controlled phase, amplitude, and/or polarization modulating material (e.g., electro-optic). In the reflective configuration illustrated, there is a mirror and light-blocking layer at the center of the sandwich, which allows the written input information to be read out by reflection from the modulating-material side of the SLM. As seen in the first part of Table II, the modulating materials which have been employed in this configuration include electro-optic crystals (e.g., KD*P) in the Phototitus (Donjon *et al.*, 1973; Casasent, 1978b;

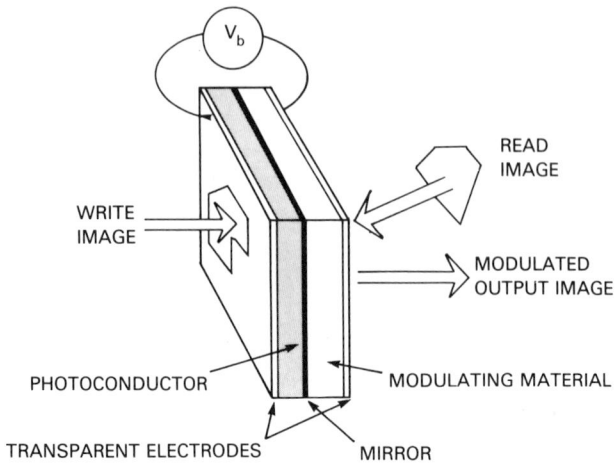

FIG. 8. Generic sandwich construction for an optically addressed 2D spatial light modulator.

Armitage et al., 1985), liquid crystals in the liquid crystal light valve (LCLV) (Bleha et al., 1978; Augbourg et al., 1982) and variable grating mode (VGM) devices (Tanguay et al., 1983), and ferroelectric ceramics in the PLZT devices (Land, 1978).

In the slightly modified sandwich structure of Fig. 9, the photoconductor transfers the bias voltage to a mechanical material which is deformed by the applied electric field. Reflecting a read beam from the deformed surface thus produces a phase-modulated output beam, which can be viewed as an intensity image with suitable interferometric or Schlieren optics. Flexible materials which have been employed in this basic configuration include: bulk deformable elastomers in the Ruticon devices (Sheridon, 1972; Lakatos, 1974; Ralston and McDaniel, 1979), plastics which deform only when heated in the thermoplastic (TP) device (Colburn and Chang, 1978), a thin film of an insulating liquid (Schneeburger et al., 1979) or gel (Hess and Danliker, 1985), thin conducting membranes stretched over an array of insulating holes in the membrane light modulator (MLM) (Preston, 1969; Reizman, 1969) and in the deformable membrane device (DMD) (Pape, 1985a), and arrays of tiny metallic or $SiO_2$ diving-board-like structures in the cantilevered beam devices (Brooks, 1985) and the DMD. In most of these structures electrical contact is directly made to a reflecting metallic coating on the deformable surface; however, in the β-Ruticon (Ralston and McDaniel, 1979) and thermoplastic devices the deformable surface is contacted by an ionized gas. A conducting liquid is employed for contact in the α-Ruticon (Sheridon, 1972).

Materials which have commonly been employed as the photosensor layer in the sandwich-type SLMs of Figs. 8 and 9 include the photoconductors PVK:TNF, amorphous Se, CdS, ZnS, and ZnCdS. More recently, sensitivity,

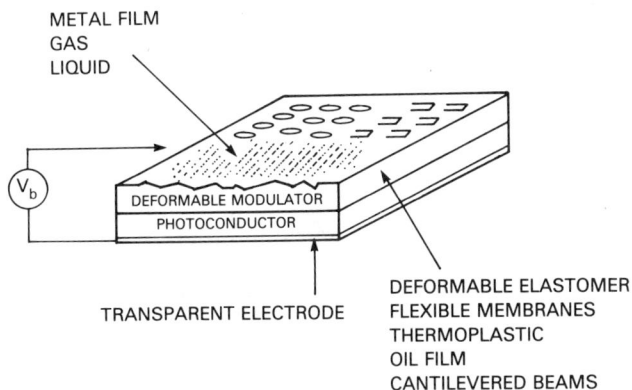

FIG. 9.   Construction of 2D spatial light modulator employing mechanical deformation of a material.

speed, and resolution improvements have been obtained by replacing the photoconductor in the LCLV and Phototitus sandwiches by a large planar silicon photodiode junction (Efron *et al.*, 1983; Armitage *et al.*, 1985). In a related development, new hybrid planar 2D SLM structures are emerging which are compatible with planar silicon microcircuits such as arrays of phototransistors. For instance, one version of the DMD device (Pape, 1985a) is fabricated directly on a standard Si wafer by etching holes in a thick polysilicon layer over an array of Si phototransistors, and then covering the device with a reflecting flexible membrane. The phototransistor in each resolution cell controls the potential applied to an electrode below the membrane modulator element formed at each hole. In another device (Lee *et al.*, 1986), islands of recrystallized polysilicon are deposited on a PLZT substrate to produce an array of modulator cells; each Si island contains a phototransistor which controls the potential applied across the surface of an adjacent region of the PLZT. A read beam transmitted through this PLZT region, and on through the substrate, is thus modulated by the transverse electro-optic effect in the PLZT. The addition of a reflecting coating on the PLZT surface between the Si islands results in a reflective configuration, where the readout beam enters through the substrate and the output beam is reflected back out. This configuration offers lower operating voltages and better isolation between the write and read beams. Another proposed improved version of this device is built on a commercially available, silicon-on-sapphire substrate with islands of single-crystal PLZT deposited through openings etched in the Si.

By utilizing a modulating electro-optic crystal which is also photo-conductive, the Pockels readout optical modulator (PROM) (Horwitz and Corbett, 1978; Owechko and Tanguay, 1982) and PRIZ (Petrov, 1981) devices can assume a very simple sandwich structure, which eliminates the separate photosensor. These devices consist merely of two transparent electrodes which are isolated from a photorefractive crystal such as BSO ($Bi_{12}SiO_{20}$) or BGO ($Bi_{12}GeO_{20}$) by thin dielectric layers; in some instances even the dielectric layers can be eliminated. Different write-beam and read-beam wavelengths are generally employed in these devices. A short-wavelength write beam generates charge carriers, and by properly applying an external bias voltage these charges can be made to drift through the crystal and produce spatial electro-optic refractive index changes as a function of the input imagery. These index changes phase modulate a read beam transmitted through the crystal; the read beam is at a longer wavelength where additional carriers are not generated. Reflective readout, similar to the other modulators discussed above, can be obtained by coating one surface with a reflective layer.

Even simpler structures, with many important SLM properties, are possible with such photorefractive ferroelectric materials as BSO, BGO,

$Fe:LiNbO_3$, $BaTiO_3$, KTN, or SBN. Employing one of these materials with no additional structure or only an applied bias voltage, one can perform such operations as (Burke et al., 1978; Pepper, 1982; Huignard et al., 1985): image multiplication and amplification by two-wave mixing (TWM) or by degenerate four-wave mixing (FWM), real-time holography, and long-term storage of high-resolution holograms. A relatively new operating mode for these materials, called the photorefractive incoherent-to-coherent optical converter (PICOC) (Marrakchi et al., 1985) fills in most of the missing functions of a general SLM including wavelength conversion and incoherent-to-coherent conversion. This mode of operation involves writing a spatial photorefractive index grating in the material with two coherent light beams and then selectively erasing the grating with the image information in a third beam which need not be coherent with the other beams. The modulated grating can be read out with a fourth read beam. The volume-holographic photorefractive operations mentioned here involve separation of the photoinduced charges over the wavelength-scale distances of interference fringes. The PROM and PRIZ discussed above generally involve much larger charge separations, which restricts the number of suitable photorefractive materials and reduces their image resolution by a factor of more than one hundred, due largely to fringing electric fields. On the other hand, the PROM/PRIZ structure is generally easier to use, being a bulk modulator which does not require coherent auxilliary beams or spatial carriers.

Electroabsorption, or the Franz–Keldysh effect, provides the basis for another type of single-material modulator (Kingston et al., 1984). The active material GaAs is photoconductive, and a device can be configured where the optically generated carriers produce fields which in turn change the hole–electron bandgap and thereby modify the optical absorption at wavelengths near the band edge. (GaAs can also potentially be employed as a photorefractive modulator.)

A significantly different SLM structure, where photoemission is employed as the light sensing mechanism, is illustrated in Fig. 10. In operation, an incident write image is converted by a photocathode into an electron image, which is amplified by a device called a microchannel plate (MCP). The electrons are then deposited onto a mirror on the back of a light modulating material. The MCP is an array of tiny tubes ($\sim 10$ $\mu$m diameter), each acting as a continuous-dynode electron multiplier, where electrons "bouncing" down the tubes generate additional electrons by collisions with the walls. There is also a grid mesh between the MCP and modulating material to facilitate the active removal of charge by excess secondary emission of electrons from the modulating material. The grid controls the energy of the primary electrons and collects the emitted secondary electrons (Fisher, 1981). The modulating material can be an electro-optic crystal, such as $LiNbO_3$

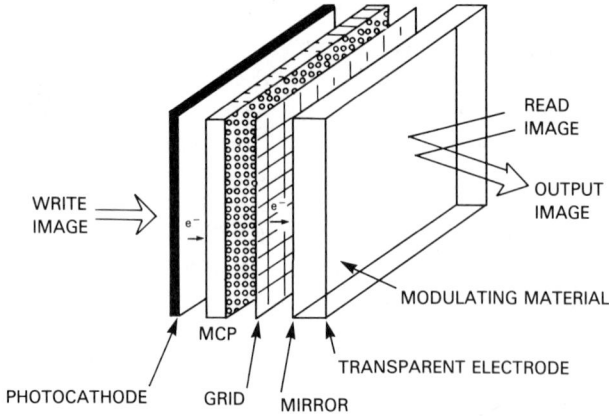

FIG. 10.   2D spatial light modulator employing photoemission as the light-sensing mechanism and including electron amplification.

or $LiTaO_3$ in the microchannel spatial light modulator (MSLM) (Warde et al., 1981; Warde and Thackara, 1983; Hara et al., 1985), or a reflective, flexible membrane array covering the MCP in the photoemitter membrane light modulator (PEMLM) (Somers, 1972; Fisher et al., 1986). These devices offer storage of images as a net positive or negative charge distribution on the modulating material. Storage times of more than six months have been observed in some MSLM devices.

Heat generated by absorption of the writing image intensity is an alternative optical sensing method which results in fairly simple device structures. The $VO_2$ (Eden, 1979; Strome, 1984) and smectic LCLV devices consist simply of a single thin film of the active material. Optical heating of the material causes phase transitions that modify its optical properties. In some instances an auxiliary heating means is provided for active temperature control and erasure purposes. In another variant (O.39 in Table II), a reflective oil film is deformed by local heating from an underlying, light-absorbing, plastic substrate (Schneeberger et al., 1979).

Bistable optical devices (BODs), which have been receiving increasing attention in recent years, can also perform many of the functions of an optically addressed spatial light modulator. (References providing an overview of recent developments in this field include: Gibbs, 1985; Peyghambarian and Gibbs, 1985; Smith et al., 1985; Dove, 1985; Chemla et al., 1985; Gibbs et al., 1986; Dagenais and Sharfin, 1986.) Bistable optical devices often take the form of an optical etalon, such as a Fabry–Perot cavity or multilayer interference filter, containing a nonlinear material whose optical absorption, refractive index, and/or physical length are a function of the incident optical intensity. Nonlinear materials which have been utilized in the Fabry–Perot cavities include Na vapor, $CS_2$, nitrobenzene, CdHgTe, CdS, CuCl, ZnS, ZnSe, InAs, InSb, GaAs, and multiple quantum well (MQW) structures consisting of very thin (0.5–100 nm) layers of GaAs and $Al_{1-x}Ga_xAs$. Recent attention has focused on the last three materials and on interference color filters with layers of ZnS or ZnSe, all of which offer fairly large nonlinearities at room temperature. In operation, part of the incident intensity on the bistable device is transmitted or reflected to form an output beam. The nonlinear material provides a feedback mechanism, whereby the intensity in the etalon changes the effective cavity length (and/or absorption), which in turn modifies the cavity intensity. The net transmission or reflection follows a nonlinear transfer function of the general form shown in Fig. 11, which is characterized by hysteresis with two possible bistable outputs for input intensities near $I_2$. The feedback required for bistability has also been realized in such other configurations as: the self-electro-optic effect device (SEED), which places an MQW structure inside a $p-i-n$ photodiode (Miller et al., 1984); hybrid structures where the output intensity from an etalon is incident on a detector which in turn drives an electro-optic (e.g., $LiNbO_3$) modulator in the cavity (Smith and Turner, 1977); laser diode cavities where the external intensity tunes the cavity resonance (Dagenais and Sharfin, 1986); a conventional SLM with its output beam fed back to its input (Collins, 1980; Fisher, 1981); or reflection at an interface with a nonlinear material.

The general bistability characteristic of Fig. 11 offers switching and/or memory functions. Thresholding can be implemented by operation near $I_3$ in Fig. 11, and it is possible to configure devices with only threshold and no hysteresis; i.e., $I_1 = I_3$ in Fig. 11. The input intensity can also be a summation of multiple input beams, in which case nonlinear beam interactions such as binary logic or amplification of a weak beam (accompanying a biasing-to-threshold beam) can be implemented. Similarly, switching between output

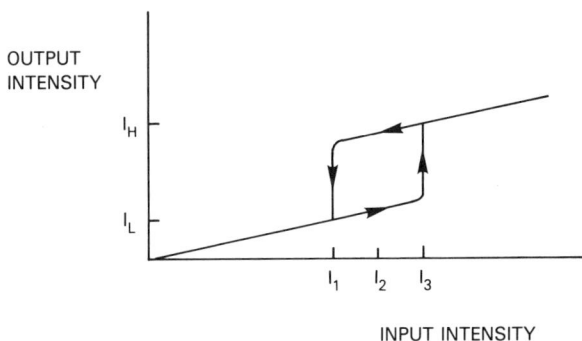

FIG. 11.   The nonlinear transfer function generally obtained with a bistable optical device.

hysteresis states can be induced by a weak beam accompanied by a bias beam. The relatively flat regions below and above the hysteresis region provide an optical limiting function, for example, to supress intensity fluctuations in a noisy beam.

To be useful in SLM applications these bistable devices must be able to operate on 2D information or image fields. To date, simultaneous parallel operation of only 2–25 elements has been demonstrated (private communications with H. M. Gibbs and S. D. Smith). Low contrast, with $I_L$ large relative to $I_H - I_L$ in Fig. 11, and large input-offset or holding intensity, $I_2$, also limit the utility of many of the bistable technologies. High speed is a potential strong point of these devices; some of the bistable etalon devices offer picosecond switching times. Sustained operation of most of the bistable devices has generally required 1–10 mW of optical power per pixel (for switching times in the ns to ps range), which implies that sustained operation of a large array may demand impractically large optical intensities. For example, a $10^6$ element array constructed in 1 cm$^2$ (Dove, 1985) would require 1–10 kW/cm$^2$ of optical intensity, with 1 kW/cm$^2$ or more being dissipated as heat. An alternative is a low duty cycle operating mode, with occasional transient high-speed decisions being made.

A closely related SLM device is the image-emission platelet laser, which is essentially a GaAs etalon, but it is pumped by a write-beam image of one wavelength to produce a self-luminous output image at another wavelength by lasing (Seko and Nishikata, 1977). Operation on imagery, with a resolution of 50 lp/mm, has been demonstrated.

## 2. Electronically Addressed Devices

Many electronically addressed SLMs (E-SLMs) are closely related to a corresponding optically addressed device sharing the same modulator material and hence material-dictated performance properties; in many instances both devices were developed by the same research group. This is reflected in Table II, where the electronically addressed devices are grouped among the optically addressed SLMs, mostly according to the modulating material. An alternative classification is shown in Fig. 12 according to addressing mechanism, which tends to dictate the overall physical structure. The following discussion of E-SLMs is according to the addressing mechanisms listed in Fig. 12.

In the electron-beam devices, the modulating material is placed in a vacuum envelope and written on by a scanned electron beam, much like the phosphor screen in a cathode ray tube. Specific examples of electron target materials include: electro-optic crystals in the Titus (Groth and Marie, 1970; Casasent, 1978c) and MSLM (Schwartz et al., 1985), electrostatically de-

**Electron Beam**

| NAME | MOD. MATERIAL | DEVELOPER |
|---|---|---|
| Titus | E-O | LEP, CMU |
| MSLM | E-O | MIT, Hamamatsu |
| PEMLM | membrane | NRL, England |
| Ruticon | Elastomer | IBM |
| Micro-mech. | Cant. beams | Westinghse, RCA |
| Eidophor, Talaria | Oil film | GE, Swiss |
| Lumatron | Thermoplas. | CBS, Erim |
| $VO_2$ | $VO_2$ | Vaught, USC |

**Electrode Matrix**

| NAME | MOD. MATERIAL | DEVELOPER |
|---|---|---|
| PLZT | PLZT | Bell, Sandia |
| MLM | Membrane | Perkin Elmer |
| L/SIGHTMOD, | YIG | Litton |
| LISA | (Magnetoopt.) | Philips |
| Liq. Film | Thermal $I^2R$ | Cannon |
| Opt. Tun. Arr. | Anisotropic part. sus. | Holotronics |

**Semiconductor**

| NAME | MOD./ADDR. MATERIAL | DEVELOPER |
|---|---|---|
| LCLV | Liq. Xtal/ Si CCD | Hughes |
| PLZT | PLZT/ Si Transist. | UCSD |
| DMD | Membrane/ Si CCD, Transist. | TI, RCA |
| Micro-mech. | Cant. Beams/ Si Circuits | TI, Telesens., TRW, IBM |
| TIR | $LiNbO_3$/Si Circuits | XEROX |
| Electro-abs. | GaAs/GaAs CCD | Lincoln |

**Other**

| Multichan. Bragg | Acoustooptic | Harris |
|---|---|---|
| Display Dev. | Misc. | Many |

FIG. 12. Categories of electronically addressed spatial light modulators according to addressing mechanism.

formable materials in the MLM (Van Raalte, 1970; Thomas *et al.*, 1975), an oil film in the Talaria and Eidophor (Mol, 1974) or thermally modifiable materials in the $VO_2$ (Eden, 1979) and thermoplastic devices. Since there is no optical sensor to be isolated from the read beam, readout by either transmission or reflection is often possible. An additional flood electron gun is sometimes included for electron erasure by secondary emission. In the PEMLM and MSLM devices, an MCP in front of the membrane or electro-optic crystal intensifies the electron beam for faster charge writing, and the incorporation of a photocathode results in a dual-function device having either electrical or optical addressing (Schwartz *et al.*, 1985).

Individual pixels are addressed in the electrode matrix devices at the intersections of two perpendicularly crossing linear arrays of electrodes. The crossed electrodes can either be on opposite faces of the modulating material, as in the optical tunnel array (OTA) device, or in a single plane, as in the LIGHTMOD/SIGHTMOD (Ross et al., 1983) or LISA magneto-optic devices. The OTA device contains a dielectric with pores or "tunnels" filled with a particle-in-liquid suspension which changes its opacity as a function of the voltage between the opposing transparent electrodes. The SIGHTMOD/LIGHTMOD and LISA devices consist of a 2D array of small (76 × 76 μm) mesas of the magneto-optical material yttrium iron garnet (YIG). Each mesa is a single magnetic domain which can be in one of two opposite magnetic orientations; the two orientations produce different Faraday rotation in the polarization of a transmitted readout beam. Operation between crossed polarizers thus yields two binary amplitude modulation levels. The conjunction of currents in both crossing matrix electrodes at the ion-implanted corner of a mesa induces switching in the SIGHTMOD/LIGHTMOD. In the LISA there is a small resistive element at each mesa corner to thermally induce domain flipping. The Canon liquid film device also employs local heating by a resistive element at each matrix pixel (Minoura et al., 1984).

Simple electrode matrix addressing schemes are generally restricted to modulating mechanisms with a threshold-like or rectifying nonlinear drive characteristic that allows discrimination of the drive signal of two intersecting electrodes from that of one electrode. An alternative is to add an active device such as a thin-film transistor at each intersection, which places the device in the semiconductor-addressed category and can greatly increase the device's complexity, fabrication/yield difficulty, and expense. Matrix addressing often offers the option of loading an entire row of information simultaneously to obtain a frame update time equal to the material response time multiplied by the number of rows. However, in a device with storage, such as the LIGHTMOD/SIGHTMOD, random pixel-by-pixel addressing can be advantageous for some applications.

Many of the newer additions to the repertoire of E-SLMs employ semiconductor addressing. As mentioned above, some of these employ semiconductor devices at the junctions of a crossed matrix to provide nonlinearity. Others implement specialized serial addressing structures such as shift registers or strings of charge coupled devices (CCDs). For example, the CCD-addressed LCLV (Efron et al., 1983) contains an array of parallel 1D analog CCD shift registers. In operation, imagery is first formatted into lines by a master serial-input CCD register, and is then shifted line by line into the array via parallel operation of all the other CCD shift registers. After $N$ shifts a full $N \times N$ frame has been loaded, and the charge image is then dumped through the silicon substrate to the mirror on the back of a liquid crystal modulator cell. Another silicon-addressed liquid crystal modulator is under development, with each pixel containing an $n$MOS memory element (Underwood et al., 1985). In an electronically addressed variant of the DMD (Hornbeck, 1983), the pixels are arranged in an $x-y$ matrix, with each pixel containing a transistor and storage element. Each line of imagery is serial-to-parallel converted in a master shift register and then latched and parallel transferred into a line of the 2D matrix which is accessed by an address decoder. Other proposed variants of the DMD employ full CCD addressing, similar to the LCLV, and more recent versions of the DMD deflect diving-board-like cantilevered overhangs of $SiO_2$ instead of the flexible membrane elements mentioned above. The previously mentioned TIR device (Johnson et al., 1983) also places an electro-optic material ($LiNbO_3$) in direct contact with silicon addressing circuitry. To date, the TIR device has been fabricated only in a 1D format. However, with a suitably large crystal, it should be possible to apply a full read beam with multiple rows of imagery at the required oblique angle. The GaAs electroabsorption device (Kingston et al., 1984) has also been implemented using direct electronic access, with the modulating fields being provided by charge packets in a high-speed GaAs CCD. The silicon plus PLZT technology (Lee et al., 1986) discussed above is also amenable to implementing silicon addressing circuitry.

In a significant departure from the above electronic-addressing schemes, the multichannel Bragg cell (VanderLugt et al., 1983; Pape, 1985b) addresses a 2D format by stacking an array of 1D acousto-optic Bragg cells of the type discussed in Section III,B. In practice, a single slab of acousto-optic material is employed, with a linear array of transducers launching parallel acoustic beams across the readout beam aperture. The multichannel Bragg approach can potentially offer a larger, more spatially uniform, dynamic range of modulation than most other SLM technologies, and promises a respectable number of resolution elements, particularly along the direction of propagation. However, lack of storage is a severe limitation, unless the requisite multichannel high-bandwidth electronic data happen to be available as part of the application. The Bragg cells must generally be operated with very low duty cycle illumination, using only a very short flash of light at the instant all the information has propagated into the cell. In some applications, the optical processor sits dormant most of the time, and its capabilities are thus not fully utilized. In addition, the short light pulses must be very intense to provide sufficient energy to a subsequent integrating detector, and optical power is wasted if the pulses are created by a shutter, with the source remaining continuously on.

The rapidly developing field of display devices can also provide some potential candidates for an electronically addressed SLM. (Heynick et al.,

1973; Kmetz and Von Williams, 1976; Hirshon, 1984; Perry, 1985; Apt, 1985; and Blechman, 1986, provide an overview of display technology. The yearly digests of the Society for Information Display International Symposia (SID) and regular display-devoted Proceedings of the Society of Photo-Optical Instrumentation Engineers (SPIE) are also excellent sources for additional information.) Many of the addressing schemes employed in E-SLMs were first pioneered in a display context, such as scanned electron beams, electrode matrices, and more recently, arrays of thin-films transistors in amorphous Si and direct Si addressing. In fact, the development of some SLM technologies was initially motivated by flat screen and projection display applications. Most display devices have tended to lack the cosmetic quality and flatness required for use as an E-SLM with coherent light. Furthermore, the fundamental SLM function of multiplying the electrical write image by a read-beam image to obtain a product output beam demands a phase, amplitude, and/or polarization modulating material. However, many of the display technologies currently being most actively pursued are self-luminous, for example, electroluminescent, vacuum fluorescent, gas plasma panel, LED, and, of course, CRT phosphor. Of course, there are some E-SLM roles, such as input to an incoherent processor, which can be served by a self-luminous display. In addition, there have been some very promising developments in such nonluminous display technologies as electrophoretic, electrochromic, and, in particular, liquid crystal. In fact, good-quality, low-cost ($\approx$ \$100), nematic liquid crystal television displays with up to 240 × 240 resolution elements are now commercially available (Blechman, 1986) and are beginning to be exploited by optical researchers (Lui et al., 1985; McEwan et al., 1985b).

*3. 2D Spatial Light Modulator Functions*

An appreciation of the basic functions performed by a spatial light modulator provides an understanding the issues which have motivated their development and set their performance requirements, and offers a perspective for evaluating and comparing the large variety of alternative devices discussed above. The most basic functions of an optically addressed SLM can be seen by looking back to Fig. 8 as a somewhat generic view of a spatial light modulator. Notice that this device is fundamentally a real-time parallel multiplier of 2D image fields. In the reflective configuration of Fig. 8, the intensity of a write image incident from the left controls the amplitude (and/or phase or polarization) changes imposed on a read image incident from the right. Each point in the read image is in effect *multiplied by a reflectivity* which is a function of the corresponding point in the write image to obtain the output product image exiting to the right in Fig. 8. With suitable optics, this basic image multiplication operation can be exploited for applications ranging from actual analog numeric multiplication of vectors and matrices (Tamura and

Wyant, 1976; Goodman *et al.*, 1978) to matched filtering, wave-front conjugation (Fisher, 1981), reconfigurable interconnect between 2D fields (Goodman *et al.*, 1984), and programmable template-masking operations. Also note that Fig. 8 is essentially a three-port device and as such can be loosely viewed as a two-dimensional "optical transistor." For example, with an intense spatially uniform read beam, the output image can become an amplified version of a weak write image. The desirable feature of decoupling or isolation of the optical system producing the weak write beam from the amplified output beam is usually also obtained. When the uniform read beam has different properties than the write beam, the SLM can perform very useful conversion functions, such as from incoherent to coherent light or from one wavelength to another.

The electronically addressed devices generally serve as interface transducers between electronic and optical processing subsystems. Many implement a parallel image multiplication function analogous to the O-SLMs, with an incident read image being multiplied by an effective reflectivity which is a function of the electronically written input information. This multiplication operation can be applied to such applications as analog multiplication by matrices, matched filters, or masks stored or generated by a digital computer, or implementing control or programmable interconnects under the direction of an electronic sequencer. The E-SLMs also serve to input and/or format data into an optical processor, for example, from such nonoptical sources as multisensor arrays or the results of digital manipulations in a hybrid electro-optical processor. The formatting role depends on the specific addressing mechanism, with sequential electronic to the 2D parallel-optical conversion being common. The input and format functions usually employ a uniform read beam and hence do not demand the multiplication capability. Once the transduction to optics is accomplished it is appealing to remain in the optical processing domain by exploiting the functional capabilities of optically addressed SLMs, which are emphasized in much of the remainder of this section.

Many SLMs also have storage capabilities, with their output image becoming the product of a previously stored write image and the current read image. These devices can perform memory or information latching functions, and optically addressed SLMs with storage are in general a reusable replacement for photographic film. Storage is particularly important in the electronically addressed devices, because continuous updating of a 2D data array places huge information bandwidth demands on a driving electronic processor. Modulators in Tables II and III which can store images for on the order of an hour, and in some instances much longer, include the Phototitus, Titus, smectic or ferroelectric liquid crystal, PLZT, PROM/PRIZ, MSLM, PEMLM, thermoplastic, photorefractive, PICOC, L/SIGHTMOD, LISA, bistable, and $VO_2$ devices. Storage capabilities are often accompanied by the ability to perform numeric addition between successively stored images and also to detect very low light level input imagery with an O-SLM by integration over an extended period.

Virtually all SLMs can also perform addition between simultaneously applied incoherent image intensities. Some SLMs can also directly perform image subtraction. For example, the previously mentioned ability of the MSLM and PEMLM devices to either add or remove electrons to/from the stored-charge image distribution on the modulating material can implement algebraic subtraction between the stored image and new image written with opposite polarity (Warde et al., 1981; Warde and Thackara, 1983; McEwan et al., 1985a). A few other SLMs, such as the Phototitus, can also be operated in an analogous subtraction mode. With the inclusion of additional optics, the arithmetic operations of parallel subtraction (Marom, 1986) and division (Efron et al., 1985) of image fields can also be implemented with most SLMs.

Another major additional class of SLM functions is in the performance of nonlinear operations on each point in the input write-image intensity. In practice, most modulators are intrinsically nonlinear; for example, many modulators are used in conjunction with crossed-polarizer or interferometric readout, which produces an output beam amplitude proportional to the sine of the write-beam intensity, or output intensity proportional to the sine squared of the input intensity. Modulation approximately linearly proportional to the input intensity can be obtained by operation near a $\sin 2n\pi$ point of the modulation characteristic. For example, the PEMLM and MSLM are easily biased by erasure to a uniform stored-charge distribution corresponding to the $\sin 2n\pi$ point. On the other hand, it is possible to obtain reversed-contrast modulation by operation about a $\sin(2n - 1)\pi$ point. This contrast reversal operation can often be implemented on a previously stored image by manipulating crossed polarizers or Schlieren optics in the readout system. A few SLMs can also perform contrast reversal by alternate means, for instance, by switching between the electron accumulation and depletion modes in the MSLM or PEMLM. Other examples of intrinsic nonlinearities are output intensity proportional to $\sin^2(\text{input intensity}^2)$ in some configurations of the mechanical modulators (e.g., cantilevered beam, PEMLM, MLM, or DMD), or exp(input intensity) in the electroabsorption devices. (A thin electroabsorption device can offer an output amplitude approximately proportional to the input intensity.) Other technologies, such as photo/cathodochromic, can also provide output amplitude directly proportional to input intensity.

In image thresholding, another important category of SLM nonlinear operations, there is little or no change in the output intensity until the input write-beam intensity exceeds a specific threshold level. In analog thresholding, after the input intensity has exceeded the threshold, the output intensity varies as a function of the input intensity. Thresholding behavior can be obtained by employing specific modulator materials, such as liquid crystals, YIG, PLZT, or $VO_2$ or by using particular aspects of the device physics as in the MSLM and PEMLM, the platelet laser, and most of the bistable optical devices. In binary or "hard-clip" thresholding there are two fixed output levels, corresponding to input intensities below or above threshold, respectively. Many of the bistable optical devices can approximate this behavior. The MSLM and PEMLM devices have a very useful intrinsic hard-clipping mode where all three critical parameters, input threshold level, below-threshold output level, and above-threshold output level, are fully adjustable, including inversion with the above-threshold output less intense than the below-threshold output. Hard-clip image thresholding allows decisions (e.g., yes/no, true/false, exists/doesn't exist) to be made in an optical processor in terms of the two output intensity levels and can also provide a "Schmitt trigger" function to regenerate binary optical signals in cascaded systems. In addition, binary logic operations can be performed between multiple input beams; for example, setting the input threshold below the logical "1" intensity level produces the OR operation between two input beams, and setting the input threshold at $\frac{3}{2}$ of the logical "1" level produces an AND gate between two input beams. With inversion, NOT, NAND, and NOR, all possible Boolean logic operations can be implemented. Hard-clipping can also be utilized to build an analog-to-digital converter for optical image fields or can form the nucleus of an optical half-tone system which can implement virtually arbitrary nonlinear intensity functions (Dashiell and Goodman, 1975). Particularly useful nonlinear functions include log for transforming multiplicative signals and/or noise into additive signals, and square root for obtaining output amplitude modulation directly proportional to the write-beam amplitude.

Some SLMs also have intrinsic abilities to directly perform more advanced processing functions. For example, the variable grating mode (VGM) liquid crystal device produces an output beam modulated by local spatial gratings proportional to the local intensity of the input write beam (Tanguay et al., 1983). Almost any nonlinear function of the input intensity can be implemented by passing the output beam through an appropriate spatial filter in a subsequent Fourier plane. The programmable hard-clipping mode of the MSLM and PEMLM devices can be used to implement edge detection between regions of the input image having intensities above and below the threshold level (Warde et al., 1981; McEwan et al; 1985a). Essentially, both output levels are set to produce the same output intensity (e.g., white), but separated by one full cycle of the device's sinusoidal modulation characteristic. Due to finite device resolution, a line of the intermediate half-cycle level (e.g., black) is produced along portions of the input image divided by the threshold. Other devices such as the PRIZ (Petrov, 1981; Owechko and Tanguay, 1982) and photorefractive configurations (Feinberg, 1980) also offer intrinsic spatial differentiation or edge detection capabilities. The MSLM and PEMLM devices can also intrinsically implement the more exotic function of operating as a pixel array of lock-in amplifiers or synchronous detectors (McEwan et al., 1985a). This function is implemented by oscillating the device between its charge deposition and removal modes at a rapid rate. Portions of the write image which are flashing on in synchrony with, for example, the charge-deposition cycles, result in net charge accumulation over many cycles which eventually integrates up to a visible modulation level. However, portions of the write image which are continuously illuminated or oscillating at a slightly different frequency result in no net charge accumulation. Applications of this lock-in amplifier array include discrimination from the ambient background of a structured light pattern projected into a scene for robot vision, removal of the dc bias from a heterodyne interferometric image, and target designation.

Some of the technologies mentioned in Sections III,C,1 and 2 do not completely fall into the basic SLM functional framework outlined at the start of this section and are probably more accurately classed as "SLM-related" devices. For example, most of the bistable optical devices are fundamentally two-terminal devices, with only a write beam and an output beam, and thus do not directly offer the basic real-time image multiplication and image-transistor functions outlined above. They are more analogous to electronic diodes in that their output intensity is a nonlinear interaction function of the beam intensities applied to their input port. Furthermore, they require the continuous application of illumination to store information, unlike many SLMs which have a passive storage mode requiring no power. However, some of the three-port operations can be approximated with bistable devices by superimposing two beams at the input, a weak "write" beam accompanied by a stronger "read" beam which provides power for the "output" beam. The bistable devices also intrinsically provide such specialized nonlinear functions as thresholding, switching, or logic on multiple simultaneous input beams. However, as mentioned previously, their major potential strength is speed— offering nanosecond to picosecond switching times, as opposed to the milli-second operation projected for most SLMs.

As discussed earlier, such self-luminous technologies as CRTs, other display devices, arrays of LEDs, or laser diode arrays do not provide the fundamental multiplication SLM function. They are of course still useful for input and formatting functions. The lack of versatility of such a device may explain why there are very few examples of optically addressed emissive modulators. However, there was some work in the 1960s and early 1970s on photoconductor-accessed electroluminescent devices (Pollock et al., 1963;

Tippett *et al.*, 1965; Shaefer and Strong, 1975). Currently, there is interest in platelet lasers (Seko and Nishikata, 1977), and semiconductor integrated circuits incorporating photodiode-activited LEDs or surface-emitting 2D planar arrays of injection lasers (Uchiyama and Iga, 1985; Liau *et al.*, 1984; Goodman *et al.*, 1984).

The photorefractive volume-holographic devices can in principle perform most SLM functions, including multiplication and amplification, but their use is generally subject to a variety of additional constraints (Pepper, 1982; Huignard *et al.*, 1985; Tanguay, 1985)]. For example, in four-wave mixing the write (or "pump"), read (or "probe"), and output (or "conjugate") beams must have specific geometric, polarization, and intensity relations to each other and must be mutually coherent. The coherence condition generally means that the write and read beams must come from a common laser source, which precludes wavelength conversion or incoherent-to-coherent conversion. The coherence constraints are relaxed in the previously mentioned PICOC configuration (Marrakchi *et al.*, 1985).

### 4. Spatial Light Modulator Performance Issues

There is a large number of parameters for quantitatively evaluating and comparing how well the various SLM technologies perform their basic functions. Due to space limitations, only spatial resolution and/or total number of pixels, optical-addressing sensitivity, and time response characteristics are given in Table III. Readout dynamic range, e.g., in phase, amplitude, and/or polarization, is another critical parameter which is coupled to such additional issues as: contrast, signal-to-noise ratio, input–output non-linearity, and the practical consideration of the required operating voltages. Optical quality is also extremely important, involving such issues as scattering, cosmetic defects, spatial variations in response uniformity, and optical flatness. The total spatial–temporal throughput, i.e., area × resolution/frame time, is sometimes used to provide a single measure for crude comparison of the various technologies. Other important parameters include fabrication difficulty, lifetime, reproducibility of behavior, specialized capabilities, and fundamental performance limitations.

Assigning realistic, comparable numeric values to SLM performance is difficult. Performance numbers are critically dependent on measurement conditions, but the conditions are often not given in the literature or are not consistently related to actual device applications. Resolution, sensitivity, and speed parameters, for instance, can be made arbitrarily good by making the measurement with an arbitrarily small depth of modulation. As another example, dynamic range is closely linked with spatial uniformity in these 2D format devices. While the measured dynamic range at a single point may be very large, this is useless for most applications if a given analog level is not the same from point to point across the device. In fact, very few 2D devices can claim a "spatial dynamic range" better than 20 levels (5% uniformity). Performance for new device technologies is also sometimes confused with overly optimistic projections based on perceived limitations. In addition, research groups often concentrate on one or a few performance characteristics at the expense of others.

The performance parameters in Table III should be interpreted with care: they are generally best-reported values or near-term, low-risk projections. They are sometimes not all simultaneously achievable in one version of a given device. The sensitivity and speed parameters in Table III are generally given for a response of 90% of maximum-contrast amplitude modulation and/or 90% of half-wave phase modulation (i.e., $0.9\pi$ radians). The time-response data are generally for updating a full frame in the optically addressed devices, and are footnoted as to whether they are for a full frame or an individual pixel response in the electronically addressed devices. In the latter case, the full frame write time is usually the pixel response time multiplied by the total number of pixels, or by the total number of lines for line-by-line addressing. The resolution data are footnoted as to whether they are for 50% or 10% contrast expressed in line pairs/mm, or for pixel density of discrete-array devices expressed in pixels/mm. When available, the contrast numbers are derived from the modulation transfer function, which is essentially a plot of contrast of the output beam as a function of the spatial frequency of a sinusoidal fringe pattern applied to the write image; the contrast being measured as (max amplitude − min amplitude) divided by (max amplitude + min amplitude). A footnote in Table III also designates that some of the devices have a spatial-bandpass MTF, which falls off at both low and high spatial frequencies; either the upper and lower 50% points or the total range of the 50% passband is given. For device comparison purposes, the 50% resolution of a discrete-pixel device can be taken as approximately $\frac{1}{2}$ of the actual pixel density. In applications, the total number of available pixels is often of greater import than the actual resolution, particularly with discrete-pixel devices.

The improvement of one SLM performance parameter at the expense of another is often unavoidable as a result of fundamental physical tradeoffs. In many electro-optic modulators, for example, a thinner slab of electro-optic material offers higher resolution, but at the same time also increases the device capacitance, which in turn decreases the device speed. A similar tradeoff is also seen with some mechanical modulators; as the modulator elements become smaller, more charge is required to deflect them and the device is slowed (Fisher et al., 1986). There is also a fundamental trade-off between resolution and sensitivity. As the resolution elements become smaller, more optical energy per unit area (i.e., $\mu J/cm^2$) must be collected to obtain enough photons in each pixel to maintain a given quantum-limited signal-to-noise ratio (S/N). The MCP-intensified PEMLM and MSLM modulators can be operated in this quantum-limited regime, and it is often desirable to optimize the MCP gain so that the number of photons required to obtain the desired S/N also drives the device to full-depth modulation (e.g., $\pi$ phase shift).

Many aspects of SLM performance arise from an interplay between limits imposed by material properties and device geometries. Using time response as an example, the typical response times of a variety of modulating materials are given in Table IV; however, most SLMs operate considerably more slowly than these ultimate limits. For instance, the maximum speed of the MCP and silicon-addressed devices is generally limited by thermal dissipation constraints on the maximum current available to charge the modulating material. However, this is impacted by the amount of charge required to obtain full-depth modulation (e.g., $\pi$ phase shift), which is a function of both modulation material properties and the geometric configuration of the modulator (for instance, the thickness and orientation of the crystal in an electro-optic modulator). The time response of photoconductive modulators tends to be limited by the material properties of the photoconductor, particularly carrier decay lifetimes and carrier mobility.

## 5. Current Status and Future Directions

Most of the 2D spatial light modulators listed in Table II exist only as laboratory prototypes and some are not even being actively developed any longer. Only the LCLV, MSLM, Librascope, PROM, and thermoplastic optically addressed devices and LIGHTMOD/SIGHTMOD, Talaria/

TABLE IV

RESPONSE TIMES OF MODULATOR
MATERIALS

| | |
|---|---|
| Electro-optic crystals | 1 ps |
| GaAs (in MQW or etalon) | 1 ps |
| PLZT (single crystal) | 1 ns |
| Deformable membranes | 0.5 $\mu$s |
| Cantilevered beams | 1 $\mu$s |
| Magneto-optic switching | 1 $\mu$s |
| Liquid crystal (ferroelectric) | 1 $\mu$s |
| Acousto-optic (Bragg cell) | 1 $\mu$s |
| PLZT (polycrystalline) | 10 $\mu$s |
| Particle suspensions | 1 ms |
| Deformable elastomers | 1 ms |
| Liquid crystal (nematic) | 10 ms |

Eidophor, and Titus electronically addressed devices are commercially available. The DMD, optical tunnel array, PLZT, Phototitus, PEMLM, Ruticon, and/or $VO_2$ modulators could potentially become available within the next 5–10 years, depending on corporate and funding agency decisions.

While some researchers feel that none of these existing SLMs performs sufficiently well for serious application, the actual performance levels required are, of course, application specific. It is possible at present to prototype simple cases of quite a few "unrealizable" sophisticated applications using many of the 2D SLMs available today—including even some display devices. This currently achievable performance level is on the order of 100 × 100 resolution elements, 10 Hz framing rates, 1 s storage, less than 50 $\mu J/cm^2$ sensitivity (O-SLM), 5 level dynamic range, a few wavelengths flatness, and 10% spatial uniformity. In practice most applications do not require all of these features simultaneously. This initial prototyping activity can serve an important function in identifying critical parameters and defining directions for future device research.

There are a number of promising, but less developed, modulator technologies which may have an important impact in the future; some of these are new organic crystal and polymer electro-optic materials (Garito and Singer, 1982; Williams, 1983), ferroelectric liquid crystals (Clark et al., 1983; Armitage et al., 1986), 2-D arrays of diode lasers (Uchiyama and Iga, 1985; Liau et al., 1984), and stacks of integrated-optical modulators. Other modulation mechanisms which are also currently being explored include photochromic, cathodochromic, electrophoretic, photodichroic, electrowetting, electrocapillary (Lea, 1984), and various magneto-optical effects. Some of the newer semiconductor-compatible technologies offer the exciting prospect of bypassing many of the limitations imposed by material properties and the simple sandwich-type device geometries. One can conceive of a hybrid 2D SLM built on a semiconductor substrate with simple electronics between a phototransistor and modulator in each resolution cell. A reprogrammable SLM could result which accurately implements such useful nonlinear intensity functions as log, square root, inversion, and hard-clip thresholding, as well as optical logic, image storage, switching functions, and general arithmetic operations (i.e., $+, -, \times, /$). It should also be possible to include connectivity between pixels (McAulay, 1983; Dove, 1985), for example to implement arithmetic and/or logic operations between adjacent pixels, spatial image shifting operations, or structures for more sophisticated algorithms. Such a device could play an important role in optical interconnect and SIMD (single instruction, multiple data) parallel processing applications. Candidate silicon-addressed technologies include cantilevered beam, deformable membrane, PLZT, liquid crystal, and electro-optic crystal devices, and also the electroabsorption and surface-emitting injection laser devices constructed with GaAs.

Current SLM research is moving in the direction of a device with better than 1000 × 1000 resolution elements, kilohertz framing rates, quantum-limited sensitivity (O-SLM), hour storage time, 100 levels of dynamic range, $\frac{1}{5}$ wavelength flatness, and 3% spatial uniformity. There are also a variety of practical issues which should guide future SLM development; these include ease of use, reproducibility and reliability of operation, minimal and unsophisticated support electronics and other equipment, simple low-loss optical readout system, small footprint on the optical table, and nonintimidating cost ( < $5000).

## IV. PHOTODETECTORS

Next to the unavailability of good light-control devices, a major impediment to the implementation of optical processors has been the difficulty in accurate, high-speed detection of optical data fields and the inability to rapidly and effectively extract the pertinent information from such fields. A large variety of photodetectors are available commercially in several basic types: single-element photodetectors, 1D arrays, and 2D arrays. However, the photodetector characteristics required in optical processors are often not available in commercial devices, particularly in the areas of dynamic range and access speed. Large dynamic range is needed because most optical processors are analog in nature; high-speed access is required to match the speeds at which computations are performed in the optical domain. Large numbers of pixels are also desirable in photodetector arrays; applications other than optical processing also require large arrays, but readout/access times are a particular problem for large arrays. Those aspects of photodetector design and physics that relate to specific optical processor needs will be emphasized here. Many references exist with more complete descriptions of various photodetectors and their operation, including some specifically constructed for acousto-optic processors (Borsuk, 1981).

The degree of shortfall in commercial detectors depends somewhat on the type of detector and the particular optical processor configuration. The categories of detectors can be divided into either vacuum tube or semiconductor types. Vacuum tube types include photomultiplier tubes which consist of a photoemissive cathode followed by amplification stages such as discrete electrodes (often called dynodes) or microchannel plates; the former are employed for single-element detectors, while the latter are generally employed in imagery. Single-element photodetectors are well developed;

a variety of detector types can be obtained for photon-limited dynamic range performance or for high speed (several GHz). Single-element vacuum tube photodetectors can achieve single-photon, shot-noise-limited performance. However, the use of photoemissive cathodes in such devices restricts the attainment of such performance to visible light wavelengths or shorter.

Semiconductor-type photodetectors (Si, GaAs, etc.) employ photosensing elements such as $p-n$ and $p-i-n$ photodiodes, charge-coupled devices, and charge-injection devices. Semiconductor photodetectors have the advantages of ruggedness, smaller size, longer life, and lower operating voltages than vacuum tube devices. Arrays of detectors are available either as large numbers of single-element devices or on a single monolithic chip. One-dimensional monolithic arrays can be obtained with almost 2000 pixels, and two-dimensional arrays are now approaching sizes of $1000 \times 1000$. In general, 1D arrays offer greater dynamic range than 2D arrays. Readout of such detector arrays can be either serial, fully parallel, combined serial–parallel, or random-access matrix type.

An optical processor can be configured for either incoherent or coherent light detection. With incoherent detection, i.e., detection of optical power only, a square-law relationship exists between the optical power ($I$) and the electrical power from the detector, i.e.,

$$P_e = Ri_d^2 = cI^2 \tag{12}$$

where $i_d$ is the detector signal current, $R$ is the load resistance for the detector, and $c$ is a constant. The electrical output power $P_e$ will range over twice the number of dB as the range of the optical power. With coherent detection, where both the amplitude and phase of the optical beam are preserved by combining a reference light beam with the signal beam on the photodetector and detecting the beat frequency, the amplitude of the signal beam $\bar{A}$ and the amplitude of the reference beam $\bar{B}$ are summed before square-law detection occurs, i.e.,

$$i_d = |\bar{A} + \bar{B}|^2 = |\bar{A}|^2 + |\bar{B}|^2 + 2|\bar{A}||\bar{B}|\cos\phi \tag{13}$$

where $\phi$ is the phase difference between the two beams (a constant in the usual case of coherent beams). Hence, disregarding the bias terms $|\bar{A}|^2$ and $|\bar{B}|^2$, a linear relationship exists between the optical power in the signal beam and electrical power from the detector. Coherent detection may be performed either in the time domain or spatial domain—i.e., the "beat" frequency may be either temporal or spatial.

### A. Photodetector Dynamic Range—General Considerations

The total span of optical intensities that can be detected unambiguously may be considered one definition of the dynamic range of a photodetector.

However, many factors enter into the determination of dynamic range. These include noise sources and signal-accumulation levels possible with a given photodetector bandwidth or integration time, and the presence of spurious signals, especially in arrays. Due to the photon nature of light, there exists a fundamental quantum noise limit in the detection of very low light levels. The photon noise limit may be expressed as a noise current

$$i_n = (2ei_dB)^{1/2} \tag{14}$$

where $e$ is the electronic charge, $i$ the average photodetector current, and $B$ the bandwidth of the detector circuit. Note the trade-off between bandwidth, or speed of detector response, and the noise level.

Semiconductor detectors can operate over broader wavelength bands than photomultipliers. For example, intrinsic (undoped) Si photodetectors operate out to about 1100 nm in the infrared, with peak response in the 800–900 nm region; this is a distinct advantage when employing laser diode light sources. Since the response is over a wider wavelength band and the photocharge consists of carriers traveling through the semiconductor, the noise limits in semiconductor detectors can be different from the photon-noise limit. As one goes beyond 1500 nm into the infrared wavelengths, noise becomes dominated by background blackbody radiation at 300 K. However, as noted earlier, longer wavelengths tend to be less useful for optical processing purposes and are therefore rarely encountered. The ultimate quantum noise–current limit for the semiconductor detector may be expressed in a form analogous to Eq. (14):

$$i_n = (4Gei_dB)^{1/2} \tag{15}$$

where $G$ is an additional intrinsic gain factor offered by a semiconductor detector.

The noise limits of Eqs. (14) and (15) are often not attained due to noise in the amplifier that follows the photodetector. This noise is thermal in origin and can be expressed as a noise current

$$i_n = (4k_BTB/R)^{1/2} \tag{16}$$

where $k_B$ is the Boltzmann constant, $T$ is the temperature in degrees kelvin, and $R$ is the resistance through which the photocurrent flows. The noise power $(k_BTB)$ over a 1 MHz bandwidth into a 50 Ω load at a temperature of 300 K is $-114$ dBm.

For maximum dynamic range, not only must the noise floor be low, but the photodetector current over an integration time $\tau$ must be as large as possible with a given light intensity. The current will scale with detector area, but generally detector speed and signal-to-noise (e.g., due to leakage currents) are reduced as detector area increases. This is particularly true for intrinsic semiconductor detectors where current versus light level is controlled by basic physical material properties. With large bandwidths $B$ it is difficult to obtain high dynamic range, since the signal level decreases as $1/B$ ($\tau \approx 1/B$ is the integration time) while the noise current increases as $B^{1/2}$. The signal level actually delivered also depends on the characteristics of the circuit following the detector. While this circuit itself might be high bandwidth, the impedance seen by the photodetector can limit the amount of charge actually delivered to this amplifier.

Semiconductor photodetectors can be obtained either with or without internal gain. With certain diode structures avalanche mechanisms exist that produce amplification of the photocurrent; this amplication can be as high as several hundred in these so-called avalanche photodiodes (APDs). This internal gain can be of advantage if the system-limiting noise is not amplified, or equivalently, the noise sources that are amplified do not determine the noise floor. This condition will generally be true where amplifier noise dominates. However, APDs are not available in large monolithic arrays, and they require higher bias voltages than nonavalanche diodes.

While the above noise considerations should hold for any detector configuration, arrays of photodetectors have additional sources of noise. Among these are (1) clock reset noise during readouts employing serial shifting of the charges; these often consist of voltage spikes occurring between individual pixel readouts or at the beginning of each pixel readout, (2) internally generated fixed pattern noise varying from pixel to pixel, (3) unequal sensitivities of pixels to the same light flux, and (4) crosstalk between pixel elements. The first two are independent of the light signal and in principle may be corrected with subsequent subtraction of fixed values from the output; the third is a function of the light flux, and requires corrective gain factors for each pixel. Without such corrections the usable dynamic range of the array is reduced. Crosstalk is the generation of signals at pixels other than the one being illuminated. It is most prominent when some pixels are illuminated with high optical intensities, and it interferes with the detection of low light intensities at the other pixels. Crosstalk mechanisms are numerous and include migration of photocarriers to other pixels, light scattering by the photodetector structure, and electrical coupling between different pixels. The electrical coupling depends on both the photodetector array and readout circuit configuration and may be capacitive, ohmic, or inductive; the first tends to be the most and the last tends to be the least significant.

## B. General Speed/Throughput Considerations

In optical processing applications it is generally desired to manipulate 2D fields that are spatially as large as possible. This will usually mean that photodetector throughputs must be commensurately large, even if the data array to be detected has been reduced in dimension to 1D or even a single channel. Optical processing architectures determine the format of the output data array, but the performance of photodetectors determines the efficacy of alternative candidate architectures. Large throughput can be obtained by either employing output channels of high bandwidth or by employing many more lower-bandwidth channels. Use of single-element, high-bandwidth detectors would seem attractive because of the potential for photon-noise-limited performance; however, signal levels (i.e., dynamic range) may not be high due to small integration time. A high-bandwidth serial output may also be obtained from a 1D array connected to a serial shift-register line. If the array data can be shifted in parallel to the shift register, one obtains a higher integration time (for continuous operation by up to a factor equal to the number of elements in the array). In these serial readout schemes further postprocessing of the output must be performed on the high-bandwidth data stream. Parallel output channels require lower bandwidth by a factor equal to the number of channels, but difficulties exist in producing arrays with large numbers of parallel channels and in interfacing the channels to postprocessors.

The size of semiconductor photodetector elements impacts speed in several ways. Generally, the smaller the photodetector element, the faster the response; typically, the capacitance of the element becomes smaller. Further, the total charge accumulated in a given integration time will decrease with detector size, so less energy is required to transport the signal, or a given voltage will more rapidly transport the signal. However, as mentioned, there is some loss in dynamic range.

## C. Design Approaches

Various photodetector designs have evolved in order to meet the specific needs of optical processors. For each of the specific needs the major approaches will be summarized here.

### 1. Speed/Throughput

For both single-element detectors and arrays, the use of III–V semiconductor compounds such as GaAs provides higher speed because of the higher photocarrier mobility. Single-element detector/amplifier packages have been demonstrated at up to 4 GHz bandwidth, employing InGaAs APD structures having separate absorption and multiplication regions and a graded-composition interface layer; $p-i-n$ and photoconductor structures and materials such as Ge have also been demonstrated with lower speed and sensitivity (Brain and Lee, 1985). These single-element photodetectors have been developed for fiber communications applications, and are generally optimized for near-infrared wavelengths such as 1300 and 1500 nm.

GaAs can be employed in detector arrays both as the sensing elements and as the material for a high-speed serial shift-register readout line. 1D arrays have been demonstrated with 1 GHz readout rates (Sahai et al., 1984).

High throughput for 1D arrays can also be obtained by employing parallel output channels. To obtain a large number of such channels, with reasonable dynamic range, VLSI techniques must be utilized. Often the preamplifiers for the detectors are placed onto the same chip to increase dynamic range by minimizing loading effects on the signal (Boling and Dzimianski, 1984).

If 2D arrays are used it is not easy to simply extend all the approaches used for 1D arrays. Fully parallel readout is not feasible for large arrays because of the complexity of wiring and pin-output constraints for VLSI chips. If only signal peaks need to be located within a 2D output field (a situation often encountered in optical processing), several approaches may be taken for such a search. A serial–parallel readout of the 2D array may be performed, whereby one identifies in parallel the rows in which high-intensity signals occur, followed by a serial search of the active rows. If only one signal is to be located, it is possible to use a matrix-address technique, where both the rows and the column are identified in parallel. This technique has been demonstrated with a microchannel-plate addressed matrix of electrodes (Timothy, 1985).

The general necessity to perform a 2D search in parallel for signal detection and to perform postprocessing of the detected signals (such as for parameter determination) might be satisfied by further stages of optical processing. In the context of photodetectors, one could consider light detection to be one of the functions to be performed by 2D spatial light modulators. 2D modulators such as the MSLM and the PEMLM discussed earlier have sensitivities comparable to most detector arrays— $< 1$ nJ/cm$^2$. If appropriate optical-processing algorithms exist for handling the detected optical field, parallel readout schemes are not needed.

## 2. Dynamic Range

The use of on-chip amplifiers to improve dynamic range in 1D semiconductor arrays has been mentioned. The other major approaches are to reduce the leakage or dark currents in the photodetector arrays, reduce crosstalk between elements, and to increase the charge storage capability for the elements. If long integration times are permissible, commercial 1D silicon $p-n$ diode arrays with about 40 dB in power detection dynamic range are available (Talmi and Simpson, 1980). For extremely long integration times, it is possible to perform analog-to-digital conversion periodically and to accumulate these results in an external digital memory. Different detector structures may also result in better dynamic range. A structure using the source region of a MOSFET transistor as the light-sensitive region has demonstrated a dynamic range of greater than 70 dB, with logarithmic output relative to input, and has been fabricated into 1D arrays (Chamberlain and Lee, 1984).

Since single-element detectors can achieve larger dynamic range than attainable with the elements in a monolithic array, one can attempt to construct an array of single-element detectors. Packing the detectors as closely as possible is a major problem; an alternative is a closely packed array of optical fibers or channel waveguides that fan out to large single-element detectors. This approach can also reduce crosstalk (the actual photodetectors may be separated at arbitrarily large distances from each other), giving an effectively larger dynamic range.

Methods for reducing crosstalk in semiconductor detector arrays include (1) deep grooves between elements to prevent charge migration, (2) design of pixel size/shape to minimize inter-pixel capacitance, (3) masking around the pixels with opaque dielectric to prevent entry of light, and (4) use of a high-resistivity semiconductor to reduce ohmic coupling and carrier migration. Use of a high-resistivity semiconductor in $p-n$ structures may also allow higher bias voltages to be employed; this results in deeper depletion regions, reducing ohmic coupling and direct light leakage, especially at the near-infrared wavelengths.

Use of coherent detection provides greater dynamic range theoretically, since the output current of a photodetector is proportional to the light amplitude rather than the intensity. Quantum-limited sensitivity is obtainable, and use of single-detector coherent schemes have long been employed in applications such as coherent laser communications. It is also desirable that the detector arrays used in optical processors be structured to perform coherent detection. However, coherent detection in the time domain requires that every pixel in the array be sampled at a rate greater than twice the beat frequency. This places an intolerable burden on purely sequential readout schemes; fully parallel readout of all pixels is desirable but has not yet been accomplished for large arrays. Coherent detection in the spatial domain requires significant amounts of postprocessing of the detector outputs to obtain optimum performance (e.g., background subtraction and gain equalization).

## 3. Resolution

The highest resolution currently available is obtained with vacuum tube devices such as vidicons; the equivalent of 1400 × 1400 resolution elements has been obtained. However, such devices are bulky, have high power consumption, and have limited dynamic range. Much effort has gone into the development of high-density semiconductor photodetector arrays, to fit the maximum number of pixels (picture elements) onto a limited-sized semiconductor (usually silicon) chip. However, for optical-processing purposes, high-density, large arrays must be achieved without sacrificing speed/throughput and dynamic range.

The largest semiconductor arrays have been charge-coupled devices (CCDs). Imagers with $1024 \times 1024$ pixels without on-chip storage have been fabricated. High-resolution CCD TV cameras have become available with formats of $600 \times 450$ pixels (4:3 aspect). 1D arrays with $>2000$ pixels are available, and are fabricated on silicon chips which are close to the present size which can be made defect-free. Hence, the technology to produce large semiconductor photosensing arrays with relatively few defects is now available. However, dynamic range continues to be a problem. With some sacrifice in ultimate array size, additional structures can be incorporated to eliminate or minimize effects that reduce dynamic range. A serious problem in CCDs is blooming at high light intensities, where charge spreads from illuminated cells to many neighboring cells. Specific CCD sensor structures have been developed with drains to remove excess charge before blooming can occur (Borsuk, 1981).

Large arrays may also be fabricated using charge-injection devices (CIDs). While CCDs accumulate charge in potential wells, CIDs employ metal–oxide–silicon (MOS) capacitors to store charge. When the charge is injected back into the bulk silicon, it is sensed either via the recharging currents or the changes in capacitance. By using two CID devices per pixel it is possible to construct devices with random-access, matrix addressing and to compensate partially for unequal bias levels at various pixels. However, use of two CID devices per pixel limits the array size compared to CCD arrays, and the use of matrix readout requires more complex circuitry compared to sequential readout.

Ultimately, the best solution to detector limitations may have to involve the optical-processing architectures. As mentioned earlier, optical architectures should ideally result in only a few output pixels of information. Large intermediate arrays of information in optical form can be reduced by further processing in the optical domain, involving the manipulation of these arrays with high-resolution light control devices.

## V. Summary

The performance of light sources, modulators, and detectors drives progress in the utilization of optical techniques in signal processing and computing. Of the three device categories, optical sources are the most developed. Most developments of relevance are now pointed toward improvements in semiconductor laser diodes to achieve better optical beam outputs (e.g., power, mode shape and stability, and shorter wavelengths), since the advantages of such lasers in size, weight, and power efficiency are well known. In the critical area of light modulators, single-point modulation techniques are well developed into the multi-gigahertz regime, and the acousto-optic 1D modulators have been developed for gigahertz bandwidths, reasonable time bandwidths, and high efficiency. The acousto-optic performance parameters are at present sufficiently demonstrated and detailed to serve as reliable guides to the effectiveness of acousto-optic architectures. While substantial improvement is still possible in the performance of both the single-point and acousto-optic devices, these devices have made optical approaches attractive in many applications such as spectrum analysis and correlation. However, more extensive applications require architectures involving two-dimensional spatial light modulators. A small number of 2D SLMs are beginning to emerge from laboratories into manufacture by commercial sources, however these devices fall short of attaining the overall features and performance levels required to be generally useful. Nevertheless, many may perform well enough for use in a few practical applications and for prototyping simple cases of more sophisticated concepts, particularly when the specific strengths of a given technology are matched with the requirements of the application. This often implies that current optical architectures be designed to employ the strengths of particular devices while de-emphasizing the weaknesses. Optical-processing techniques generally require photodetectors with much higher dynamic range and effective speeds than most other photodetector applications. While substantial progress has been achieved, much remains to be done, especially in the area of 2D arrays. High output data rates from detector arrays are often undesirable since further processing in the electrical domain is usually required, and this can prove to be the bottleneck in system throughput. Ideally, optical architectures should ensure that significant optical-domain processing has occurred, so that the resulting output consists of a small number of data channels or is sparsely populated if in a large spatial-array format.

REFERENCES

Apt, C. M. (1985). *IEEE Spectrum* **22**, p. 60.
Armitage, D., Anderson, W. W., and Karr, T. J. (1985). *IEEE J. Quant. Electron.* **QE-21,** 1241.
Armitage, D., Thackara, J. I., Clark, N. A., and Handschy, M. A. (1986). *Dig. Conf. Lasers Electro-Opt. (CLEO)* p. 366.

Athale, R. A., and Collins, W. C. (1982). *Appl. Opt.* **21**, 2089.

Athale, R. A., and Lee, J. N. (1984). *Proc. IEEE* **72**, 931.

Augbourg, P., Huignard, J. P., Hareng, M., and Mullen, R. A. (1982). *Appl. Opt.* **21**, 3706.

Bartolini, R. A. (1977). *Proc. SPIE* **123**, 2.

Berg, N. J., and Lee, J. N., eds. (1983). "Acousto-optic Signal Processing: Theory and Implementation." Dekker, New York.

Blauvelt, H., Margalit, S., and Yariv, A. (1982). *Appl. Phys. Lett.* **40**, 1029.

Blechman, F. (1986). *Radio Electron.* **57**, 39, 47.

Bleha, W. P., Lipton, L. T., Wiener-Avnear, E., Grinberg, J., Reif, P. G., Casasent, D., Brown, H. B., and Markevitch, B. V. (1978). *Opt. Eng.* **17**, 371.

Boling, E. J., and Dzimianski, J. W. (1984). *Proc. SPIE* **477**, 174.

Borsuk, G. M. (1981). *Proc. IEEE* **69**, 100.

Brain, M., and Lee, T. P. (1985). *J. Lightwave Tech.* **LT-3**, 1281.

Brooks, R. E. (1985). *Opt. Eng.* **24**, 101.

Burke, W. J., Staebler, D. L., Phillips, W., and Alphonse, G. A. (1978). *Opt. Eng.* **17**, 308.

Casasent, D. (1977). *Proc. IEEE* **65**, 143.

Casasent, D. (1978a). Guest ed., Special issue on spatial light modulators. *Opt. Eng.* **17**, 307.

Casasent, D. (1978b). *Opt. Eng.* **17**, 365.

Casasent, D. (1978c). *Opt. Eng.* **17**, 344.

Casasent, D. (1978d). *In* "Applied Optics and Optical Engineering" (R. Kingslake, ed.), Vol. 6, p. 143. Academic Press, New York.

Chamberlain, S. G., and Lee, J. P. Y. (1984). *IEEE Trans. Electron Devices* **ED-31**, 175.

Chang, I. C. (1983). *Proc. IEEE Ultrasonics Symp. IEEE No. 83CH 1947-1*, p. 427.

Chemla, D. S., Miller, D. A. B., and Smith, P. W. (1985). *Opt. Eng.* **24**, 556.

Clark, N. A., Handschy, M. A., and Lagerwall, S. T. (1983). *Mol. Cryst. Liq. Cryst.* **94**, 213.

Colburn, W. S., and Chang, B. J. (1978). *Opt. Eng.* **17**, 334.

Collins, S. A. (1980). Guest ed, Special issue on feedback in optics. *Opt. Eng.* **19**, 441.

Cutrona, L. J., Vivian, W. E., Leith, E. N., and Hall, G. O. (1961). *IRE Trans. Mil. Electron.* **Mil-5**; Leith, E. N. (1970). *Proc. IEEE* **59**.

Dagenais, M., and Sharfin, W. F. (1986). *Opt. Eng.* **25**, 219.

Dashiell, S. R., and Goodman, J. W. (1975). *Appl. Opt.* **14**, 1813.

Dean, P. J., and Thomas, D. G. (1966). *Phys. Rev.* **150**, 690.

Dieulesaint, E., and Royer, D. (1980). "Elastic Waves in Solids." Wiley, Chichester.

Dixon, R. W. (1967). *IEEE J. Quantum Electron.* **QA-3**, 85.

Donjon, J., Dumont, F., Grenot, M., Hazan, J.-P., Marie, G., and Pergrale, J. (1973). *IEEE Trans. Electron. Devices* **ED-20**, 1037.

Dove, B. L. (1985). Ed., *Digital Opt. Circuit Technol. AGARD Conf. Proc. Springfield, VA. (362) NTIS.*

Eden, D. D. (1979). *Proc. SPIE* **185**, 97.

Efron, U. (1984). Ed., *Proc. SPIE* **465**.

Efron, U., Bratz, P. O., Little, M. J., and Schwartz, R. N. (1983). *Opt. Eng.* **22**, 682.

Efron, U., Marom, E., and Soffer, B. H. (1985). *Tech. Dig. OSA Top. Meet. Opt. Comput.* TuF2.

Feinberg, J. (1980). *Opt. Lett.* **5**, 330.

Fisher, A. D. (1981). "Techniques and Devices for High-Resolution Adaptive Optics." Ph.D. thesis, MIT, Cambridge, MA.

Fisher, A. D. (1985). *Tech. Dig. OSA Top. Meet. Opt. Comput.* p. TuCl.

Fisher, A. D., and Giles, C. L. (1985). *Proc. IEEE COMPCON Spring Meet.* **CH135-2**, 342.

Fisher, A. D., Giles, C. L., and Lee, J. N. (1984). *J. Opt. Soc. Am.* **A1**, 1337.

Fisher, A. D., Ling, L.-C., Lee, J. N., and Fukuda, R. C. (1986). *Opt. Eng.* **25**, 261.

Flannery, J. B. (1973). *IEEE Trans Electron Devices* **ED-20**, 941.

Forkner, J. F., and Kuntz, D. W. (1983). *Proc. SPIE* **390**.

Garito, A. F., and K. D. Singer. (1982). *Laser Focus*, **Feb.**, 59.

Gaylord, T. K., and Moharam, M. G. (1985). *Proc. IEEE* **73**, 894.

General Electric Corp. (1984). TN2500 Solid State Video/Digital Camera-Operating Manual.

Gibbs, H. M. (1985). "Optical Bistability: Controlling Light with Light." Academic Press, New York.

Gibbs, H. M., Mandel, P., Peyghambarian, N., and Smith, S. D. (1986). "Optical Bistability III." Springer-Verlag, Heidelberg.

Goldberg, L., Taylor, H. F., and Weller, J. F. (1985). *Appl. Phys. Lett.* **46**, 236.

Goldstein, B., Ettenberg, M., Dinkel, N. A., and Butler, J. K. (1985). *Appl. Phys. Lett.* **47**, 655.

Goodman, J. W. (1968). "Introduction to Fourier Optics." McGraw-Hill, New York.

Goodman, J. W., Dias, A., and Woody, L. M. (1978). *Opt. Lett.* **2**, 1.

Goodman, J. W., Leonberger, F. J., Kung, S. Y., and Athale, R. A. (1984). *Proc. IEEE* **72**, 850.

Gordon, E. I. (1966). *Proc. IEEE* **54**, 1391.

Groth, G., and Marie, G. (1970). *Opt. Commun.* **2**, 133.

Haney, M., and Psaltis, D. (1985). *Appl. Opt.* **24**, 1926.

Hara, T., Sugiyama, M., and Suzuki, Y. (1985). *Adv. Electroni. Electron Phys.* **64B**, 637.

Hess, K., and Danliker, R. (1985). *Dig. Conf. Lasers Electro-Opti. (CLEO)* p. 44.

Heynick, L. N., Reingold, I., and Sobel, A. (1973). Guest eds., Special issue on display devices. *IEEE Trans. Electron Devices* **ED-20**.

Hirshon, R. (1984). *Electron. Imag.* **Jan.**, 40.

Hornbeck, L. J. (1983). *IEEE Trans. Electron Devices* **ED-30**, 539.

Horwitz, B. A., and Corbett, F. J. (1978). *Opt. Eng.* **17**, 353.

Huang, A. (1985). *Tech. Dig. OSA Top. Meet. Opt. Comput.* WA2.

Huignard, J. P., Rajbenbach, H., Refregier, Ph., and Solymar, L. (1985). *Opt. Eng.* **24**, 586.

Ikeda, M., Mori, Y., Takiguchi, M., Kaneko, K., and Watanabe, N. (1985). *Appl. Phys. Lett.* **45**, 661.

Ishikawa, M., Ohba, Y., Sugawara, H., Yamamoto, M., and Nakanisi, T. (1986). *Appl. Phys. Lett.* **48**, 207.

Johnson, R. V., Hecht, D. L., Sprague, R. A., Flores, L. N., Steinmetz, D. L., and Turner, W. D. (1983). *Opt. Eng.* **22**, 665.

*J. Lightwave Tech.* (1983 +). IEEE-OSA.

Kawaguchi, T., Adachi, H,, and Setsune, K. (1984). *Appl. Opt.* **23**, 2187.

Kingston, R. H., Burke, B. E., Nichols, K. B., and Leonberger, F. J. (1984). *Proc. SPIE* **465**, 9.

Kitamura, M., Yamaguchi, M., Murata, S., Mito, I., and Kobayashi, K. (1985). *IOOC-ECOC Conf. Dig., Italy*.

Klein, W. R., and Cook, B. D. (1967). *IEEE Trans. Son. Ultrason.* **SU-14**, 123.

Kmetz, A., and Von Willisen, F. K., eds. (1976). "Non-emissive Electro-optic Displays." Plenum, New York.

Knight, G. R. (1981). *In* "Optical Information Processing" (S. Lee, ed.), p. 111. Springer-Verlag, New York.

Kobayashi, K., and Mito, I. (1985). *J. Lightwave Tech.* **LT-3**, 1202.

Kogelnik, H. (1969). *Bell Syst. Tech. J.* **48**, 2909.

Kressel, H., and Butler, J. (1975). "Semiconductor Lasers and Heterojunction LED's." Academic Press, New York.

Lakatos, A. I. (1974). *J. Appl. Phys.* **15**, 4857.

Land, C. E. (1978). *Opt. Eng.* **17**, 317.

Lau, K. Y., Yariv, A., Eshi, I., and Mito, I. (1985). *J. Lightwave Tech.* **LT-3**, 1202.

Lea, M. (1984). *Proc. SPIE* **465**, 12.

Lee, S. H., Esner, S. C., Title, M. A., and Drabik, T. J. (1986). *Opt. Eng.* **25**, 250.

Liau, Z. L., Walpole, J. N., and Tsang, D. Z. (1984). *Tech. Dig. Top. Meet. Integrated Guided-Wave Opt. OSA, 7th, Washington, DC.* p. TuC5.

Lindley, J. P. (1983). *In* "Acousto-Optic Signal Processing: Theory and Implementation" (N. J. Berg and J. N. Lee, eds.), p. 87. Dekker, New York.

Lipson, S. (1979). *In* "Advances in Holography" (N. Farhat, ed). Dekker, New York.

Lockwood, H. F., and Kressel, H. (1975). *Tech. Rep. AFAL-TR-75-13*, Wright-Patterson AFB, Dayton OH.

Lohmann, A. W., and Wirnitzer, B. (1984). *Proc. IEEE* **72**, 889.

Lui, H.-K., Davis, J. A., and Lilly, R. A. (1985). *Opt. Lett.* **10**, 635.

McAulay, A. D. (1983). *Proc. SPIE* **431**, 215.

McEwan, J. A., Fisher, A. D., and Lee, J. N. (1985). *Dig. Conf. Lasers Electro-Opt. (CLEO)* p. PD-1.

McEwan, J. A., Fisher, A. D., Rolsma, P. B., and Lee, J. N. (1985b). *J. Opt. Soc. Am.* **A2**, 8.

Marcuse, D. (1973). "Integrated Optics." IEEE Press, New York.

Marom, E. (1986). *Opt. Eng.* **25**, 274.

Marrakchi, A., Tanguay, A. R., Yu, J., and Psaltis, D. (1985). *Opt. Eng.* **24**, 124.

Mergerian, D., Marlarkey, E. C., Pantienus, R. P., Bradley, J. C., Marx, G. E., Hutcheson, L. D., and Kellner, A. L. (1980). *Appl. Opt.* **19**, 3033.

Miller, D. A. B., Chelma, D. S., Damen, T. C., Gossard, A. C., Wiegmann, W., Wood, T. H., and Burrus, C. A. (1984). *Appl. Phys. Lett.* **45**, 13.

Minoura, K., Usui, M., Matsouka, K., Baba, T., Suzuki, M., and Asai, A. (1984). *Dig. Conf. ICO, 13th* **ICO-13**, 154.

Mol, J. (1974). *Dig. Int. Opt. Comput. Conf.* p. 34.

Owechko, Y., and Tanguay, A. R. (1982). *Opt. Lett.* **7**, 587.

Paoli, T. L., Streifer, W., and Burnham, R. D. (1984). *Appl. Phys. Lett.* **45**, 217.

Pape, D. R. (1985a). *Opt. Eng.* **24**, 107.

Pape, D. R. (1985b). *Tech. Dig. OSA Top. Meet. Opt. Comput.* p. TuC6.

Pepper, D. M. (1982). Guest ed., Special Issue on nonlinear optical phase conjugation. *Opt. Eng.* **21**, 156.

Perry, T. S. (1985). *IEEE Spectrum* **22**, 53.

Petrov, M. P. (1981). *In* "Current Trends in Optics" (F. T. Arecchi and F. R. Aussenegg, eds.), p. 161. Taylor Francis, London.

Peyghambarian, N., and Gibbs, H. M. (1985). *Opt Eng.* **24**, 68.

Pollock, D. K., Koester, C. J., and Tippett, J. T., eds. (1963). "Optical Processing of Information." Spartan Books, Baltimore.

Preston, K. P. (1969). *Opt. Acta* **16**, 579.

Psaltis, D., and Farhat, N. (1984). *Dig. Congr. ICO, 13th* **ICO-13**, 24.

Psaltis, D., and Farhat, N. (1985). *Opt. Lett.* **10**, 98.

Ralston, L. M., and McDaniel, R. V. (1979). *Proc. SPIE* **185**, 86.

Reizman, F. (1969). *Proc. Electroopt. Syst. Design* p. 225.

Ross, W. E., Psaltis, D., and Anderson, R. H. (1983). *Opt. Eng.* **22**, 485.

Sahai, R., Pierson, R. L., Martin, E. H., and Higgins, J. A. (1984). *Proc. SPIE* **477**, 165.

Sawchuk, A. A., and Jenkins, B. K. (1984). *Tech. Dig. OSA Top. Meet. Opt. Comput.* p. TuA2.

Schneeberger, B., Laeri, F., Tschudi, T., and Mast, F. (1979). *Opt. Commun.* **31**, 13.

Schwartz, A., Wang, X.-Y., and Warde, C. (1985). *Opt. Eng.* **24**, 119.

Scifres, D. R., Lindstrom, C., Burnham, R. D., Streifer, W., and Paoli, T. (1983). *Electron. Lett.* **19**, 169.

Seko, A., and Nishikata, M. (1977). *Appl. Opt.* **16**, 1272.

Shaefer, D. H., and Strong, J. P. (1975). Tse Computers. NASA Report X-943-75-14. Goddard SFC, Greenbelt, MD.

Sheridon, N. K. (1972). *IEEE Trans. Electron Devices* **ED-19**, 1003.

Smith, P. W., and Turner, E. H. (1977). *Appl. Phys. Lett.* **30**, 280.

Smith, S. D., Janossy, I., MacKenzie, H. A., Mathew, J. G. H., Reid, J. J. E., Taghizadeh, M. R., Tooley, F. P. A., and Walker, A. C. (1985). *Opt. Eng.* **24**, 569.

Somers, L. E. (1972). *Adv. Electroni. Electron Physi.* **33A**, 493.

Strome, D. H. (1984). *Proc. SPIE* **465**, 192.

Talmi, Y., and Simpson, R. W. (1980). *Appl. Opt.* **19**, 1401.

Tamura, P. N., and Wyant, J. C. (1976). *Proc. SPIE* **83**, 97.

Tanguay, A. R. (1983). Guest ed., Special issue on SLMs: Critical Issues. *Opt Eng.* **22**, 663.

Tanguay, A. R. (1985). *Opt. Eng.* **24**, 2.

Tanguay, A. R., and Warde, C. (1985). Guest eds., Special issue on optical information processing components. *Opt. Eng.* **24**, p. 91.

Tanguay, A. R., Wu, C. S., Chaval, P., Strand, T. C., Sawchuck, A. A., and Soffer, B. H. (1983). *Opt. Eng.* **22**, 687.

Thomas, R. N., Guldberg, J., Nathanson, H. C., and Malmberg, P. R. (1975). *IEEE Trans. Electron Devices* **ED-22**, 765.

Thompson, B. J. (1977). *Proc. IEEE* **65**, 62.

Timothy, J. G. (1985). *Opt. Eng.* **24**, 1066.

Tippett, J. T., Berkowitz, D. A., Clapp, L. C., Koester, C. J., and Vanderburgh, A., Jr., eds. (1965). "Optical and Electro-optical Information Processing." MIT Press, Cambridge, MA.

Turpin, T. M. (1981). *Proc. IEEE* **69**, 79.

Uchida, N., and Niizeki, N. (1973). *Proc. IEEE* **61**, 1073.

Uchiyama, S., and Iga, K. (1985). *Dig. Conf. Lasers Electro-Opt. (CLEO)*, p. 44.

Underwood, I., Sillitto, R. M., and Vass, D. G., (1985). *IEEE Colloq. Opt. Tech. Image Signal Process.*

VanderLugt, A. (1963). Radar Lab Report No. 4594-22-T, Institute of Science and Technology, U. Michigan, Ann Arbor.

VanderLugt, A., Moore, G. S., and Mathe, S. S. (1983). *Appl. Opt.* **22**, 3906.

Van Raalte, J. A. (1970). *Appl. Opt.* **9**, 2225.

Verber, C. M. (1984). *Proc. IEEE* **72**, 942.

Verber, C. M., Kenan, R. P., and Busch, J. R. (1981). *Appl. Opt.* **20**, 1626.

Warde, C., and Efron, U. (1986). Guest eds., Special issue on materials and devices for optical information processing. *Opt. Eng.* **25**, 197.

Warde, C., and Thackara, J. I. (1983). *Opt. Eng.* **22**, 695.

Warde, C., Weiss, A. M., Fisher, A. D., and Thackara, J. I. (1981). *Appl. Opt.* **20**, 2066.

Welch, D. F., Cross, P. S., Scifres, D., Streifer, W., and Burnham, R. D. (1986). *Electron. Lett.* **22**, 293.

Williams, D. J., ed. (1983). "Nonlinear Optical Properties of Organic and Polymeric Materials." American Chemical Society, Washington, D.C.

Yatagai, T. (1986). *Proc. SPIE* **625**, 54.

ADVANCES IN ELECTRONICS AND ELECTRON PHYSICS, VOL. 69

# Monte Carlo Methods and Microlithography Simulation for Electron and X-Ray Beams

## KENJI MURATA

*Electronics Department, College of Engineering*
*University of Osaka Prefecture*
*Sakai, Osaka 591, Japan*

## DAVID F. KYSER

*Philips Research Laboratories Sunnyvale*
*Signetics Corporation*
*Sunnyvale, California 94088*

## I. INTRODUCTION

The present rate of integrated circuit development is very high (see for example, Einspruch, 1981). This rate of development depends on various fields of modern science and technology. At present, the density and speed of IC devices are increasing continually. One of the important IC fields is microfabrication technology, which supports IC progress. In conventional lithographic fabrication technology, visible light is used with resolution limited to about one micrometer owing to diffraction effects. Newer technologies use shorter wavelength light, i.e., ultraviolet or deep uv light, where the resolution can be extended to the submicrometer region and is expected to produce half-micrometers resolution. However, the pursuit of much higher resolution has been directed to utilization of electron, ion and x-ray beams, which have much shorter effective wavelengths. The former two radiation beams are electronically controllable, and therefore can generate patterns through use of digital computers. An electron beam can easily be focused to a 10 nm spot with sufficient beam current to be utilized as a pattern-writing machine. This is known as electron-beam lithography (for example, Thornton, 1979, 1980; Munro, 1980), which is widely used in practice for mask making at present. The electron-beam technique is also used for direct wafer writing because it can realize fast turnaround in the fabrication process. At high accelerating voltages, a beam spot of 0.5 nm can be produced. However, one difficulty for electron-beam lithography is the scattering effect, where electrons penetrate deep into a resist–substrate sample and scatter back into the resist film over a wide region. This effect deteriorates the fidelity of the pattern writing even when the incident beam is finely focused. Much work has been done to understand this effect and to reduce it to obtain better pattern fidelity. Namely, we have to know how far the electrons spread out in the sample, and then "correct" the electron dose to obtain the desired exposure pattern in the resist film. Typically, this electron spreading is a few micrometers in a silicon target at 20 keV. In an actual lithographic process we encounter various types of samples. Therefore, it is not sufficient to know the electron range, but the electron behavior according to the sample geometry. Monte Carlo (hereafter called MC) simulation of electron scattering is a powerful tool to investigate this type of problem. Various applications of MC simulations to electron microscopy, microanalysis, and microlithography are reviewed by Kyser (1981).

Analytical models for electron scattering in a film–substrate target have been proposed by Nosker (1969), and further models have been established for studies of electron energy dissipation in a resist film by Hawryluk and Smith (1972), Heidenreich et al. (1973), Hatzakis et al. (1974), Hawryluk et al. (1974b), Greeneich and Van Duzer (1974), and Heidenreich (1977).

One noteworthy merit of these calculations is their short calculation time. However, such analytical methods cannot treat arbitrary boundary conditions, while MC simulation can handle any type of physical sample boundary just by programming the geometrical configuration, and can also give insight into the physical processes of the phenomena. Electron trajectory plotting is a typical example. In the present article we review the MC simulation of electron scattering and its various applications to microlithography. As a matter of fact, this simulation is also useful for x-ray lithography, where photoelectrons and Auger electrons are produced by x-ray beams. In the near future direct writing will become an important technology, where the technology of registration mark detection has to be established. As pattern features get smaller, mask inspection may require a finely focused electron beam because visible light cannot show details of the mask pattern. In both technologies a study of electron scattering is essential. Recent MC applications to these problems are reviewed. Ion beams also suffer scattering problems (Karapiperis *et al.*, 1981; Onga and Taniguchi, 1985; Mladenov *et al.*, 1985), but fortunately the scattering is small. With the size reduction of circuit elements and the development of ion-beam technology, the ion scattering effect may be important in the future. In this article we concentrate only on electron-beam scattering for both electron-beam lithography and x-ray lithography.

For a theoretical basis of MC simulation, see, for example, the review paper by James (1980). For a general scheme of MC calculations with combined photon–electron transport a report by Colbert (1974) is helpful.

## II. Monte Carlo Modeling of Electron Scattering

### A. Electron Scattering Modeling of Incident Electrons

An incident electron loses its energy within a solid target through inelastic scattering such as excitation and ionization of atomic electrons, and changes its direction of motion through elastic scattering with a nucleus. More strictly, energy loss can also occur through elastic scattering accompanying bremsstrahlung radiation, and a direction change through inelastic scattering between incident and atomic electrons. Among many incident electrons, some electrons lose their whole energy in the target by repeating such scattering processes, and some escape from the target with a finite energy. This scattering phenomenon can be modeled with MC methods, as shown in Fig. 1. Namely, an individual electron trajectory is divided into many segments with appropriate step lengths, where both the electron energy and direction (scattering angle) are determined. Note that crystallographic effects are

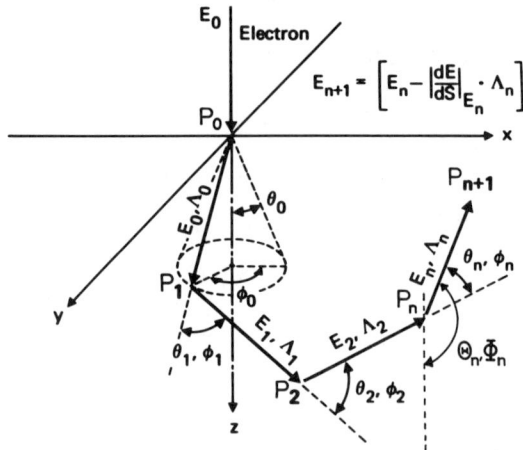

FIG. 1.    Simplified electron trajectory for the single scattering model (Kyser and Murata, 1974a).

neglected in the present discussion. Various models can be considered for calculating the step length, for example the mean free path of single scattering, or an equal step length or the length in which the multiple scattering by 20 or more collisions occurs. Modeling also depends on the energy loss process, i.e., the continuous slowing-down approximation of the Bethe law or the discrete process which takes account of each inelastic scattering event.

MC simulation of electron scattering in a thick target was started in the field of high-energy physics (Berger, 1963), where multiple scattering theory can be applied owing to the large electron range. At the beginning, experimental data for the angular distribution were used in MC applications (Green, 1963) to electron scattering phenomena in electron microprobe analysis (EPMA) and scanning electron microscopy (SEM). Subsequently, MC calculations based on purely theoretical models were performed (Bishop, 1966; Shimizu et al., 1966; Shinoda et al., 1968) through use of the multiple scattering theory (Lewis, 1950). A typical step length for multiple scattering is 0.2 μm in aluminum at 20 keV. This step length is too large for studies of electron scattering at low energies, especially for lithography applications where we have various sample structures such as a thin resist film on a substrate. Therefore, the single scattering model is more desirable since it utilizes very small step lengths.

Once the step length and the scattering angle are determined, the position of the electron is calculated. We take the $z$ axis as the direction of depth and the $x-y$ plane as the specimen surface as shown in Fig. 1. Describing the position of the endpoint, $P_n(x_n, y_n, z_n)$, at the $n$th step the position of the endpoint,

$P_{n+1}(x_{n+1}, y_{n+1}, z_{n+1})$, at the $(n+1)$th step is obtained as follows. Let $(\theta, \phi)$ be the scattering angles from and around the incident axis, respectively. The value of $\theta$ is determined by the angular distribution of scattering. The value of $\phi$ is uniformly distributed around the axis. The direction of electron motion $(\Theta_{n+1}, \Phi_{n+1})$ at the $(n+1)$th step in the $(x, y, z)$ coordinate is calculated by using the Euler equation:

$$\cos \Theta_{n+1} = \cos \Theta_n \cos \theta - \sin \Theta_n \sin \theta \cos \phi \tag{1}$$

$$\sin \Phi_{n+1} = U \sin \Phi_n + V \cos \Phi_n \tag{2}$$

$$\cos \Phi_{n+1} = U \cos \Phi_n - V \sin \Phi_n \tag{3}$$

where

$$U = \frac{\cos \theta - \cos \Phi_n \cos \Theta_{n+1}}{\sin \Theta_n \sin \Theta_{n+1}} \tag{4}$$

$$V = \frac{\sin \phi \sin \theta}{\sin \Theta_{n+1}} \tag{5}$$

Thus, the position $P_{n+1}$ is calculated as follows, using the step length $\Delta s$:

$$x_{n+1} = x_n + \Delta s \sin \Theta_{n+1} \cos \Phi_{n+1} \tag{6}$$

$$y_{n+1} = y_n + \Delta s \sin \Theta_{n+1} \sin \Phi_{n+1} \tag{7}$$

$$z_{n+1} = z_n + \Delta s \cos \Theta_{n+1} \tag{8}$$

When only the penetration depth is desired, Eqs. (1) and (8) are sufficient. Based on the above structured model, the scattering angle and energy loss are determined by using a computer-generated random number. One trajectory is composed of many step lengths by the repetitive calculation of steps until the whole electron energy is spent. This is only one typical trajectory sampled from infinitely possible trajectories. A few thousand to a few tens of thousands trajectories are sampled, depending on the type of calculation. There are also various computational procedures to realize the above models.

## B. Single Scattering Model

The single scattering model of MC simulation has been reported by Reimer et al. (1968, 1970), Curgenven and Duncumb (1971), McDonald et al. (1971) and Murata et al. (1971) in the field of microprobe analysis. The simulation has been applied successfully to many problems of electron scattering in EPMA and SEM targets. The models used are somewhat different from each other, but we shall not go into details of these models. A model is described which has been applied to electron-beam lithography. This model utilizes the mean free path, which expresses the distance between two successive collisions as a step length. In Fig. 1, strictly speaking, the first step takes a straight line without scattering because the first scattering event occurs after a finite penetration depth. Therefore, both values of $(\theta, \phi)$ have to be set to zero at the first step. Three factors, (1) angular distribution of electron scattering, (2) step length, and (3) energy loss required in the simulation are described in the following.

### 1. Angular Distribution of Electron Scattering

As is well known, the differential scattering cross section when an electron interacts with a nucleus of an atom is given by the classical Rutherford equation. When electrostatic screening by shell electrons is taken into consideration, the Schrödinger equation has to be solved using an atomic potential. The solution with an exponential field based on the Born approximation yields the following expression (Wentzel, 1927):

$$\frac{d\sigma_i^{el}}{d\Omega} = \frac{e^4 Z_i(Z_i + 1)}{m^2 v^4 (1 - \cos\theta + 2\chi_i)^2} \tag{9}$$

where $m$(g) is electron mass, $v$(cm/sec) is electron velocity, $e$(cgs esu) is electron charge, $Z_i$ is the atomic number of the $i$th element, and $\chi_i$ is the screening parameter, which depends on the atomic potential utilized. The following equation for $\chi_i$ was derived by Nigam $et\ al.$ (1954) using a Thomas–Fermi-type potential:

$$\chi_i = \frac{1}{4}\left(1.12 \frac{h\lambda_i}{2\pi mv}\right)^2 = \tfrac{1}{4}(4.67 E^{-1/2} Z_i^{1/3})^2 \tag{10}$$

where $\lambda_i = Z_i^{1/3}/0.885 a_0$, $a_0$ is the Bohr radius, $h$ is Planck's constant, and $E$ is the kinetic energy of the incident electron. The value 1.12 is an adjustable parameter. Shinoda $et\ al.$ (1968) reported that this value, when used with multiple scattering theory, gave reasonable agreement with experimental data of Cosslett and Thomas (1964) for the angular distribution of transmitted electrons through a thin aluminum film. McDonald $et\ al.$ (1971) also investigated this parameter in comparison with experimental results of the saturation backscattering and showed that Mott and Massey's expression for $\chi_i$ [Eq. (11)]

$$\chi_i = \tfrac{1}{4}(2.75 E^{-1/2} Z_i^{1/3})^2 \tag{11}$$

gave good agreement. The values differ from each other. However, the difference does not significantly influence the final results of the MC calculation.

In Eq. (9) the term $Z_i(Z_i + 1)$ is substituted for $Z_i^2$ in the classical theory (Kulchitsky and Latyshev, 1942; Bethe, 1953). This is introduced to take account of the cross section for electron–electron collisions by simply assuming the same equation as that for electron–nucleus scattering. This effect becomes significant with low atomic number materials, but the accuracy is not established.

Integration of Eq. (9) gives the following total elastic scattering cross section for element $i$:

$$\sigma_i^{el} = \frac{\pi e^4 Z_i(Z_i + 1)}{\chi_i(1 + \chi_i)m^2 v^4} \quad (\text{in cm}^2) \tag{12}$$

In a compound element target the collision atom has to be determined by prorating a computer-generated pseudo-uniform random number $R(0 \leq R \leq 1)$ according to the fraction of the cross section as follows:

$$p_i = \frac{\rho N_A C_i \sigma_i^{el}/A_i}{\sum_i \rho N_A C_i \sigma_i^{el}/A_i} = \frac{n_i \sigma_i^{el}}{\sigma_{tot}^{el}} \tag{13}$$

where $\sigma_{tot}^{el}$ is the sum of the elastic scattering cross sections per unit volume, $C_i$ the weight fraction of the $i$th element, $A_i$ is atomic weight, $\rho$ is mass density, $N_A$ is Avogadro's number, and $n_i$ is the number of $i$ atoms per unit volume. Namely, the region where $R$ applies is checked in the order of the following equation:

$$0 < R \leq p_1, \quad p_1 < R \leq p_1 + p_2, \quad p_1 + p_2 < R \leq p_1 + p_2 + p_3, \cdots \tag{14}$$

Finally, the scattering angle is determined in each collision with the $i$th atom. The probability $P(\theta)\,d\Omega$ for scattering into a small solid angle $d\Omega$ ($= \sin\theta\,d\theta\,d\phi$) is given by

$$P(\theta)\,d\Omega = \left(\frac{d\sigma_i^{el}/d\Omega}{\sigma_i^{el}}\right)d\Omega \tag{15}$$

The zenith angle $\phi$ is distributed uniformly from 0 to $2\pi$. Therefore, Eq. (15) is only a function of $\theta$. To evaluate the scattering angle $\theta$ according to the probability function $P(\theta)$, we integrate Eq. (15) from 0 to $\theta$ and obtain the following equation (Fig. 2):

$$\cos\theta = 1 - \frac{2\chi_i F(\theta)}{1 + \chi_i - F(\theta)} \tag{16}$$

where $F(\theta)$ is the accumulated function of $P(\theta)$. It can be easily shown that the angle $\theta$ is distributed in accordance with $P(\theta)$ when $F(\theta)$ is substituted by a uniform random number $R$. In this case the integration is carried out analytically. But if this is not the case, a numerical integration must be adopted.

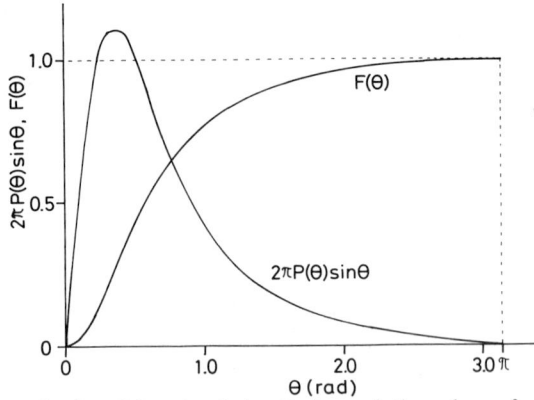

FIG. 2.   Determination of the azimuthal scattering angle through use of a generated uniform random number from the integrated function of the probability density function of angular deflection.

## 2. Step Length

The mean free path for elastic scattering, $\Lambda_{el}$(cm) is given by

$$\Lambda_{el}^{-1} = \sigma_{tot}^{el} = \frac{\pi e^4}{m^2 v^4} \sum_i \frac{n_i Z_i (Z_i + 1)}{\chi_i (1 + \chi_i)} \tag{17}$$

Since $\chi_i \propto 1/(mv)^2$ and usually $\chi_i \ll 1$, $\Lambda_{el}$ is proportional to $v^2$; i.e., $\Lambda_{el} \propto E$ for nonrelativistic electrons. When the step length is $\Lambda_{el}$ (fixed step length), one scattering event must occur during this step.

In contrast with this fixed step length, a variable step length can be adopted. This yields the so-called path length straggling effect. Namely, since collisional phenomena have a Poisson distribution, the probability that an electron takes a free path $\Delta s$ is given by

$$p(\Delta s) = \frac{1}{\Lambda_{el}} \exp(-\Delta s / \Lambda_{el}) \tag{18}$$

In a similar way to the previous sampling of $\theta$, a value for $\Delta s$ is obtained by the accumulated function $H(\Delta s)$ as follows:

$$\Delta s = -\Lambda_{el} \ln[1 - H(\Delta s)] \equiv -\Lambda_{el} \ln H'(\Delta s) \tag{19}$$

when a generated uniform random number $R$ is substituted into $H(\Delta s)$ or $H'(\Delta s)$.

It should be noted here that Eq. (9) is derived based on the Born approximation, which assumes that scattered electron waves are much weaker than an incident electron wave. The condition for validity of the Born approximation is given by $Z_i/137 \ll \beta$ (electron velocity/light velocity).

The electron energies for equality are about 2.7 and 115 keV for Si($Z = 14$) and Au($Z = 79$), respectively (for example, Motz et al., 1964; Mott and Massey, 1965). Therefore, at low energies, typically lower than 10 keV or for heavy elements, the accuracy becomes worse. Kyser and Murata (1974a) found that there were some discrepancies between experiments and MC calculations based on Eq. (9) in quantitative microprobe analysis and that the discrepancies increased with increasing atomic number $Z$. This arises probably from the failure of the Born approximation. Thus, they introduced a new multiplication factor $\mu_s = (1 + Z/300)$ to increase the mean free path, so that a better agreement is obtained between theory and experiment. This means that Eq. (12) overestimates the total cross section. Recent papers by Krefting and Reimer (1973), Ichimura et al. (1980), and Kotera et al. (1981) show that utilizing Mott cross sections (which are derived by partial wave expansion for each angular momentum) gives an improvement over the Born approximation. Therefore, this newer cross section is recommended in any future work.

## 3. Energy Loss

The energy loss process is complicated when an electron travels in a solid, which is an assembly of atoms. It is almost impossible to calculate exactly the individual discrete energy loss processes. In MC calculations, the Bethe law (Bethe, 1933) is commonly used, which gives the continuous slowing-down approximation. This is given by

$$-\frac{dE}{ds} = \frac{2\pi e^4}{E} \sum_i n_i Z_i \ln \frac{\gamma E}{J_i} \tag{20}$$

where $J_i$ is the mean ionization potential. The value of $\gamma$ is 2 or 1.166, depending on nonrelativistic or relativistic electrons, respectively. Usually a value of 1.166 is used for $\gamma$. The Bethe equation is derived from a quantum-mechanical theory, but in the final result a value for $J_i$ is empirically determined as a function of atomic number $Z_i$. Typical proposals for $J_i$ are those by Berger and Seltzer (1964) and Duncumb and Reed (1968). The former value is given by

$$J_i = 9.76Z_i + 58.8Z_i^{-0.19} \quad \text{eV} \tag{21}$$

The values are based on the analysis of proton stopping data. This equation loses validity for elements whose atomic number are smaller than 13. The values for these elements are given separately. The values of Duncumb and Reed are obtained so that calculated data with those values agree with the experimental data in quantitative microprobe analysis. Examples of $J_i$ values from Berger and Seltzer and Duncumb and Reed are listed in Table I for

TABLE I

THE MEAN IONIZATION POTENTIAL $J_i$ IN
THE BETHE EQUATION. TWO TYPICAL
VALUES ARE COMPARED

| | $J_i$ | |
|---|---|---|
| Elements | BS[a] | DR[b] |
| H | 18.7 | |
| C | 78 | 146 |
| O | 89 | 127 |
| Si | 172 | 154 |
| Cu | 314 | 377 |
| Mo | 439 | 567 |
| Au | 797 | 1071 |

[a] From Berger and Seltzer (1964).
[b] From Duncumb and Reed (1968).

typical elements. Relatively large discrepancies are seen between both values, which produce a difference of about 10% in the calculated electron range. The detailed accuracy is not clear for both proposals. When $J_i$ is divided by the atomic number $Z_i$, the proportions are around 11.5, which has been suggested by Wilson (1964).

Another difficulty of the Bethe equation is that Eq. (20) takes negative values for energies lower than $J_i/1.166$. To avoid this failure, Rao-Sahib and Wittry (1974) proposed the following equation:

$$-\frac{dE}{ds} = \frac{2\pi e^4}{1.26\sqrt{E}} \sum_i \frac{n_i Z_i}{\sqrt{J_i}}, \qquad E \leqq 6.338 J_i \tag{22}$$

This equation has a smooth continuation from the Bethe equation. Love et al. (1978) proposed a single equation which is applicable to a wide region of energy.

According to Spencer and Fano (1954), the Bethe law derived for an infinite target has to be modified as shown in Eq. (23) when the boundary condition is taken into consideration that electrons are incident on the surface of a semi-infinite target:

$$-\frac{ds}{dE} = (\kappa M)^{-1}[1 - (\pi^2/6)M^{-2}]$$

$$\kappa = \frac{\pi e^4 \rho N_A}{AE} \tag{23}$$

$$M = 1 + \ln[4E(E_0 - E)/J^2] - \ln(E_0/E)$$

where $E_0$ is the initial energy. This theory is based on the physical mechanism that discrete processes with a long mean free path for inelastic scattering have to be subtracted from all existing processes for electrons just after incidence on the target. The theory has no relativistic correction. Brown *et al.* (1969) first applied this correction to microanalysis with the EPMA (electron probe microanalyzer), by using the following artifice for a relativistic correction:

$$\frac{ds}{dE} = \frac{\text{nonrelativistic expression of Eq. (20)}}{\text{relativistic expression of Eq. (20)}} \tag{24}$$

This theory has also been utilized in a study of secondary electron emission by Shimizu (1974). Results of numerical calculations with Eq. (24) are shown in Fig. 3 for a target of polymethyl methacrylate (PMMA, $C_5H_8O_2$) at energies of 10, 20, and 30 keV. As seen from the figure, there is a peaking near the initial energy. That means the Bethe equation may overestimate the energy loss near the incident point. A typical calculation for an electron beam of 20 keV shows that the Spencer–Fano theory reduces by a factor of 1.5 the estimate of the absorbed energy density with the Bethe theory in the top layer of 500 Å thickness in an 8000 Å PMMA resist film and that both theories predict about the same energy density at the bottom (Murata *et al.*, 1978a). This has not yet been verified experimentally.

Once the energy loss rate at the $n$th step is found, the energy $E_{n+1}$ at the next step is simply calculated by the following equation:

$$E_{n+1} = E_n - |dE/ds|_{E=E_n}\Delta s \tag{25}$$

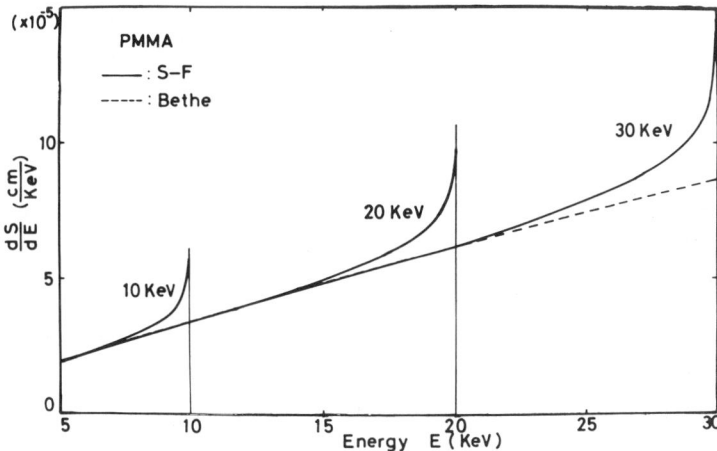

FIG. 3.   A comparison in the reciprocal stopping power for PMMA at 10, 20, and 30 keV between the theories of Spencer–Fano (———) and Bethe (– – –) (Murata, 1982).

TABLE II

THE MEAN FREE PATH FOR ELASTIC SCATTERING, THE BETHE RANGE, THE NUMBER OF STEPS, AN
ENERGY LOSS RATE, AND AN ENERGY LOSS FOR THE FIRST STEP AT 20 keV FOR VARIOUS
MATERIALS

| Material (atomic number) | $\Lambda$ (Å) | $R_B$ ($\mu$m) | No. of steps | $dE/ds$ (eV/Å) | $E$ (eV) |
|---|---|---|---|---|---|
| PMMA (6.2) | 773 | 7.31 | 212 | 0.152 | 117 |
| Si (14) | 363 | 5.08 | 327 | 0.224 | 81.3 |
| Ga (31) | 128 | 2.66 | 503 | 0.438 | 56.1 |
| Mo (42) | 68.7 | 1.69 | 606 | 0.697 | 47.9 |
| Au (79) | 32.5 | 1.19 | 923 | 1.027 | 33.4 |

## 4. Examples of Physical Quantities Used in the Simulation

Examples of the mean free path, an energy loss rate, the energy lost in one step, the Bethe range, and the number of steps at an incident energy of 20 keV are listed in Table II for various materials. The Berger and Seltzer values are utilized for the mean ionization potential $J_i$. The correction factor $\mu_s$ is not included for the mean free path, but the $Z(Z + 1)$ correction in Eq. (9) is included. The Bethe range $R_B$ is calculated by summing all the step lengths until an electron slows to 100 eV. As seen in the table, the number of steps is only about 5 within a 0.4 $\mu$m PMMA resist film. The number of steps increases with increasing atomic number, resulting in an increase in computational time.

### C. Hybrid Model for the Discrete and Continuous
### Energy Loss Processes

Various types of discrete processes are involved in the actual energy loss process. This yields the so-called energy straggling effect. Sometimes atoms are ionized and generate secondary electrons with low energy. These electrons often deviate from the traveling direction of the incident electron, and give a further electron spreading which is not considered in the single scattering model. The inclusion of the secondary electron generation may be important for a study of, for example, the ultimate spatial resolution in electron lithography. Murata *et al.* (1981) developed a new MC program by including the fast secondary electron production, based on a model by Berger (1963, 1971) and Seltzer (1974). Originally the idea came from a paper by Schneider and Cormack (1959). The model is described in the following.

*1. Energy Loss*

In Fig. 4 the primary electron energy degradation in this hybrid model is shown schematically. There is an abrupt change in energy at an arbitrary traveling distance, generating a secondary electron whose energy and generation frequency depend on the inelastic cross section to generate it. The Bethe law is shown by a dashed line. From this complete continuous slowing-down approximation, some of the discrete energy loss processes are drawn out. However, we limit the theory to collisions with an energy transfer larger than, say, $\Delta E_c$. For an energy loss resulting from a smaller energy transfer than $\Delta E_c$, a continuous slowing-down approximation is assumed. In the figure a slowly decreasing solid curve corresponds to this continuous energy loss. We can see a smaller decreasing energy loss rate for the hybrid model than that for the Bethe law. The energy loss rate caused only by knock-on collisions is calculated by using the differential inelastic cross section $d\sigma/d(\Delta E)$ as follows:

$$\left(\frac{dE}{ds}\right)_{\text{single}} = \sum_i n_i Z_i \int_{\varepsilon_c}^{1/2} E\varepsilon \left(\frac{d\sigma}{d\varepsilon}\right) d\varepsilon \tag{26}$$

where $\varepsilon$ is energy transfer $\Delta E$ normalized by the kinetic electron energy $E$ and $\varepsilon_c$ is a cutoff energy $\Delta E_c$ normalized by $E$. The continuous energy loss caused by the energy transfer less than $\varepsilon_c$ is obtained by subtracting the energy loss by knock-on collisions from the Bethe equation:

$$\left(\frac{dE}{ds}\right)_{\varepsilon_c} = \left(\frac{dE}{ds}\right)_{\text{Bethe}} - \left(\frac{dE}{ds}\right)_{\text{single}} \tag{27}$$

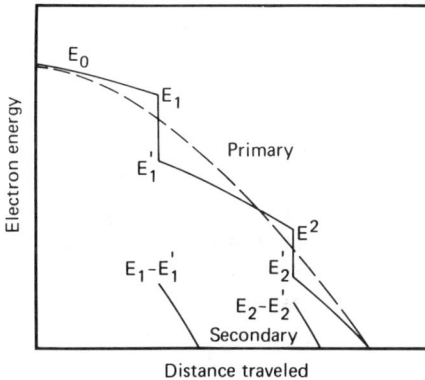

FIG. 4.   Schematic illustration of an energy loss process in the hybrid model. Fast secondary electrons are generated at catastrophic collisions.

Now the explicit expression of the differential inelastic cross section is required. One candidate is the Moller equation for free electrons in the following (Moller, 1931):

$$\left(\frac{d\sigma}{d\varepsilon}\right)_{\text{M}} = \frac{B}{E}\left[\frac{1}{\varepsilon^2} + \frac{1}{(1-\varepsilon)^2} + \left(\frac{\tau}{\tau+1}\right)^2 - \frac{2\tau+1}{(\tau+1)^2}\frac{1}{\varepsilon(1-\varepsilon)}\right] \quad (28)$$

$$B = 2\pi e^4/mv^2$$

where $\tau$ is the kinetic energy of an electron normalized by the rest mass energy of the electron. For small values of $\tau$ the equation reduces to the following:

$$\left(\frac{d\sigma}{d\varepsilon}\right) = \frac{B}{E}\left(\frac{1}{\varepsilon^2} + \frac{1}{(1-\varepsilon)^2} - \frac{1}{\varepsilon(1-\varepsilon)}\right) \quad (29)$$

The third term in Eq. (29) does not appear in the classical theory. Since the Moller theory is based on the assumption of free electron scattering, the effect of the binding energy of atomic electrons is not included. Eggarter (1975) proposed a semiempirical form for the cross section, including the ionization potential $J_i$, as follows:

$$\left(\frac{d\sigma}{d\varepsilon}\right)_{\text{E}} = \frac{B}{E}\left(\frac{1}{\varepsilon^2} + \frac{1}{(1-\varepsilon+J_i/E)^2} - \frac{1}{\varepsilon(1-\varepsilon+J_i/E)}\right) \quad (30)$$

Although the effect of $J_i$ is not included in the paper by Murata et al. (1981), the assumption of free electron scattering is not so fatal for light elements, which comprise the usual resist polymer film in microlithography.

According to Berger and Seltzer (1964), based on the Moller equation for $d\sigma/d\varepsilon$, a value for $(dE/ds)_{\varepsilon_c}$ is given by

$$-\left(\frac{dE}{ds}\right)_{\varepsilon_c} = \sum_i n_i Z_i B\left(\ln\frac{E^2(\tau+2)}{2J_i^2} + f^{-1}(\tau,\varepsilon_c) - \Delta\right)$$

$$f^{-1}(\tau,\varepsilon_c) = -1 - \beta^2 + \left(\frac{\tau}{\tau+1}\right)^2\frac{\varepsilon_c^2}{2} + \frac{2\tau+1}{(\tau+1)^2}\ln(1-\varepsilon_c) \quad (31)$$

$$+ \ln[4\varepsilon_c(1-\varepsilon_c) + (1-\varepsilon_c)^{-1}]$$

where $\beta = v/c$ and the parameter $\Delta$ is the density effect correction for ionization loss due to the polarization of the target material. This effect can be neglected for nonrelativistic electrons.

Values of $\varepsilon_c$ are usually 0.01–0.001. When the value of $\varepsilon_c$ becomes very small, then Eq. (31) will be negative. This is caused by the fact that, in the calculation of $(dE/ds)_{\text{single}}$ in Eq. (26), the Moller theory overestimates the energy loss due to electron and free-electron collisions with small energy transfer. Ideally, $(dE/ds)_{\varepsilon_c}$ should be zero in an extreme case of $\varepsilon_c = 0$. Namely, from Eq. (27) $(dE/ds)_{\text{Bethe}} = (dE/ds)_{\text{single}}$, which means that the whole energy loss is replaced by discrete processes. Thus there exists a minimum energy $E_{min}$ for the primary electron so that $-(dE/ds)$ stays positive for a specified value of $\varepsilon_c$. Assuming that $\tau$, $\beta^2$, and $\varepsilon_c$ are much smaller than unity, then

$$E_{min} = J_i/2\sqrt{\varepsilon_c} \tag{32}$$

Note that $E_{min}$ depends on the ionization potential. Therefore, one must choose the largest value among those of $E_{min}$ for all elements comprising the sample. Typical values of $E_{min}$ with $\varepsilon_c = 0.001$ and $J_i$ of Berger and Seltzer are 0.30, 1.23, 1.40, and 2.72 keV for elements of H, C, O, and Si, respectively.

## 2. The Angular Distribution of Elastic and Inelastic Scattering

In this model both elastic scattering events between electron and nucleus and inelastic scattering events between electron and electron are associated with angular deflection in the penetration process. For elastic scattering the screened Rutherford equation [Eq. (9)] is used. In inelastic electron–electron scattering both the primary and secondary electrons have angular deflections, $\theta$ and $\varphi$, respectively. According to Moller (1931), they are described as follows:

$$\sin^2 \theta = 2\varepsilon/(2 + \tau - \tau\varepsilon) \tag{33}$$

$$\sin^2 \varphi = 2(1 - \varepsilon)/(2 + \tau\varepsilon) \tag{34}$$

It can easily be shown that these equations approach the result for a classical binary collision of nonrelativistic electrons ($\tau \ll 1$).

## 3. Step Length

In the model, there are two mean free paths for elastic and inelastic collisions. For elastic scattering the mean free path $\Lambda_{el}$ is already given by Eq. (17). For inelastic scattering the mean free path $\Lambda_{in}$ is given by

$$\Lambda_{in} = \left(\sum_i n_i Z_i \sigma^{in}\right)^{-1} \tag{35}$$

using a value of $\sigma^{in}$:

$$\sigma^{in} = \frac{\pi e^4}{E^2} \left[\frac{1}{\varepsilon_c} - \frac{1}{1 - \varepsilon_c} - \ln\left(\frac{1 - \varepsilon_c}{\varepsilon_c}\right)\right] \tag{36}$$

Then, the total mean free path for both scatterings is given by the harmonic mean as follows:

$$\Lambda_{tot}^{-1} = \Lambda_{el}^{-1} + \Lambda_{in}^{-1} = \sigma_{tot} \tag{37}$$

Similar to Eq. (18), the variable step length is calculated as follows using a generated uniform random number $R$:

$$\Delta s = -\Lambda_{tot} \ln R \tag{38}$$

## 4. Selection of Elastic or Inelastic Scattering

The probabilities of an elastic ($p^{el}$) or inelastic ($p^{in}$) collision are calculated as follows:

$$p^{el} = \sum_i n_i \sigma_i^{el} / \sigma_{tot} \tag{39}$$

$$p^{in} = \sum_i n_i Z_i \sigma_i^{in} / \sigma_{tot} \tag{40}$$

Their selection is determined by prorating a computer-generated uniform random number $R$ according to each probability. Namely, when $0 < R \leq p^{el}$, an elastic event occurs and when $p^{el} < R \leq p^{el} + p^{in} = 1$, an inelastic event occurs.

## 5. Calculation Procedure

The flow chart for MC simulation with fast secondary electron production is shown in Fig. 5. The general flow is the same as that for the single scattering model. When the primary electron has a catastrophic collision, the secondary electron is produced. Then, storing the characteristic data for the primary electron, the secondary electron is tracked until its energy is dissipated. Then, the primary electron is recessed, and the same procedure is repeated. For the secondary electron the single scattering model is applied. The knock-on model can be extended to the production of tertiary and higher-order knock-on electrons if necessary.

### D. Discrete Model for Energy Loss

## 1. Discrete Energy Loss Processes

In order to investigate the ultimate spatial resolution of electron-beam lithography, Samoto et al. (1983) treated the electron energy loss by two discrete processes: inner- and outer-shell electron excitations following the proposal of Shimizu and Everhart (1978). They utilized the cross section of Gryzinski (1965) for both inner- and outer-shell electron excitations as described in the following:

$$\left(\frac{d\sigma}{d\varepsilon}\right)_G = \frac{\pi e^4}{\varepsilon^2} \frac{1}{E^2} (1 + \varepsilon_{nl})^{-3/2} (1 - \varepsilon)^{\varepsilon_{nl}/(\varepsilon_{nl} + \varepsilon)}$$

$$\times \left\{ (1 - \varepsilon_{nl}) + \frac{4}{3} \frac{\varepsilon_{nl}}{\varepsilon} \ln \left[ 2.7 + \left( \frac{1 - \varepsilon}{\varepsilon_{nl}} \right)^{1/2} \right] \right\} \tag{41}$$

where $\varepsilon_{nl}$ is the ionization energy of the $(n, l)$ shell electron, normalized with $E$. Since the binding energies are well known for inner-shell electrons, the cross

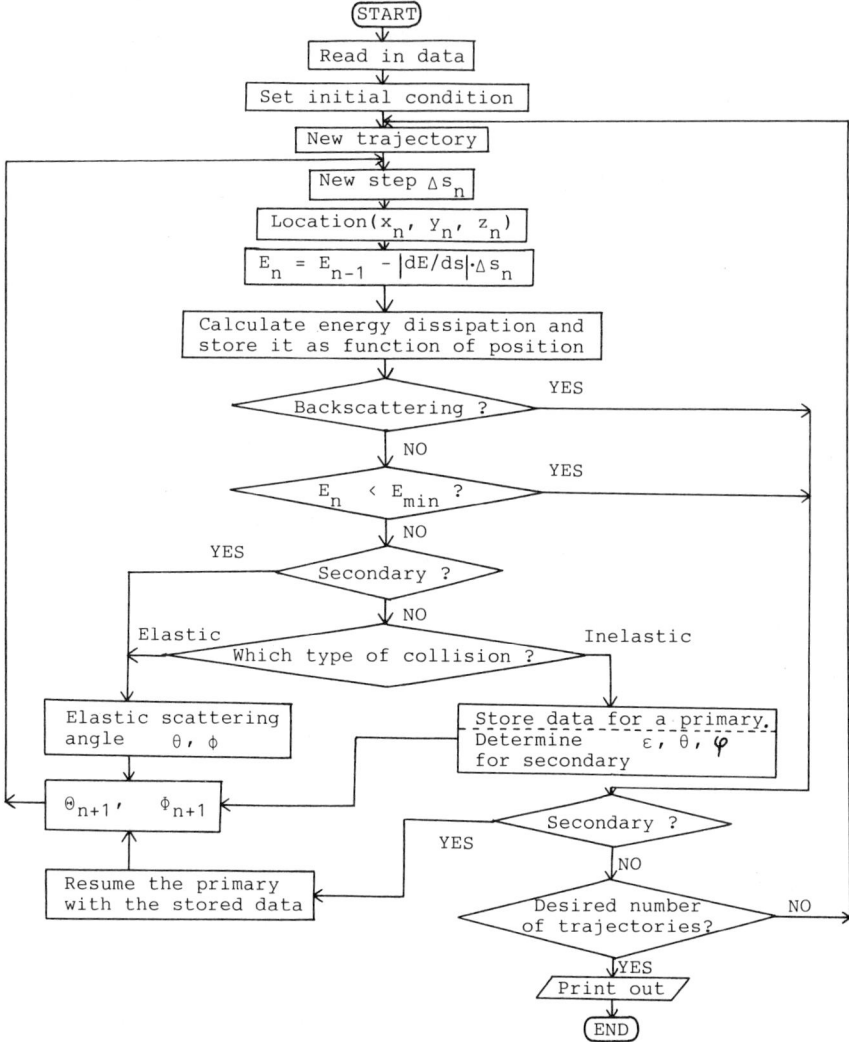

FIG. 5.  Flow chart of the Monte Carlo calculation with the hybrid model (Murata, 1982).

section can be evaluated numerically. For outer-shell electrons, i.e., valence electrons, they introduced the expedient mean binding energy $\bar{E}_b$ in the following way. They divided the Bethe equation into two categories of core and valence as given by

$$\left(\frac{dE}{ds}\right)_{\text{Bethe}} = \left(\frac{dE}{ds}\right)_{\text{core}} + \left(\frac{dE}{ds}\right)_{\text{valence}} \tag{42}$$

The value of $(dE/ds)_{core}$ can be calculated using Gryzinski's equation. If one assumes an appropriate value for $\bar{E}_b$, $(dE/ds)_{valence}$ can be calculated based on the Gryzinski equation. They tried to find a value of $\bar{E}_b$ so that both terms agree on the right- and left-hand sides of Eq. (42). They obtained $\bar{E}_b = 10$ eV for PMMA and $\bar{E}_b = 4$ eV for Si.

## 2. Tracking of Secondary Electrons

Shimizu and Everhart (1978) used the inelastic mean free path for PMMA proposed by Seah and Dench (1979) to track generated slow secondaries as follows:

$$\lambda_s = \frac{1}{\rho}(49/E^2 + 0.11E^{1/2}) \quad \text{nm} \tag{43}$$

where $E$ is in eV and $\rho$ is the material density. They adopted the variable step length, and also assumed experimental energy loss spectra by Ritsko et al. (1978), which are caused by inelastic collisions. Actually they approximated the spectrum by a Gaussian form with a standard deviation of 6.6 eV around a mean value of 20 eV.

## E. Handling of Boundary Conditions

One characteristic feature of MC calculations is the ease of setting geometrical boundary conditions just by changing input data such as incidence angle, or by programming a special specimen geometry. In electron-beam lithography we can encounter various types of geometry depending on fabrication process technologies. One of the most important geometries is that of a thin resist film on a thick substrate. MC programs have been developed to take account of this special geometry by many authors (for example, Saitou, 1973; Hawryluk et al., 1974a,b; Kyser and Murata, 1974b). A typical modeling approach is shown in Fig. 6. There are six possibilities for the location of the step as shown on the right side of the figure. A problem is how to handle electron transport behavior when electrons cross the boundary. The first approximation is to use the physical quantities at the starting point of the step to calculate the mean free path and scattering angle (Kyser and Murata, 1974b). It is easy to extend this model to a multiple layer structure. However, when the initial step position comes close to an interface, the accuracy of this approximation becomes worse. Or in the case that there is a large difference between the mean free paths on both sides of the interface, larger errors are induced. A new model to improve this deficiency has been published by Horiguchi et al. (1981). They introduced a differential equation for the

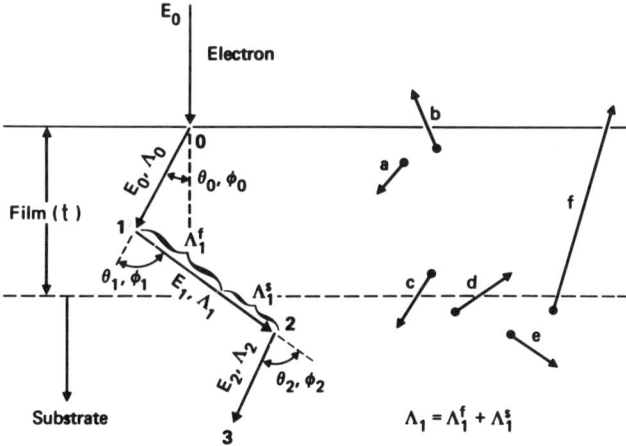

FIG. 6.   Electron trajectory modeling for a sample of a thin resist film on a substrate (Kyser and Murata, 1974a).

probability $P_m(\Delta s)$ that electrons travel a distance $z = \Delta s$ in the $m$th layer without being scattered after one scattering event:

$$dP_m(\Delta s) = -\sigma^{el}_{tot,m} P_m(\Delta s)\, d(\Delta s) \tag{44}$$

where $P_1(\Delta s = 0) = 1$, $P_{m+1}(\Delta s = a_m) = P_m(\Delta s = a_m)$, and $a_m$ is the maximum depth in the $m$th layer. The total cross section $\sigma^{el}_{tot,m}$ in the $m$th layer is given by

$$\sigma^{el}_{tot,m} = \sum_i n_{im}\sigma^{el}_{im} = \Lambda_m^{-1} \tag{45}$$

The solution of Eq. (44) gives the following probability $P_m(\Delta s)$ for the distribution of the variable step length:

$$P_m(\Delta s) = \frac{1}{\Lambda_1}\exp\left(-\frac{\Delta s}{\Lambda_1}\right), \qquad \text{for}\quad 0 \leq \Delta s \leq a_1 \tag{46}$$

$$= \frac{1}{\Lambda_m}\exp\left[-\sum_{k=2}^{m}\left(\frac{1}{\Lambda_{k-1}} - \frac{1}{\Lambda_k}\right)a_{k-1} - \frac{1}{\Lambda_m}\Delta s\right],$$

$$\text{for}\quad a_{m-1} < \Delta s \leq a_m, \quad m = 2, 3, \ldots, n$$

$$= 0, \qquad \text{for}\quad a_m < \Delta s$$

The accumulated function of $P_m(\Delta s)$ is used to determine the variable length $\Delta s$ in the usual manner.

According to the new model, the mean free path for scattering depends on the initial position of the electron. Let us take the example of a PMMA resist film on a Si substrate. A mean free path $\Lambda$ can be calculated as follows for an

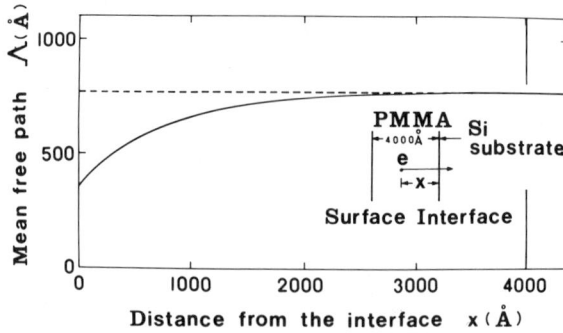

FIG. 7. The mean free path calculated with a different treatment at the boundary by Horiguchi *et al.* (1981). The result is shown for a 4000 Å PMMA film on a Si substrate at 20 keV. The mean free path changes from 773 Å for PMMA to 363 Å for Si depending on the position (Murata, 1982).

electron at a distance $X$ from the interface:

$$\Lambda = \int_0^X \Delta s \, P_1(\Delta s) \, d(\Delta s) + \int_X^\infty \Delta s \, P_2(\Delta s) \, d(\Delta s)$$

$$= \Lambda_{PMMA} - (\Lambda_{PMMA} - \Lambda_{Si}) \exp(-X/\Lambda_{PMMA}) \tag{47}$$

The equation reduces to

$$\Lambda = \Lambda_{Si}, \qquad \text{for } X = 0$$

$$= \Lambda_{PMMA}, \qquad \text{for } X = \infty \tag{48}$$

The calculated result is shown in Fig. 7 at 20 keV for a film thickness of 4000 Å. Horiguchi *et al.* (1981) have shown that the new model predicts very well the experimental data for backscattering yield from the sample with the three layer structure of CMS (chloromethylated polystyrene) resist–Mo(0.3 μm)–Si. A similar procedure has already been incorporated in the work of Hawryluk *et al.* (1974b) and was discussed recently by Hawryluk *et al.* (1982).

## F. Calculations of the Absorbed Energy Density

### 1. Calculation Configuration

It is important to obtain the absorbed energy per unit volume as a function of position in a resist film, as described in a later section. For this purpose the resist film is divided into many elemental volumes. In Fig. 8a the resist is divided into paralleelpipeds with a square section ($\Delta x \times \Delta z$) along the $y$ axis. An electron is incident at the origin. Since the electron location is always known, one can calculate the path length within this parallelepiped, $l$. Then the

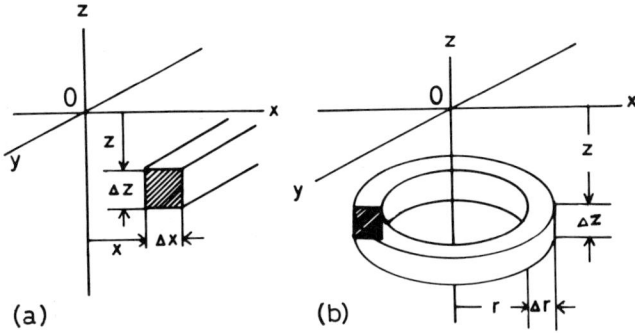

FIG. 8.   Elemental volumes of a resist film to calculate the spatial distribution of energy absorption. (a) Line source: (b) point source.

electron loses an energy of $\Delta E = |dE/ds|l$ in this volume. The parallelepipeds extend from $y = -\infty$ to $+\infty$. The geometric energy loss of any and all electrons which traverse these volumes is stored as a histogram of $\Delta E$(total) versus $x_i$ and $z_i$. As far as electrons are incident anywhere on the $y$ axis, the calculation condition is the same as that at the origin. Therefore, this calculation gives us the result for a line-source electron beam. In Fig. 8b the resist space is divided into concentric donut-shaped volumes with inner radius $r_i$, outer radius $r_i + \Delta r$, and thickness $\Delta z$ at depth $z_i$. Similarly, the total energy is stored as a function of $r_i$ and $z_i$. This is the case of a point-source electron beam. Actually the absorbed energy is stored in the computer as a matrix $(I, J)$.

## 2. Convolution of the Beam Size

In reality an electron beam has a finite size. A Gaussian form is usually assumed as follows for a point source:

$$g(r) = \frac{1}{\pi\delta^2} \exp\left(-\frac{r^2}{\delta^2}\right) \tag{49}$$

where $\delta/\sqrt{2}$ is the standard deviation. The beam diameter of $2\delta$ is defined here as the FWHM. Since $r^2 = x^2 + y^2$, integration of Eq. (49) over one direction $y$ yields the following lateral distribution (a line source):

$$g(x) = \frac{1}{\sqrt{\pi}\delta} \exp\left(-\frac{x^2}{\delta^2}\right) \tag{50}$$

There are two ways to take account of the beam size in an MC simulation. One is to begin from the MC calculation of the absorbed energy distribution $f_0(r)$ or $f_0(x)$ for an ideal point or line source, respectively. This is like a Green's function. Then, the real beam size is convoluted with the Green's function distribution as given by

$$f(r') = \int_0^\infty \int_0^{2\pi} g(r)f_0(u)r\, dr\, d\theta, \qquad u^2 = r'^2 + r^2 - 2r'r\cos\theta \qquad (51)$$

or

$$f(x) = \int_{-\infty}^\infty g(x_0)f_0(x - x_0)\, dx_0 \qquad (52)$$

for the radial and lateral distributions, respectively.

The other way is to include the beam distribution at the initial stage of MC simulation. Namely, the beam distribution is formed, using a generated uniform random number $R$. The pole $(r, \phi)$ or lateral $x$ coordinate is determined for an incident point. The value for $\phi$ is distributed uniformly. The selection of $r$ or $x$ is made by using the accumulated function of $g(r)$ or $g(x)$ over the whole area in a similar way as before in the following:

$$r = \delta\sqrt{-\ln R} \qquad (53)$$

or

$$\frac{1}{\sqrt{\pi}}\text{erfc}\left(\frac{\delta x}{2}\right) = R \qquad (54)$$

The initial position of an incident electron is specified by $(r, \phi)$ or $(x, 0)$ for a point source or line source with a finite beam size, respectively. A general pattern exposure is then achieved by superposing these convoluted distributions.

## 3. Accuracy Evaluation

Generally, it is not easy to deduce the accuracy of MC results. Apart from the randomness of generated pseudo-random numbers, statistical errors can be estimated simply by $\sqrt{N_b}$ for results such as the number, $N_b$, of backscattered electrons. Accordingly, the backscattering coefficient $\eta$ has the following fluctuation:

$$\pm\Delta\eta = \pm\sqrt{N_b}/N_0 = \pm\sqrt{\eta/N_0} \qquad (55)$$

where $N_0$ is the total number of incident electrons simulated. For example, with $N_0 = 10^4$ and $\eta = 0.30$, the percentage of $\Delta\eta/\eta$ is about 2%. A more detailed representation of the accuracy is the confidence interval of the probability for occurrence of backscattering according to the central limit theorem. This analysis cannot be applied to the quantity of the absorbed energy density as it is. The accuracy depends on the number of electrons entering the elementary volumes whose capabilities to deposit energy are different. Some weighting factor has to be introduced in applying the statistical theory. A simple way is to see how the fluctuation of energy deposited decreases with the number of incident electrons (Murata, 1982). Note that the smaller the elementary volume, the higher the spatial resolution, but then the larger the statistical error. One way to decrease the statistical error is to use a smaller volume near the incident point, and a larger volume at distances where a small number of electrons are traveling. Another way is to take advantage of the symmetry of coaxial or biaxial geometries. For biaxial symmetry the $\Delta E$ calculated with the same lateral distance from the incident axis and the same depth can be added together to enhance the statistical accuracy. The calculation geometry in Fig. 8b already utilizes coaxial symmetry.

### III. MONTE CARLO MODELING OF THE PHOTO- AND AUGER ELECTRON PRODUCTION IN X-RAY LITHOGRAPHY

A typical target configuration for x-ray replication is shown in Fig. 9. X-ray beams with an energy of about 1–3 keV are exposed on the resist through mask patterns formed on the mask substrate. X-rays generate secondary electrons in the resist, which give rise to chemical changes in the resist polymer. Open mask regions transmit incident x-ray beams easily. One can think of various problems in this exposure to achieve better accuracy. For example, absorption by the mask substrate, secondary electron emission from the mask, absorption in the resist, spreading of generated secondary electrons in the resist and substrate, etc. Such effects have been examined experimentally (Maldonado *et al.*, 1975; Feder *et al.*, 1975; Hundt and Tischer, 1978; Saito

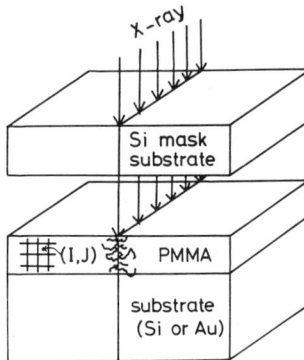

FIG. 9.   Calculation scheme for x-ray exposure on a PMMA resist film through a Si mask substrate (Murata *et al.*, 1985).

*et al.*, 1982; Okada and Matsui, 1983) and theoretically (Tischer and Hundt, 1978; Betz *et al.*, 1979). The analytical theory takes account of the spreading of generated photo- and Auger electrons, based on the depth–dose function derived by Everhart and Hoff (1971). MC simulation is more generally useful to study such electron behavior, as in electron-beam lithography.

## A. *X-Ray Absorption and Photoelectric Cross Section*

We first discuss the fundamentals of x-ray absorption in materials. Incident x-ray beams are absorbed generally through four physical events. These are (1) photoelectric effect, (2) coherent scattering effect, (3) Compton effect or incoherent scattering effect, and (4) electron pair generation. However, so far as x rays with energy less than 10 keV are concerned, only the photoelectric effect is important. The photoelectric effect is the phenomenon whereby an x-ray photon ionizes a shell electron, losing all its energy. The photoelectric cross section of this photon–electron interaction, called the ionization cross section, is obtained theoretically and experimentally (Veigele, 1973; Hubbell, 1977) and is shown as a function of photon energy *hv* in Fig. 10

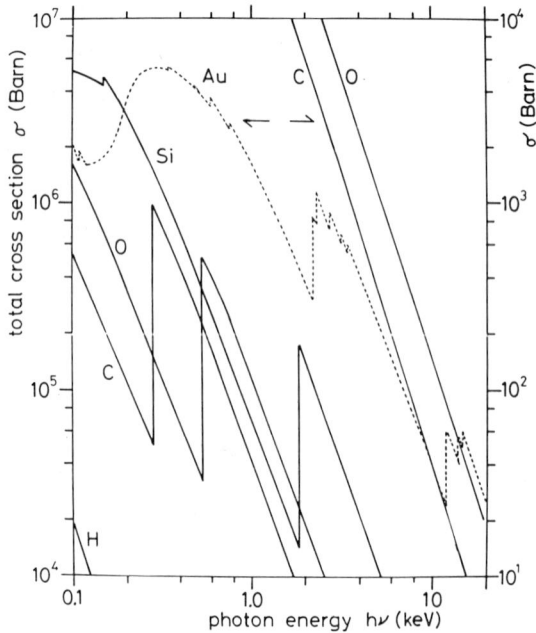

FIG. 10. The photon cross sections of H, C, O, Si, and Au as a function of photon energy (Murata *et al.*, 1985).

for typical elements. The curves are obtained from the least-squares fit of a polynomial equation through use of the numerical data in a limited energy range. We can see an abrupt change at a specific energy (called the absorption edge). There is a difference of about one order magnitude on both sides of this edge. The curves give a nearly linear variation on both the right and left sides of that energy except in the lower energy region. Since both axes in Fig. 10 are on a logarithmic scale, the cross section is approximately proportional to $(hv)^{-n}$ $(n = 3)$. For example, in the case of Si two absorption edges appear at energies of 1.83 keV ($K$ shell) and 0.149 keV ($L$ shell). In the energy range larger than $E_K$ the above cross section includes all the cross sections for $K$, $L$, and $M$ shells. Murata et al. (1985) extrapolated the curve in the lower energy region to the higher energy region, and tried to separate two cross sections for $K$ and $L$ shells by subtracting the extended curve.

Let $\sigma_i^m$ be the ionization cross section for an $m$-shell electron of element $i$. Then, the sum of the cross sections, $\sigma_{tot}$, per unit volume of absorption material is the following:

$$
\begin{aligned}
\sigma_{tot} &= \sum_i \sum_m C_i \rho N_A \sigma_i^m / A_i \\
&= \sum_i \sum_m n_i \sigma_i^m
\end{aligned}
$$

(56)

where the notation is the same as in Eq. (13) except for the difference in physical meaning of the cross section. Note that $\sigma_i^K = 0$ for $hv < E_K$, $\sigma_i^L = 0$ for $hv < E_L$, and so on.

The value of $\sigma_{tot}$ is equivalent to a linear absorption coefficient $\mu$, whose experimental values have been published by many authors (see, for example, Goldstein et al., 1981). Using a value for $\mu$, the incident x-ray intensity at a depth $z$ is expressed by

$$
I = I_0 \exp(-\mu z)
$$

(57)

where $I_0$ is the initial intensity.

On the other hand, assuming the binding energy of shell electrons is $E_b$, the energy of the photoelectron ($E_{ph}$) generated by photoionization is given by

$$
E_{ph} = hv - E_b
$$

(58)

Examples of $E_{ph}$ and $E_b$ are shown in Tables III and IV for Cu $L_\alpha$, Al $K_\alpha$, and Mo $L_\alpha$ characteristic lines when absorbed by a PMMA film on a Si substrate. In addition, when a shell electron is ionized, Auger electrons or fluorescence x rays are generated by accepting the residual energy when another shell electron of an upper energy level falls into a lower energy level shell. This probability is given by the fluorescence yield $\omega$. For light elements the value of

TABLE III

THE BINDING ENERGIES AND AUGER ELECTRON ENERGIES FOR C, O, AND Si

| | $E_b$ (keV) | | $E_A$ (keV) | |
|---------|---------|---------|---------|---------|
| Element | $K$ shell | $L$ shell | $K$ shell | $L$ shell |
| C | 0.284 | | 0.272 | |
| O | 0.532 | | 0.510 | |
| Si | 1.839 | 0.149 | 1.620 | 0.092 |

TABLE IV

THE PHOTOELECTRON ENERGIES FOR THREE TYPICAL CHARACTERISTIC LINES

| | | | $E_{ph}$ (keV) | |
|-------|-------------------|--------------------------|-----------|------------|
| Lines | Wavelength (Å) | Photon energy (keV) | Resist | Substrate |
| Cu $L$ | 13.3 | 0.93 | 0.65 (C) | 0.78 (Si $L$) |
| | | | 0.40 (O) | |
| Al $K$ | 8.3 | 1.49 | 1.21 (C) | 1.34 (Si $L$) |
| | | | 0.96 (O) | |
| Mo $L$ | 5.41 | 2.29 | 2.01 (C) | 0.45 (Si $K$) |
| | | | 1.76 (O) | 2.14 (Si $L$) |

$\omega$ is nearly zero (see, for example, Goldstein *et al.*, 1981). Namely, Auger electrons are almost always generated. In Table III are also shown the Auger electron energies.

## B. Modeling of the Photo- and Auger Electron Production

The probability $p(x)$ of photon absorption in a resist film at a traveling distance of $x$ is given by

$$p(x) = \sigma_{tot}^f \exp(-\sigma_{tot}^f x) \qquad (59)$$

Hereafter superscripts f and s mean a resist film or a substrate, respectively. In a similar manner to Eq. (19), a distance $x$ is sampled by using a generated pseudouniform random number $R$ as follows:

$$x = -\ln R / \sigma_{tot}^f \qquad (60)$$

If $x \geqq t$ (the resist thickness), the photon is not absorbed in the resist. Then, absorption in the substrate is checked for this transmitted photon. The integrated function $P(X)$ of the probability that the photon is absorbed in the substrate at a distance of $X$ from the boundary is given, according to Eq. (46)

by Horiguchi *et al.* (1981), by

$$P(X) = \int_0^t \sigma_{\text{tot}}^{\text{f}} \exp(-\sigma_{\text{tot}}^{\text{f}} x) \, dx$$

$$+ \int_0^x \sigma_{\text{tot}}^{\text{s}} \exp(-\sigma_{\text{tot}}^{\text{f}} t) \exp(-\sigma_{\text{tot}}^{\text{s}} X) \, dX$$

$$= 1 - \exp(-\sigma_{\text{tot}}^{\text{f}} t - \sigma_{\text{tot}}^{\text{s}} X) \tag{61}$$

This equation is the same as that published before (Colbert, 1974). Similarly, $X$ is sampled by

$$X = -(\sigma_{\text{tot}}^{\text{f}} t + \ln R)/\sigma_{\text{tot}}^{\text{s}} \tag{62}$$

Next, an atom and subsequently a shell that the photon interacts with are determined by prorating each cross section. The probability $p_i$ for an interaction with element $i$ is the following:

$$p_i = \sum_m n_i \sigma_i^m / \sigma_{\text{tot}}^{\text{f}} \tag{63}$$

The probability for an interaction with the $K$, $L$, or $M$ shell is the following:

$$p_K = \sigma_i^K \bigg/ \sum_m \sigma_i^m, \qquad p_L = \sigma_i^L \bigg/ \sum_m \sigma_i^m, \qquad p_M = \sigma_i^M \bigg/ \sum_m \sigma_i^m \tag{64}$$

The determination of an interaction atom and shell is done similarly for the substrate.

The angular distribution of ejected photoelectrons is calculated by the following equation (Colbert, 1974; Chapman and Lohr, 1974; Agarwal, 1979), assuming $\beta = v/c$ (electron velocity/light velocity) $\ll 1$:

$$dp_{\text{ph}}(\theta) = \tfrac{3}{4} \sin^3 \theta \, d\theta \tag{65}$$

This distribution is derived for spherically symmetric *s*-state electrons. But it is assumed also for other shell electrons. If a more exact treatment is desired, it is necessary to include the asymmetry parameter in the distribution (Chapman and Lohr, 1974). A uniform angular distribution is assumed for Auger electron ejection, namely

$$dp_{\text{A}}(\theta) = (\text{constant}) \sin \theta \, d\theta \tag{66}$$

Generated photo- and Auger electron trajectories are simulated in the same way as in the electron scattering model described in a previous section.

It should be noted that the Fresnel diffraction effect has a significant influence on the spatial distribution of exposure dose when the wafer–mask gap becomes large. Heinrich *et al.* (1981) have performed simulations of developed resist profiles by incorporating both the Fresnel diffraction effect and electron spreading based on the analytical model by Tischer and Hundt (1978). The combination with MC simulation of electron scattering has not been reported yet.

The x-ray source is also an important parameter which determines the replication quality. X-ray wavelengths of interest are in the region of a few hundred eV to a few keV, considering both resolution and sensitivity. The most popular source available at present is a characteristic x-ray line generated by electron bombardment of a solid target, or synchrotron orbital radiation SOR (for example, Neureuther, 1980). In the former source, continuous X-rays are also contained. The effect may not be neglected depending on experimental conditions. The spectral distribution of SOR is calculated by the Schwinger equation as a function of wavelength (Schwinger, 1949). Examples of the wavelength distribution are shown in Fig. 11 at electron energies of 0.7, 1.0, and 1.3 GeV with an electron orbital radius of 4 m and a current of 24 mA. The distributions are observed at a distance of 10 m. Although the intensity increases with increasing electron energy, the intensity of short-wavelength x rays also increases, which may cause undesirable effects. A similar situation is found in the characteristic x-ray source with a continuous x-ray background. The wavelengths of both sources are considered to be distributed over a wide range. Photons with energy of 0.1–10 keV can be simulated with

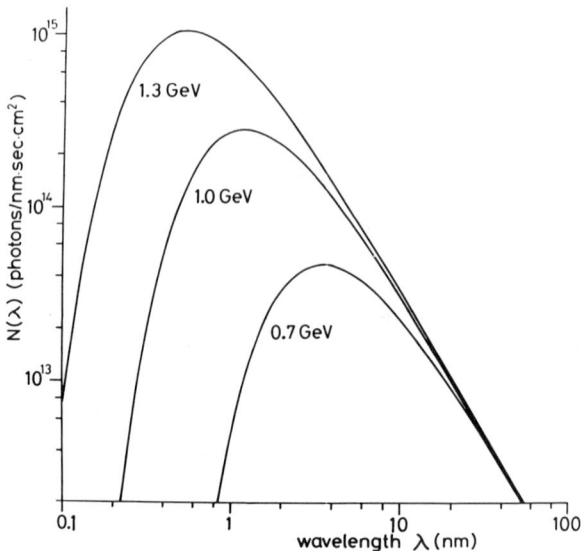

Fig. 11.   The spectral distribution of synchrotron radiation at electron energies of 0.7, 1.0, and 1.3 GeV (Murata et al., 1985). $I$ = 24 mA, distance = 10 m.

prepared cross sections. Once the spectral distribution is found for an x-ray source, the wavelength of a photon can be sampled in the same manner as in Fig. 2.

## IV. Resist Modeling

When energetic radiation is exposed onto a resist film, it collides with atoms or molecules comprising the resist material, exciting and ionizing them through inelastic events. Mostly, generated secondary electrons, tertiary electrons, or higher-order ionized electrons contribute to chemical changes in the resist material, which is usually an organic polymer with high molecular weight. The changes in the polymeric state give rise to differences in etching rate by a dry or wet developer, resulting in patterning in the resist film. Chemical changes occur in two different ways, namely molecular chain scission (positive type), which causes degradation of molecular weight, and cross linking (negative type), which causes gel formation in the polymer. It is assumed generally that the above two processes take place simultaneously in the same resist. The dominancy of one of the two processes determines positive- or negative-type behavior. In this section we discuss the relation between chemical changes and energy absorption obtained theoretically from the preceding sections.

### A. Positive Type

#### 1. Degradation of Molecular Weight and the Solubility Rate Equation

Usually a resist polymer has a molecular weight distribution. When it suffers radiation exposure, the molecular weight degrades, as shown schematically in Fig. 12. An appropriate developer can differentiate the two regions as

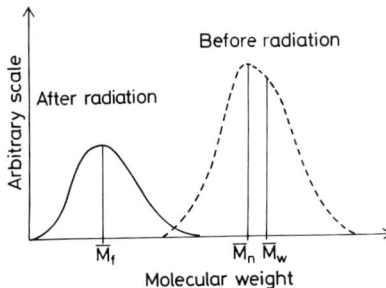

FIG. 12.   A change of the molecular weight distribution for positive electron resist before and after electron irradiation (Herzog *et al.*, 1972).

soluble or unsoluble. Separation of the two molecular weight distributions becomes larger with increasing radiation dose. This separation of the distributions can be expressed conveniently by the number average molecular weight, $\bar{M}_n$ and $\bar{M}_f$, before and after irradiation, respectively, by introducing a radiation yield $G$, which is defined as the number of bonding or scission events caused by an energy absorption of 100 eV (Charlesby, 1954; Ku and Scala, 1969; Herzog et al., 1972; Greeneich, 1973). Namely, the difference in the numbers of molecules between initial and final exposure is set equivalent to the number of scission events $(GD/100)$, where $D$ is the absorbed energy density (in $eV/cm^3$) at an arbitrary location in the resist as follows:

$$\frac{GD}{100} = \frac{\rho N_A}{\bar{M}_f} - \frac{\rho N_A}{\bar{M}_n}$$

or                                                                                      (67)

$$\bar{M}_f = \left(\frac{1}{\bar{M}_n} + \frac{GD}{100\rho N_A}\right)^{-1}$$

According to Herzog et al. (1972) and Greeneich (1973), $\bar{M}_f = 3340$ for a typical case of $G = 1.9, D = 1.1 \times 10^{22} \, eV/cm^3, \bar{M}_n = 10^5$, and $\rho = 1.2 \, g/cm^3$ for PMMA. The value of $D$ is taken from the dose required for an actual resist development, as described in a later section. The molecular size with $\bar{M}_f = 3340$ is roughly estimated to be 170–330 Å, assuming 5–10 Å for the size of a monomer of PMMA with a molecular weight of 100. Another way to estimate the molecular size is the following. The volume density of molecules with weight $\bar{M}_f$ is $\rho N_A/\bar{M}_f$. Assuming a sphere with radius $r$ for the molecule, let $(4\pi/3)r^3 = \bar{M}_f/\rho N_A$. Then the diameter of the molecule is about 20 Å for the same parameters as in the above example. Although there is no experimental verification for these estimates, we should always note that this is a fundamental limit, and will be important for studies on ultimate resolution.

On the other hand, the empirical solubility rate formula for a resist polymer with the fractional molecular weight $\bar{M}_f$ has been established as follows (Ueberreiter, 1968; Greeneich, 1974, 1975):

$$S = S_0 + S_1 \bar{M}_f^{-\alpha}$$                                                    (68)

where the constants $S_0$, $S_1$, and $\alpha$ depend on developer type and temperature, and are determined empirically. Neureuther et al. (1979) modified Eq. (68) in the following for a particular resist:

$$S = S_2(C_m + D/D_0)^\alpha$$                                                          (69)

Typical values are listed in Table V with developers of mixtures of MIBK (methyl isobutyl ketone) and IPA (isopropyl alcohol). According to Eqs. (67)

TABLE V

EXAMPLES OF SOLUBILITY RATE CONSTANTS. GREENEICH'S DATA ARE OBTAINED AT A
TEMPERATURE OF 22.8°C AND WITH $G = 0.75$

| Author | MIBK:IPA | $S_0$ (Å/sec) | $S_1$ (Å/sec)(g/mol) | | $\alpha$ |
|--------|----------|----------------|------------------------|---|----------|
| Greeneich | 1:3 | 0 | $1.555 \times 10^{13}$ | | 3.86 |
| (1975) | 1:1 | 0 | $1.117 \times 10^8$ | | 2.0 |
| | 1:0 | 1.4 | $5.233 \times 10^6$ | | 1.5 |
| | | $C_m$ | $D_0$ (J/cm$^3$) | $S_2$ (Å/sec) | |
| Neureuther | 1:1 | 1.0 | 199 | 1.0 | 2.0 |
| et al. (1979) | 1:0 | 1.0 | 325 | 8.3 | 1.40 |

and (68) or Eq. (69), we can find the solubility rate $S$ as a function of the absorbed energy density $D$. The constant $\alpha$ is an important parameter which controls developed profiles in resist films. The weaker the developer, the larger the $\alpha$ value.

## 2. Etching Model

The absorbed energy density, or the fractional molecular weight, varies with the position in the resist owing to electron scattering. Therefore, the local variation of etching rate by a developer produces a particular developed profile in the exposed region. The shape of the profile is important in microfabrication processes such as the lift-off technique (Hatzakis, 1969). Here, we describe methods to find the time dependence of developed profiles using the spatial distribution of the absorbed energy density in a resist. For this purpose the following three models have been proposed by Jewett et al. (1977).

(1)  Cell model;
(2)  String model; and
(3)  Ray-tracing model.

The cell model is applied to two-dimensional profiles, such as the case of a long line exposure. A resist film is divided into small cells. The time $\Delta T$ to dissolve cells exposed to the solvent is calculated successively as follows by using the solubility rate equation:

$$\Delta T = d/S, \qquad \text{when one face is exposed to a developer}$$

$$= d/\sqrt{2}\,S, \qquad \text{when two adjacent faces are exposed} \qquad (70)$$

where $d$ is the cell size. The total time to reach the cell $(I, J)$ is obtained by summing sequential development times. This method is easy to organize with a computer program for calculations. Recently a new calculation has been proposed for a three-dimensional analysis based on the cell model (Jones and Paraszczak, 1981).

The string model forms a profile formed with representative points in the resist. These points advance step by step during a short period $\Delta T$ at the solubility rate calculated from the absorbed energy density. The direction of motion is determined by a bisector of the angle made by the two adjacent strings.

The ray-tracing model is based on an analogy to Snell's law for refraction of light rays. The ray traces the direction of a vector $n = S_{max}/S(X, Y)$, where $S_{max}$ is the maximum value of the solubility rate, $S(X, Y)$. The line connected with the end of the ray gives an etched profile.

The latter two models are expected to give better accuracy than the first one, but the computer programming will be somewhat harder. If we simulate the etched profiles of a resist pattern with a weak developer [i.e., large values of $\alpha$ in Eq. (67)] by using the resist etching model above, we find that a weak developer has a high contrast. That means it responds sensitively to a change of the fractional molecular weight of the resist or the absorbed energy density. Namely, the front surface of the etch profile does not move farther when it meets an appropriate slope in the energy absorption distribution. Then, the etched profile is along the equienergy density contour. This is called the critical absorbed energy density model or the threshold model (Herzog et al., 1972; Greeneich and Van Duzer, 1974).

### B. Negative Type

In contrast to a positive resist, a negative resist remains insoluble in the exposed region due to gel formation caused by cross links among neighboring molecules, depending on the exposure dose. The insoluble fraction of the resist is related to the gel fraction, which determines the remaining film thickness $p = d/d_0$ normalized to the initial thickness (Atoda and Kawakatsu, 1976). Usually the relation between the dose $Q(C/cm^2)$ and the remaining thickness $p$ is given by the so-called contrast curve. A typical example of experimental data is shown in Fig. 13 for a 6000 Å PGMA (polyglycidyl methacrylate) resist film at 10 keV exposure (Nakata et al., 1981). The straight-line portion of the curve can be given by the following equation:

$$p = \Gamma \log_{10}(Q/Q_m) \tag{71}$$

where $\Gamma$ is the resist contrast ($\Gamma = 1.12$ in the figure), and $Q_m$ the minimum dose at which remaining resist starts to be observed. Heidenreich et al. (1975) and Lin (1975) have performed a developed profile analysis of negative resists based on such contrast curves. They assumed a Gaussian distribution for a

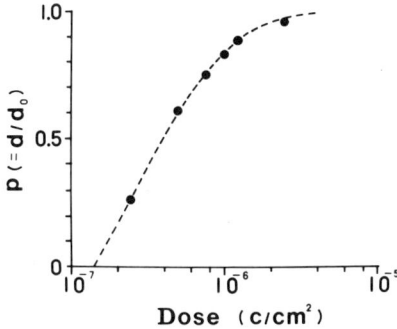

FIG. 13.    The normalized remaining film thickness as a function of dose for negative electron resist PGMA at 10 keV (Nakata et al., 1981). (●), experimental data. The cure time is 30 min, and $d_0 = 0.6$ μm. The developer is a solution of 7 parts MIBK and 1 part EtOH (ethanol). The development temperature is 22–23°C. The rinse was done with MIBK for 30 sec. No post-baking is done. (– – –) shows a calculated result based on the Charlesby theory.

line exposure dose. Although they gave an equation to take account of the lateral spreading of an electron beam, the effect is neglected in the actual calculations. Nakata et al. (1981) tried to analyze a pattern of lines and spaces by using the spatial distribution of the absorbed energy density obtained by MC calculations. The calculation procedure is described briefly. The depth distribution of the absorbed energy density is not uniform, as seen from Fig. 14, which was obtained by MC calculations for the same conditions as that in Fig. 13. For simplification a uniform depth distribution is assumed, as shown by the dotted line in Fig. 14. Then, we can replace the dose $Q(C/cm^2)$ on the abscissa by the absorbed energy density $D(eV/cm^3)$, which is obtained by the product of $(dE/dz)_{av}$ (eV/C cm) and $Q(C/cm^2)$. The theory that the gel

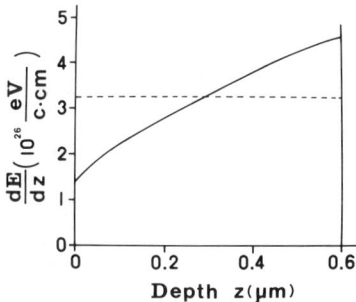

FIG. 14.    The depth variation (——) of absorbed energy in a 6000 Å thin negative resist PGMA film on Si at 10 keV with a plane source obtained by MC calculations. The averaged absorbed energy density (– – –) was determined to be $(dE/dz)_{av} = 3.24 \times 10^{26}$ eV/C cm from this result (Nakata et al., 1981).

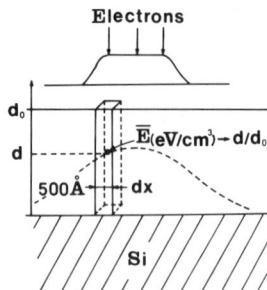

FIG. 15. The determination of the remaining film thickness with the averaged absorbed energy density within vertical parallelpipeds, the cross section of which is typically a 500 Å square (Nakata et al., 1981).

fraction is associated with the absorbed energy density has been developed based on the radiation yield $G$ (Charlesby, 1954; Atoda and Kawakatsu, 1976). In the figure a dashed line shows the theoretical curve calculated by the Charlesby theory, assuming both the Schultz–Jim function for the molecular weight distribution of polymer and the inhibitor activity $i$, using the average absorbed energy in Fig. 14.

As shown in Fig. 15, the resist film is divided into many vertical parallelpipeds. In each pipe the absorbed energy density is averaged over depths, and then from the contrast curve the gel fraction, i.e., the normalized remaining film thickness, is found, corresponding to the averaged energy density $D$. This proccess throughout all pipeds makes up the developed patterns. Subdivisions of the vertical pipeds into cells over the depth may provide more accurate analysis.

## V. APPLICATIONS TO PATTERN ANALYSES IN ELECTRON-BEAM LITHOGRAPHY

### A. Energy Dissipation Profiles in Thick Targets

It is instructive to visualize the spatial distribution of the electron energy dissipation in a specimen. Some trials have been made using, for example, cathodoluminescent targets (Ehrenberg and King, 1963). In electron-beam lithography this visualization is calculated easily with high accuracy.

As previously mentioned, developed profiles with a weak solvent, e.g., a 1:3 solution of MIBK and IPA for PMMA, can show isocontours of equal absorbed energy density which reach the threshold energy. In the earlier stage of electron-beam lithography, this type of experimental work was done by several authors (Everhart et al., 1971; Wolf et al., 1971; Hatzakis and Broers,

1971; Hatzakis, 1971; Herzog et al., 1972). The experiment is as follows. A finely focused beam (20–100 nm in diameter) is exposed on a thick PMMA sample along a long line. The experiment is done for various doses. Since the PMMA polymer is an insulator, its surface is coated with a very thin metal film. After electron exposure, the sample is developed by a weak solvent and cleaved. The cross sections are observed in a scanning electron microscope. Shimizu and Everhart (1972) and Shimizu et al. (1975) have done MC simulations of the energy dissipation in a thick PMMA target based on the single scattering model and compared with their experimental profiles. The agreement was fairly good, and the model was found to be useful.

In Fig. 16 experimental profiles of etched regions by Wolf et al. (1971) are shown at doses of $1.0q_1$, $0.7q_1$, $q_1$, and $2.0q_1$, where $q_1 = 10^{-8}$ C/cm (a line charge density) in comparison with the results of MC calculations by Kyser and Murata (1974b). The figure shows pear-shaped profiles due to electron scattering which widen with decreasing absorbed energy density. Two results are arbitrarily matched at the bottom of the isocontour of $0.3q_1$. This match gives a value of $D_c = 1.1 \times 10^{22}$ eV/cm$^3$ for the critical absorbed energy density in a threshold model. Namely, this energy density is deposited along

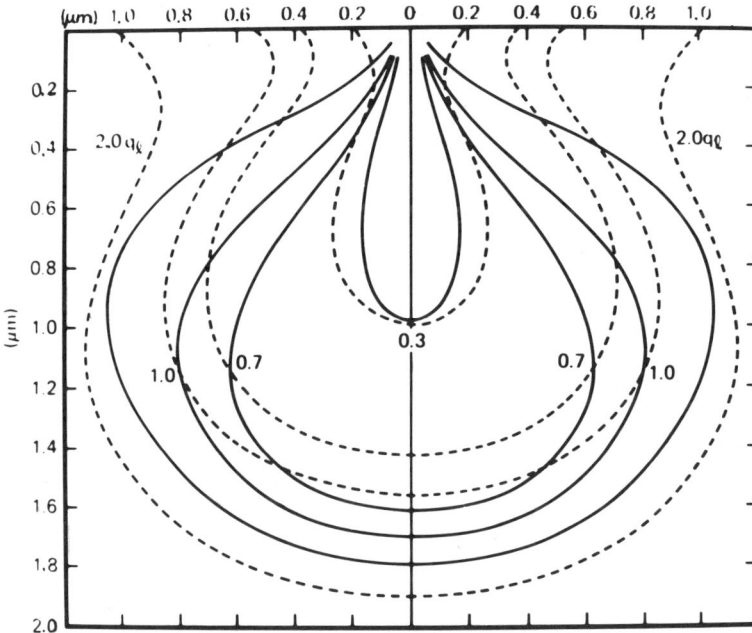

FIG. 16. Equi-energy density contours for a thick PMMA at 10 keV comparing MC simulation (——) and experiment (– – –) (Kyser and Murata, 1974b). $q_l = 1 \times 10^{-8}$ C/cm; $\Delta E/\Delta V = 1.1 \times 10^{22}$ eV/cm$^3$.

the isocontours. The same value was also used for the remaining contours. A general agreement is obtained between theory and experiment, in spite of the neglect of the time-dependent developer action. The largest discrepancy is seen near the resist surface. There are several possible causes for the large width near the surface, such as surface tension in the film after irradiation, or a finite etching rate of the polymer in the regions where absorbed energy is less than $D_c$, as discussed previously by Greeneich and Van Duzer (1974). Experimental data of Everhart et al. (1971) and Shimizu et al. (1975) at 29 and 20.7 keV, respectively, do not show the roundness near the surface. This may come from the fact that a weaker solvent of 95% ethanol was used.

## B. Exposed Intensity Distribution

The importance of electron scattering in a resist–substrate target has been shown experimentally by Wolf et al. (1971) and Hatzakis aand Broers (1971). When electrons are incident on a resist, they can give rise to chemical changes along their path, penetrating the resist film. These electrons form a designed latent image in the resist film although they spread to some extent depending on energy and resist film thickness, etc. Some electrons can reach the substrate, penetrate it, and rebackscatter into the resist. These electrons cause energy deposition in a wide region of the film, although the intensity is low.

We call the spatial distribution of energy deposition by a finely focused electron beam the "exposure intensity distribution" (EID). The electron scattering effect mentioned above is undesirable for fine lithography, and produces the so-called proximity effect, which is discussed in a later section. Greeneich and Van Duzer (1974) investigated such scattering effects by an analytical method. They divided the total contribution to energy deposition into three component parts, namely, the primary contribution, which is calculated by Lenz's plural scattering theory (1954) and the backscattered electron contribution from within the resist and from the substrate, which are calculated by the large-angle and single scattering theory. They calculated the spatial distribution of energy dissipation based on the exposure intensity distribution for a 4000 Å PMMA resist film on a Si substrate, and compared it with experimental developed profiles by Wolf et al. (1971). The theoretical prediction was reasonable. The comparison is based on a threshold energy density of $6.8 \times 10^{21}$ eV/cm$^3$. However, such analytical models cannot be readily used for general cases, as pointed out by Hawryluk et al. (1974b) based on their own analytic model.

MC calculations of energy dissipation in PMMA films on various substrates were conducted by Saitou (1973), Hawryluk et al. (1974a,b), and Kyser and Murata (1974b). Hawryluk et al. (1975) obtained good agreement between MC results and a series of experiments for energy dissipation in a PMMA film.

To illustrate electron scattering in a PMMA/Si target, trajectories simulated by an MC method are shown in Fig. 17 at 10 and 20 keV (Kyser and Murata, 1974b). The thickness of the film is 4000 Å. We can see qualitatively the extent of the forward and backward electron scattering spread. When the incident electron energy is decreased from 20 to 10 keV, the lateral spreading is reduced, but the forward electron spreading increases. A further decrease in energy can reduce the range of electron diffusion. However, the electron may not then penetrate the film completely. Moreover, electron optical systems cannot generate arbitrarily large currents for exposure with a fine spot diameter, although the development of both high brightness electron guns and fabrication technology may change this circumstance. Usually an electron energy of around 20 keV is utilized for electron-beam lithography. The use of even higher energy is discussed later. In Fig. 18 one example of the radial exposure intensity distribution is shown for a 20 keV beam incident on a 0.8 $\mu$m PMMA film on a Si substrate (Nomura et al., 1979). A Gaussian incident beam is assumed with a diameter of 0.1 $\mu$m. The film is divided into 8 layers, each with 0.1 $\mu$m thickness. The distribution is shown for the top and the bottom layers only. The curves consist of two characteristic features: a steep peak in the center caused by the forward scattered electrons and the background distribution caused by backscattered electrons. The peak for the bottom layer has a little broader distribution than that for the top layer due to the forward electron scattering. Moreover, the background for the bottom is higher than that for the top because the bottom layer has a direct influence by electrons backscattered from the substrate. This radial

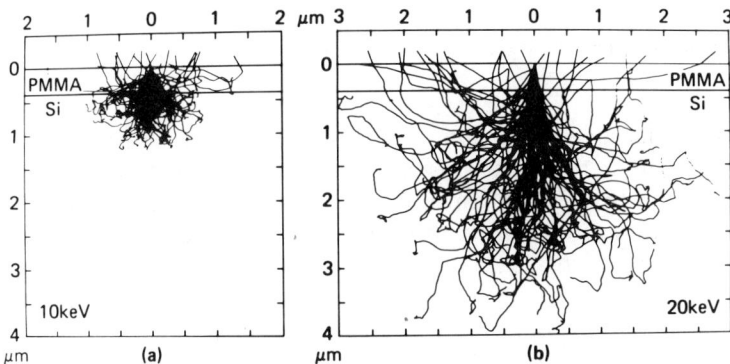

FIG. 17.    Plots of simulated trajectories in a PMMA film on a Si substrate. 100 electrons are incident at $x = 0$. Note the differences in the forward and backward electron spreading between (a) 10 and (b) 20 keV (Kyser and Murata, 1974b).

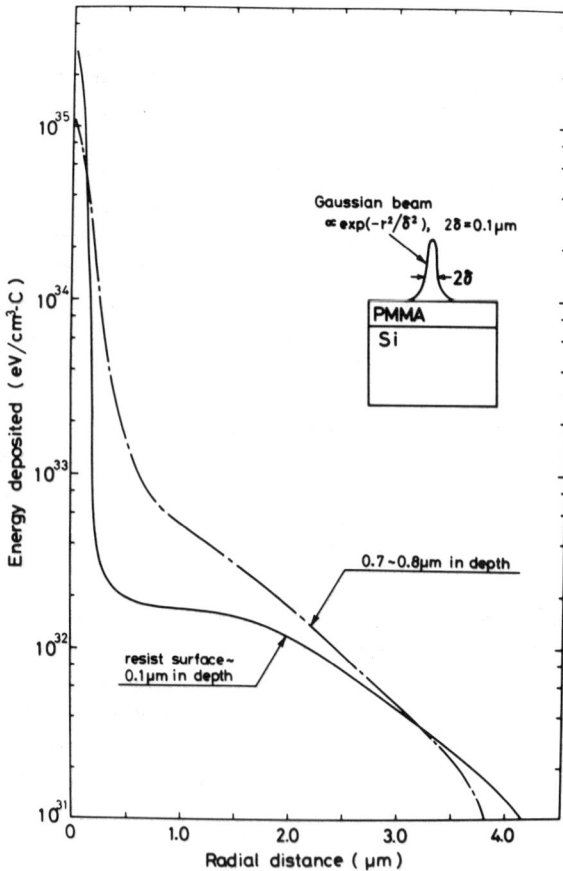

FIG. 18.    Radial exposure intensity distribution in a 0.8 $\mu$m PMMA film on Si at 20 keV. A Gaussian beam is convoluted with a diameter of 0.1 $\mu$m (Nomura et al., 1979).

exposure intensity distribution is superposed to form an arbitrary pattern analogous to a scanning-type electron-beam exposure. Usually a long line exposure is often performed for comparisons with experiments, where the lateral exposure intensity distribution is calculated by the Abel inversion, as proposed by Hawryluk et al. (1975), or originally calculated by MC calculation.

An analytic expression for EID has been proposed by Chang (1975), which consists of a coaxial double Gaussian fit in the following manner:

$$f_0(r) = \frac{F_T(E)}{\pi(1 + \eta_E)} \left[ \frac{1}{\beta_f^2} \exp\left(-\frac{r^2}{\beta_f^2}\right) + \frac{\eta_E}{\beta_b^2} \exp\left(-\frac{r^2}{\beta_b^2}\right) \right] \qquad (72)$$

where the parameter $\beta_f$ gives the forward electron spreading, $\beta_b$ the backward electron spreading, $\eta_E$ is the ratio of the backward electron contribution and the forward electron contribution to energy deposition, and $F_T(E)$ is the total absorbed energy density. Generally the expression is also a function of depth $z$, as shown in Fig. 18. The expression disregarding the depth dependence can still be useful, depending on the application. If the incident beam is a Gaussian type with a radius $\delta$, the two parameters $\beta_f$ and $\beta_b$ have to be replaced by newly introduced parameters $\beta_f^*$ and $\beta_b^*$ as follows:

$$\beta_f^* = \sqrt{\beta_f^2 + \delta^2}, \qquad \beta_b^* = \sqrt{\beta_b^2 + \delta^2} \tag{73}$$

An analytical expression is convenient for calculating the energy absorption for an arbitrary pattern exposure, and is much faster to evaluate when required for a proximity effect correction. The parameters $\beta_f$, $\beta_b$, and $\eta_E$ have been investigated experimentally and theoretically (mainly with MC calculations) by many authors. These parameters depend on various experimental factors such as resist thickness, developer type, incident beam energy, resist and substrate materials, etc.

The value of $\beta_f$ decreases with increasing beam energy and depends on depth in the resist. From MC results, the forward spreading cannot be fitted to a Gaussian form (Parikh and Kyser, 1979; Nomura et al., 1981). Nomura et al. have shown that fitted values of $\beta_f$ are proportional to $[z(\text{depth})]^{1.45}$. A typical value is around 0.1 $\mu$m at 20 keV for the bottom of a 0.5 $\mu$m thin resist film on a Si substrate. Since the value is small compared to the beam size, it is not important in general practice. The effect will be more important when the film thickness increases or when high-resolution patterns are desired.

The background distribution is expressed very well by a Gaussian expression for low atomic number substrates. The value of $\beta_b$ is comparable to the electron range in substrates. The energy dependence of $\beta_b$ is estimated to be $0.35E^{1.7}$ from MC results (Parikh and Kyser, 1979; Adesida et al., 1979; Greeneich, 1979). However, the distribution does not follow a single Gaussian for high atomic number substrates such as Cu, Mo, and Au. Note that even the distribution for the heavy substrates tends to be Gaussian at relatively high energies larger than, say, 50 keV (Y. Hirai and Y. Mano, personal communication). The difference in electron backscattering between low- and high-$Z$ substrates is shown schematically in Fig. 19 (Parikh and Kyser, 1979). Their explanation is the following: (1) The high atomic weight substrate gives rise to a greater number of large-angle elastic scattering events compared to that from silicon. Thus, the electron distribution is inherently expected to be narrower in the backward direction with increasing atomic weight. (2) The stopping power for electrons increases with increasing atomic weight of the substrates; thus the backward scattered electrons that are generated with angles to the normal greater than $\pi/4$ are likely to be attenuated in the high atomic weight

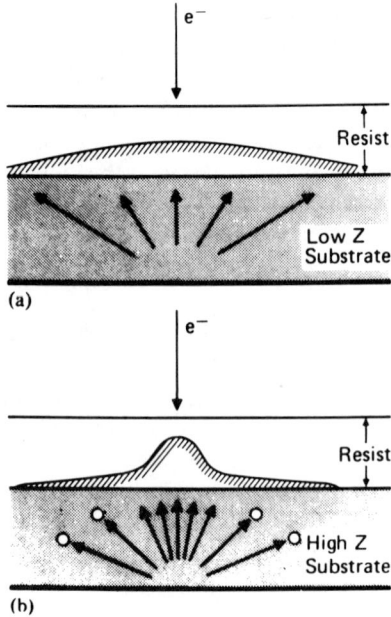

Fɪɢ. 19.   Schematic showing of the difference in the spatial distribution of energy deposition in the resist between the cases of (a) low and (b) high atomic weight substrates (Parikh and Kyser, 1979).

substrate. (3) Those backward scattered electrons which arrive at the resist will subsequently be attenuated less than they would have been, had they remained in the substrate. Thus, the range in the resist of these backward scattered electrons will define the radial extent of this component of the EID. They introduced two Gaussians into the backward component of energy deposition for high-Z substrates, resulting in three coaxial Gaussians totally. A similar explanation is given by Kato et al. (1978).

Aizaki (1979) also investigated the EID in great detail with MC calculations. He obtained approximate functions for a 0.5 $\mu$m PMMA film on Cr, Mo, and Au substrates. The function $r^{-b}$ ($r$, radial distance; $b$, constant) is utilized to fit the calculated forward distribution in his work. For the backward distribution the Gaussian and the exponential forms are found to be appropriate for Si and the above three substrates, respectively.

The value of $\eta_E$ is an important factor to determine the strength of background intensities in the proximity effect. In MC calculations it is easy to separate the absorbed energy deposited by backward scattered electrons from that by the forward scattered electrons. Although the number of backscattered electrons from light substrates such as Si is small, their contribution is enhanced both by the longer traveling path in the resist due to the wide angular distribution and by the higher stopping power due to their low energies compared to the primary electron. According to MC calculations by Parikh and Kyser (1979) and Adesida and Everhart (1980), the values of $\eta_E$ are about 0.5. The value has a weak dependence on energy and the resist film thickness. Hawryluk (1981) and Chung and Tai (1978) have reported higher values of 0.77–1.1 based on MC calculations. This difference is not clear at present. Since the MC models differ from each other to some extent, further study has to be conducted.

On the other hand, experimental measurements have been done in various ways. The simplest way to examine $\beta_f$ and $\beta_b$ is to perform point or line exposure on a positive resist by changing the exposure dose (Hawryluk et al., 1975; Adesida et al. 1979). The exposed resist is developed and observed in a scanning electron microscope. From the observed radius or the width where the absorbed energy reaches the threshold energy, the EID curve is deduced. Even with a very fine beam, however, it is not easy to obtain a value of $\beta_f$ because of shrinkage of the developed resist and the etching effect. Experimental results give larger values than the MC results. Chung and Tai (1978) prepared a special sample which had a thin gold film with an open window on the back of the resist film. When an electron beam scans on the resist film, the forward scattered electron spreading is determind from the variation of transmitted electron current through the window. A reasonable agreement was obtained between their MC calculations and the experiments.

The value of $\beta_b$ is relatively easy to obtain because it has a large value, enough to observe experimentally. It seems that MC calculations agree with experimental results for $\beta_b$. Adesida et al. (1979) have also shown the dependence of $\beta_b$ on the Si substrate thickness, and obtained a good agreement between MC calculations and experiments.

To obtain a value for $\eta_E$, areal exposure experiments are necessary. Various methods have been reported for this purpose (Chung and Tai, 1978; Jones and Hatzakis, 1978; Grobman and Speth, 1978; Stephani and Kratschmer, 1981; Kratschmer, 1981; Shaw, 1981; Jacket et al., 1984). These experiments give values between 0.72 and 1.1, which are larger than MC results by Parikh and Kyser (1979), and by Adesida and Everhart (1980). Kyser et al. (1980) attempted to explain this difference by incorporating the significant developer effects. Another reason for the difference may be in the reduction of the energy dissipation at the surface (Murata et al., 1978a) predicted by the theory of Spencer and Fano (1954).

In Fig. 20 the lateral EID curves for an isolated thin PMMA film and a PMMA film on a Si substrate (curves 5–8) are compared between the regular single scattering model and the knock-on model (Murata et al., 1981). The energy is 20 keV. The curves are shown at depths of $z = 0$ and 4000 Å. The difference is clearly seen between both models. The new model predicts much

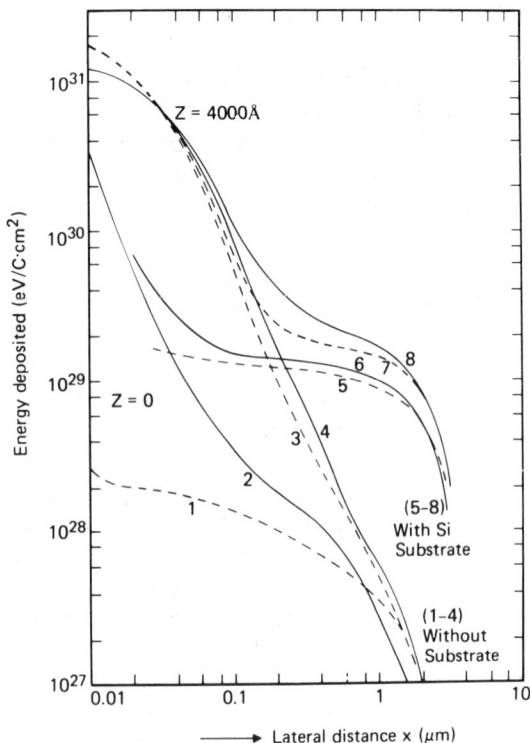

FIG. 20. Lateral distributions of the absorbed energy density from a 20 keV electron beam. The new results (knock-on or hybrid model, ——, $\varepsilon_c = 0.001$) are compared with the old results (− − −) for both an isolated thin film and a film of 4000 Å PMMA on Si (Murata *et al.*, 1981).

higher values in the case without substrates than the old model does, especially at $z = 0$. This intensity increase comes from secondary electrons generated along the primary electron, which results in an increase for the value of $\beta_f$. Also with the substrate a relatively large increase is seen, especially at the bottom layer where the backscattering contribution dominates. Thus, the backscattering factor $\eta_E$ may increase with the knock-on model. Although detailed studies have not been done yet, the new model may improve the discrepancy in $\eta_E$ between MC results and experimental ones.

## C. Proximity Effect

When a point-source electron beam is incident on a resist film on a substrate, the resist receives energy absorption in a wide range around the beam, as described in the previous section. This background builds up a large amount of energy deposition, which is comparable to that by the forward scattered electrons when the beam is exposed sequentially to form an arbitrary pattern. Then, the latent image does not replicate the designed pattern. If the effect is considered within a single pattern, this is called the "intraproximity effect" by electron scattering. If the effect extends to adjacent patterns, it is called the "interproximity effect." Fortunately these effects can be corrected to a relatively high extent. Many proximity effect correction methods have been proposed (for example, Parikh, 1979; a review paper by Hawryluk, 1981; Owen and Rissman, 1983).

In this article the details of these methods are not described, but we should point out that the EID curves previously mentioned play an important role in the correction methods. We describe the MC techniques as applied to analyses of exposed patterns and as applied directly to proximity effect corrections.

## 1. Intraproximity Effect

Typical examples of the intraproximity effect are given in Fig. 21 as equi-energy density contours for various sizes of square patterns (Nakase and Yoshimi, 1980). The contours are demonstrated at the bottom of PMMA resist films of 0.5, 1.0, and 1.3 $\mu$m thicknesses. They obtained the EID curve with MC calculations, approximated them by two Gaussian forms as a function of depth, and convoluted them with an incident Gaussian beam of 0.6 $\mu$m diameter. The energy distribution can then be calculated analytically for a finite single-line exposure. The appropriate number of distributions are added with 0.5 $\mu$m separation to form a square pattern. The dashed line shows the designed shape. The results show that the energy deposition is not uniform in the exposed pattern, mainly due to the backward electron contribution. The discrepancy of the contour shape from the designed pattern shape increases with increasing thickness. Another thickness dependence of equienergy contours is seen in the figure. The contours become dense with decreasing thickness near the exposed pattern edge, which means that the forward electron spreading decreases with decreasing thickness. These effects cause a different energy absorption within the square pattern with the same areal dose (C/cm$^2$). For the 2 $\mu$m square with $t = 1$ $\mu$m, the contour number 8 approximately follows the designed shape, while a smaller energy contour follows it for the 1 $\mu$m square and a higher energy contour for the 4 $\mu$m square.

In Fig. 22, SEM micrographs of the experimental results are shown with the same conditions as in the calculations (Nakase and Yoshimi, 1980). The developer is MIBK. Although the equienergy contours cannot be compared directly with experimental developed pattern profiles, they should reflect developed shapes. Careful observations show good correspondence with the facts derived from the theory. A similar study for rectangular patterns was reported by Murata et al. (1978b).

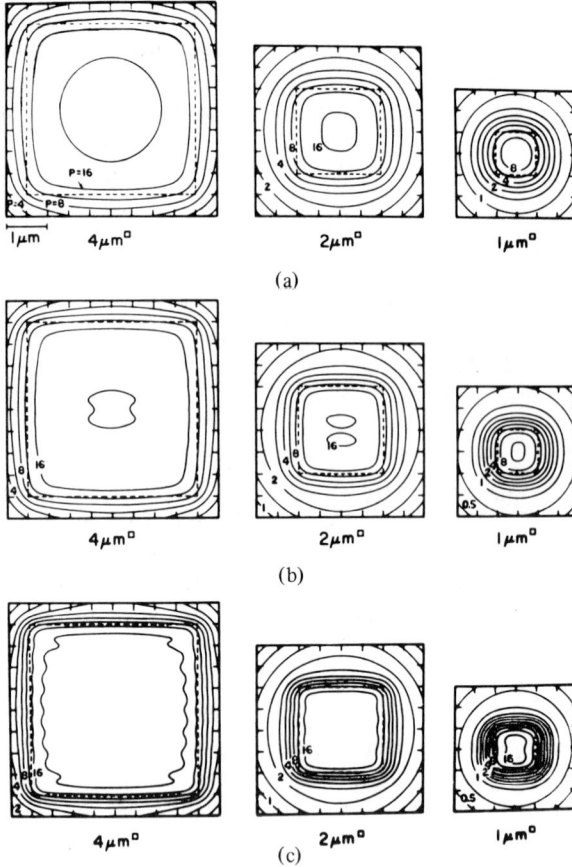

FIG. 21. Calculated equi-energy density contours in the bottom region of the resist film at 20 keV. The contours are given with relative units. The beam diameter $2\delta = 0.6$ μm. The results are shown for three different film thicknesses of (a) 1.3, (b) 1.0, and (c) 0.5 μm (Nakase and Yoshimi, 1980).

The intraproximity effect appears as a difference in the development speed. One example simulated by Kyser and Viswanathan (1975) is shown in Fig. 23a and b. Two line patterns are exposed with different beam widths. The dose is $Q = 100$ μC/cm$^2$ for both exposures. The developer is MIBK. The figure shows that the wider line (Fig. 22b) collects more energy, and thus etching reaches the Si substrate interface faster. There exist various patterns with different sizes in an actual fabrication process, but we cannot change the development time corresponding to the pattern sizes. Therefore, the dose has to be modulated so that the same absorbed energy is obtained. MC calculations

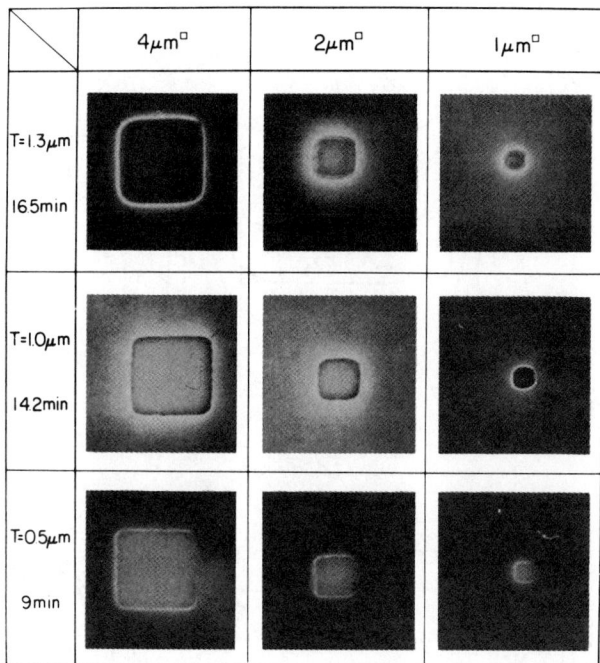

FIG. 22.    SEM micrographs of developed resist patterns under the conditions corresponding to Fig. 21. The exposure dose is 15 $\mu C/cm^2$, and the developer is MIBK (Nakase and Yoshimi,
1980).

can predict the dose modulation factor with pattern size (Neureuther *et al.*,
1979; Kyser *et al.*, 1980).

## 2. Interproximity Effect

A typical example of the interproximity effect is shown for the bottom
layer of the resist film at 20 keV in Fig. 24, which was calculated by Hawryluk
*et al.* (1974a,b). In Fig. 24(i) the lateral distribution of absorbed energy is
shown for a single-line exposure. The peak intensity of the forward scattered
electrons is higher than the background intensity by a factor of about $10^2$.
When an infinite array of lines and spaces is formed, the background piles up
as shown in Fig. 24(ii), resulting in the reduction of the peak-to-background
ratio. The effect becomes significant with decreasing interline spacing, namely
more packed lines. It seems that the background intensity is approximately
proportional to the line density in these results. Although it is not clear at this
stage what profile will result when they are combined with the development
process, the contrast of the exposure dose is certainly reduced for successful
patterning.

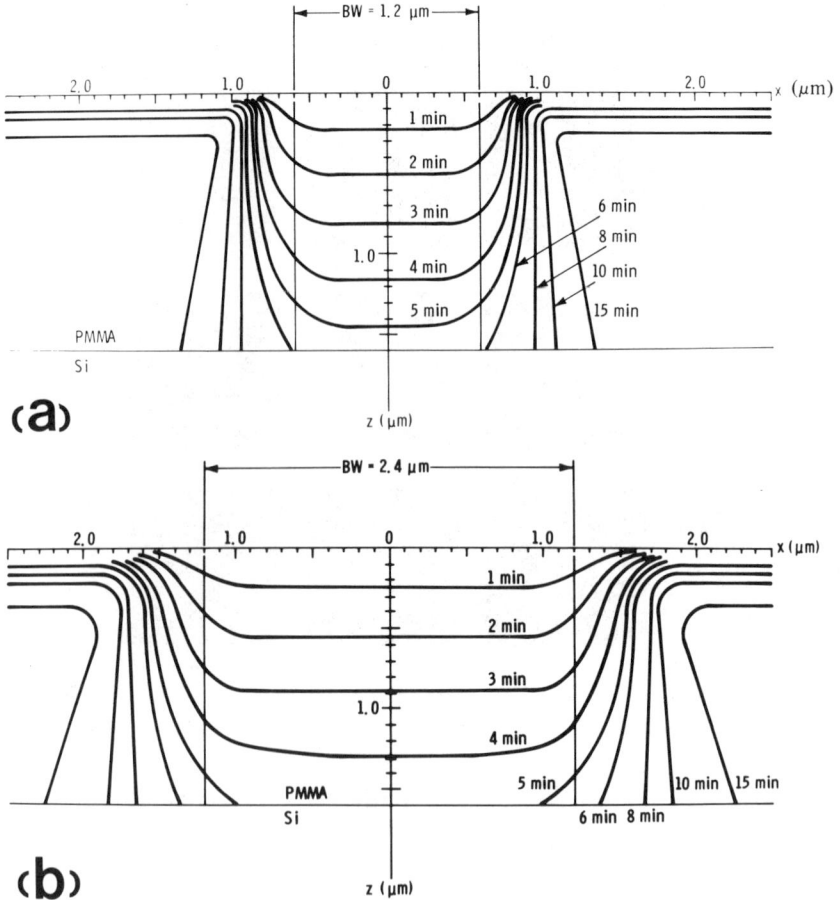

FIG. 23. Time dependence of etching fronts for line patterns. The dose is 100 $\mu C/cm^2$ and $G = 0.75/100$ eV. The results are compared between two different beam widths of (a) 1.2 and (b) 2.4 $\mu$m. In a larger pattern the front can reach the bottom faster than that in a smaller pattern due to the proximity effect (Kyser and Viswanathan, 1975).

Another example of the interproximity effect is shown in Fig. 25 using an 8000 Å PMMA resist film on a silicon substrate, which has been reported by Nomura et al. (1979). This figure shows the equienergy contours in a part of the exposed rectangular pattern ($10 \times 1.5\ \mu m^2$) adjacent to the side of a 20 $\mu$m square pad. The gap distance is 1.0 $\mu$m between the two pattern elements. The contours are calculated by superposing the EID curves in Fig. 18 obtained by MC calculations. They are normalized to an absorbed energy density of $1.12 \times 10^{26}$ eV/C cm (100%) which approximates the designed profile. As seen from the figure, the interproximity effect from the large pattern appears

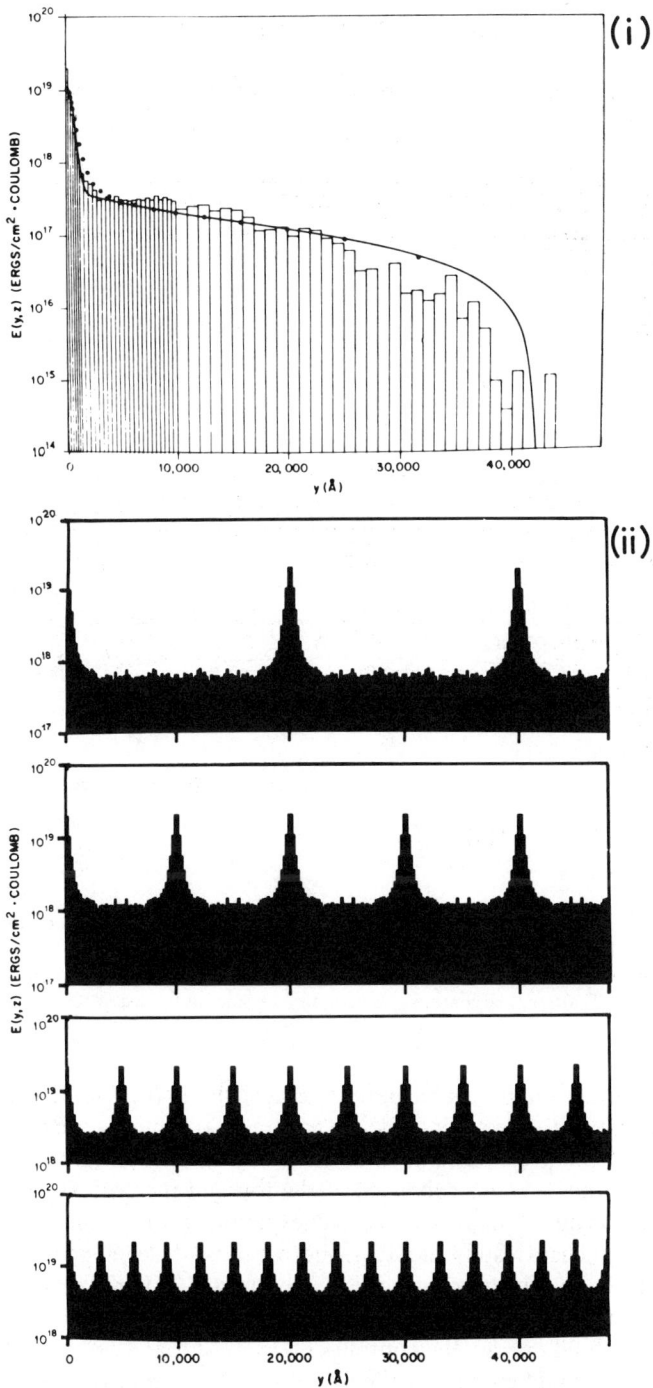

FIG. 24. (i) Lateral exposure intensity distributions in a 0.4 μm resist film on Si and (ii) spatial distributions of absorbed energy for an array of lines produced by this EID at $z = 4000$ Å. The beam energy and the beam diameter are 20 keV and 250 Å, respectively. The interline spacings are, from top to bottom, 2, 1, 0.5, and 0.3 μm (Hawryluk *et al.*, 1974a,b).

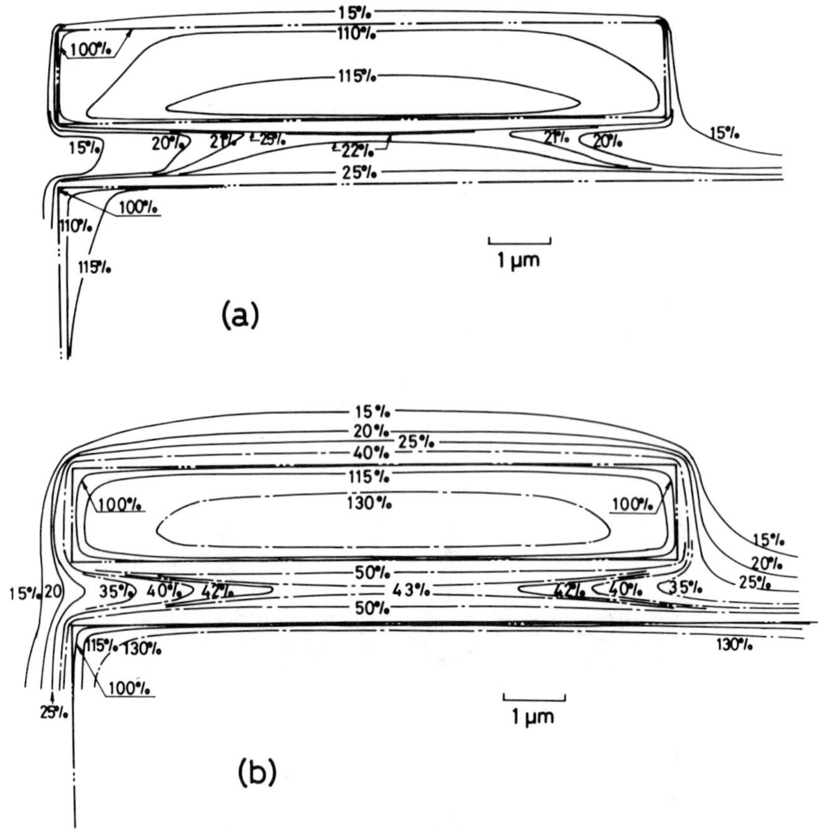

FIG. 25. Calculated equienergy density contours (a) at the top and (b) the bottom layers of the 0.8 μm resist film for a 1.5 × 10 μm² rectangular pattern adjacent to a 20 × 20 μm² square with a spacing of 1.0 μm. The beam energy is 20 keV. The contours are shown only in a region around a smaller pattern. The contour of $1.12 \times 10^{26}$ eV/cm C is labeled as 100% (Nomura et al., 1979).

in the contours for the small pattern, especially at the top, in addition to the intraproximity effect. Namely, the asymmetric contours are biased in the direction of the large pattern. In the middle of the gap region, energies of about 22% and 43% are deposited at the top and the bottom, respectively.

To see the quantitative accuracy, an experiment was planned by Nomura et al. (1979). Their results of SEM photographs of developed patterns are shown in Fig. 26. The developer is a 1:3 solution of MIBK and IPA, where a threshold model can be applied. Assuming a value of $D_c = 1.1 \times 10^{22}$ eV/cm³ for the threshold energy, the contours of 18, 34, 51, and 68% in Fig. 25 can reach the threshold energy for the four doses as given in the figure caption,

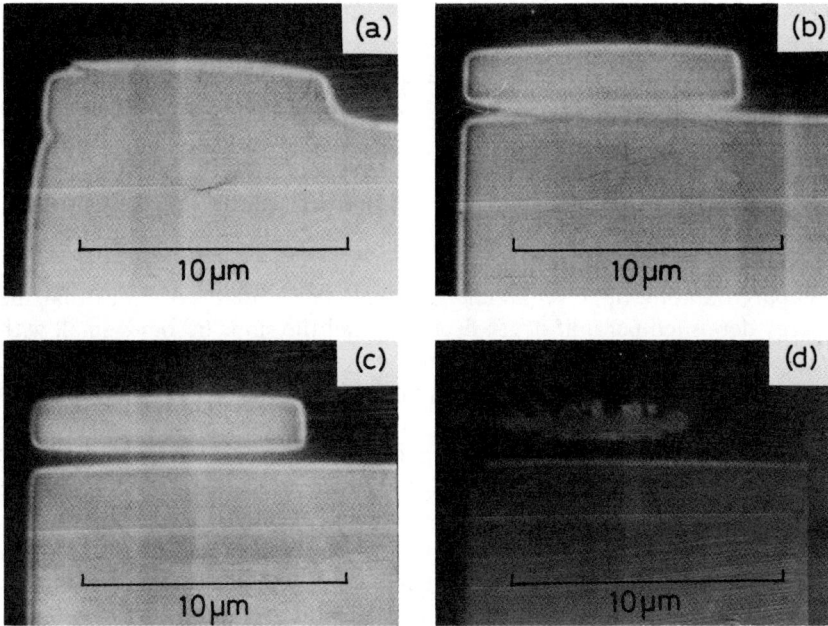

FIG. 26.   SEM photographs of developed resist patterns under the same conditions as in Fig. 25. Doses are (a) 543, (b) 289, (c) 193, and (d) 145 $\mu C/cm^2$ (Nomura et al., 1979).

respectively. Therefore, the resist is developed out along each contour. We see a relatively good correspondence between calculated contours and developed profiles. For example, at a dose of 2.89 $\times$ $10^{-4}$ $C/cm^2$ the calculated absorbed energy at the bottom in the gap region exceeds the normalized threshold energy of 34%, while it does not at the top. This is seen in the photograph of Fig. 26b. Namely, a resist width of about 0.75 $\mu$m remains with a little brighter contrast between the two pattern elements. Beneath this stripe the resist seems to be developed out and the stripe is suspended.

However, we can see some difficulties in the theory. One of them is seen at the largest dose. The experimental profile agrees well with the isocontours of 18% at the bottom, but there is poor agreement at the top. The other is observed at the lowest dose (Fig. 26d). Although the whole region of the small pattern attains the threshold energy, it is not dissolved away completely. Possible reasons for these discrepancies are the following. The threshold model cannot be applied to the low exposure dose. Thus, the time evolution may have to be evaluated. Also the energy loss calculation used in the model is important. The Bethe theory may have to be replaced by the Spencer–Fano theory. These error factors have not been studied in detail yet.

### 3. Substrate Dependence of the Proximity Effect

As mentioned in Section V,B, a heavy element substrate results in a localized energy dissipation and a large backscattered electron intensity. More details on the substrate dependence of the proximity effect are described in this section. In Fig. 27 the calculated depth variations of absorbed energy in a 0.8 $\mu$m thin PMMA film are shown with four different substrates, PMMA, Al, Cu, and Au (Kyser and Murata, 1974b). The curve is extended to deeper regions for the PMMA substrate, i.e., a thick PMMA sample. The results are compared among three initial energies. First, we should point out that the energy deposited per unit depth decreases and the slope becomes small with increasing beam energy. It seems that energy absorption becomes more uniform with increasing energy. This uniformity of the energy absorption is favorable for fabrication processes, although the absorption efficiency decreases. The increase in the absorbed energy due to a heavy element substrate depends on beam energy. At 10 keV there is almost no contribution at the surface. This is because the range of electrons reflected from the substrate is not sufficiently large to reach the resist surface. We can understand this fact from the electron range of about 1.9 $\mu$m in thick PMMA. The increase in the absorption energy with heavier element substrates is due to an increase in backscattering from the substrate. As can be seen easily from the figure at 20 keV, the additional increase is roughly proportional to the backscattering coefficient for each substrate. This is suggested by Saitou et al. (1972) from the experimental observations. Note that the coefficients are nearly 0.15, 0.30, and 0.50 for Al, Cu, and Au, respectively. Therefore, we must be careful for the apparent changes in resist sensitivity with substrate material.

Kato et al. (1978) investigated both with MC calculations and experiments the energy absorption in a sample configuration of PMMA film–metallic film–glass plate. Their results show that the developed width has an appreciable widening even with an 800 Å Ta film. They also demonstrated a large variation of two-dimensional equienergy contours with Mo and Ta thick substrates.

Neureuther et al. (1979) studied the substrate effect with a single-line exposure of 0.5 $\mu$m width on a 1.0 $\mu$m PMMA film. Their result is shown in Fig. 28 for Si, GGG($Gd_3Ga_5O_{12}$), and Au substrates. The developer and the development time are MIBK and 160 sec, respectively. The same dose of 80 $\mu$C/cm$^2$ produces different widths depending on substrate material. The width is largest for the Au substrate. The difference from a Si substrate ranges from 0.1 to 0.2 $\mu$m per edge. The difference appears also in the surface etching. For heavier element substrates, the etching rate increases at the surface. They obtained developed profiles close to that for the Si substrate by compensating

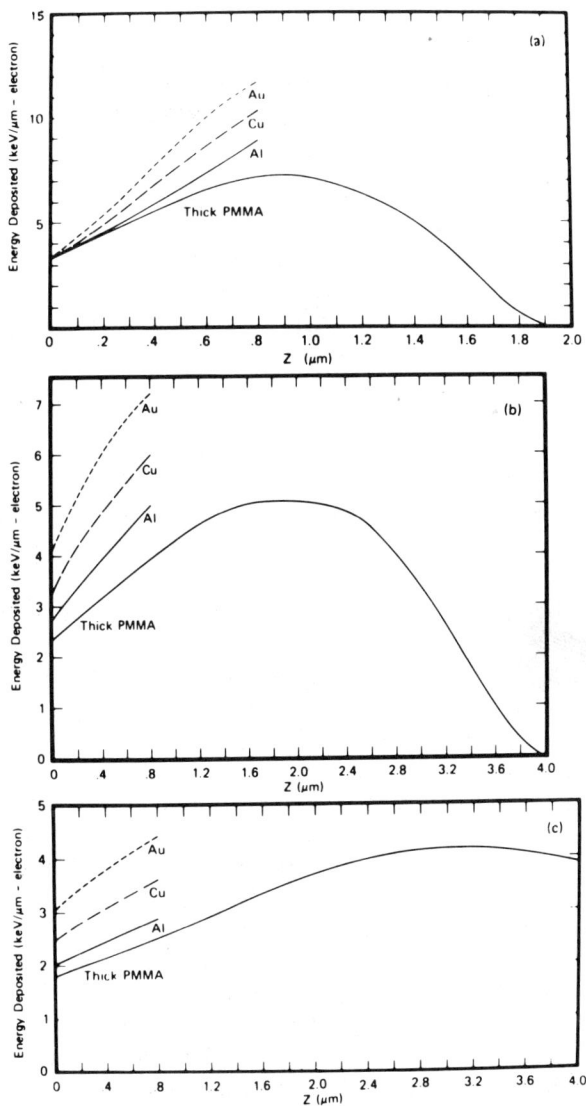

FIG. 27.    Substrate dependence of energy deposited by a plane source in a 0.8 $\mu$m PMMA film shown as a depth variation for (a) 10, (b) 15, and (c) 20 keV (Kyser and Murata, 1974b).

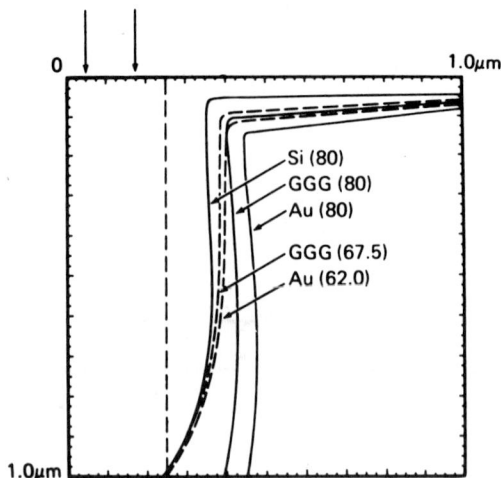

FIG. 28. The effect of substrate material on developed profiles for a 0.5 $\mu$m line in 1.0 $\mu$m PMMA films for $t = 160$ sec. An appropriate dose compensation can yield similar profiles. The dose is indicated in parentheses in ($\mu$C/cm$^2$) at 20 keV (Neureuther et al., 1979).

exposure dose by a factor of about 20%. The profile change is more significant for arrays of lines.

The increased surface etching and degradation in resolution with high atomic number substrates has been reported by Viswanathan et al. (1976) (based on MC calculations) and Todokoro (1981) (based on MC calculations and experiments with GaAs substrates).

Adesida et al. (1979) examined the substrate effect by thinning Si substrates. Their calculated results of the lateral exposure intensity distribution are shown in Fig. 29 for three different thicknesses of Si substrates. The resist thickness is 1000 Å. The spatial extent and the intensity of the backward electrons are reduced with thin substrates. In the extreme case without a substrate only the forward scattered electron spreading is important. To achieve high-resolution lithography, a thin substrate is useful. They verified this phenomenon experimentally. They also calculated the EIDs with the discrete MC model in Section II,D for energy loss and found that the model gives a larger backward contribution than that with the conventional MC model due to a larger backscattering coefficient.

## 4. Beam Energy Dependence of the Proximity Effect

As shown in Fig. 17, if the electron beam energy is too low, patterns cannot be reproduced with high fidelity due to a large forward electron scattering. On the other hand, when the beam energy increases, the electron range becomes large, the spatial extent of the backward scattered electrons increases, and the

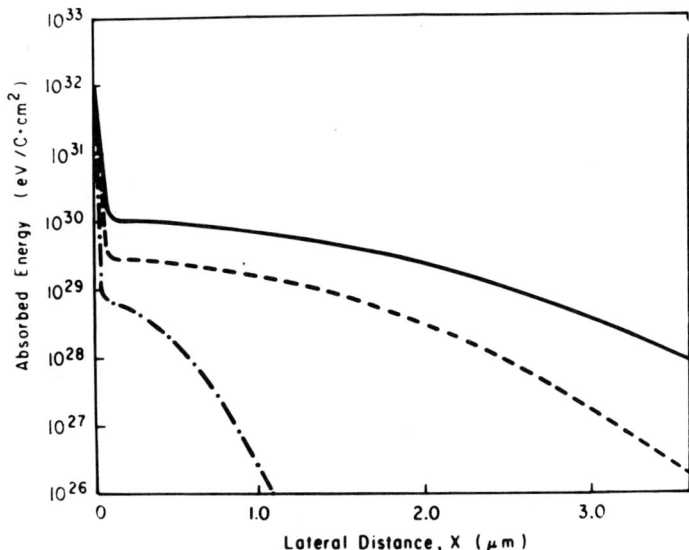

FIG. 29.    Monte Carlo results of the lateral exposure intensity distribution in a 0.1 $\mu$m PMMA resist film for substrates with various thicknesses $T_{SB}$ (= $\infty$, ——; 5400 Å, – – –; 600 Å, — · —) at 20 keV (Adesida et al., 1979). $E_0$ = 20 keV, $T_{RS}$ = 1000 Å.

proximity effect extends to large distances. The lateral EID curves obtained by MC calculations are shown in Fig. 30 at 20 and 40 kV for a 0.8 $\mu$m PMMA film on Si (Nomura, 1981). The characteristic features at both energies are specified by the numbers (1) to (4) in the figure. Each characteristic region is explained in the following.

(1) This corresponds to the electron stopping power in the vicinity of the incident point. The electron stopping power is approximately proportional to $E^{-1}$ according to the Bethe law [Eq. (20)]. Therefore, a twofold increase of energy results in the reduction of resist sensitivity by one-half. This reduction can be compensated in practice by a beam current which is proportional to the beam energy. (2) The spatial distribution extends to long distances at higher energies. (3) The background level decreases with increasing energy. A drastic estimate can be done as follows. The electron range $R_e$ is proportional to $E_0^{1.7}$ ($E_0$ is the incident energy). If it is assumed that the angular and energy distributions and the number of backscattered electrons from the substrate do not change very much, the absorbed energy density is roughly proportional to the inverse of the whole area of backscattering; i.e., $(2R_e)^{-1} = (kE_0^{1.7})^{-1}$ for a line source, where $k$ is constant. Assuming a dependence of $E_0^{-1}$ for the stopping power of the forward scattered electrons and the backward scattered electrons with an average energy of $E_0/2$, the ratio of energy density deposited

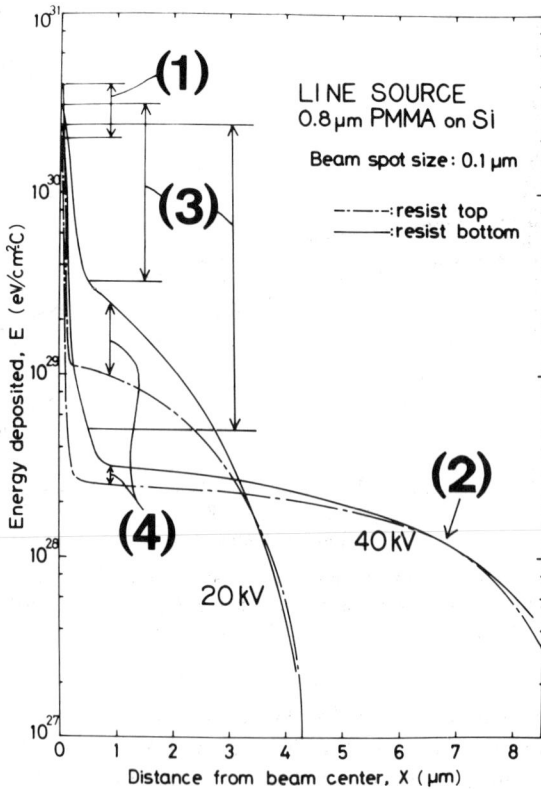

FIG. 30. Characteristic features of EID curves at low and high beam voltages for a line source, 0.8 $\mu$m PMMA on Si; beam spot size 0.1 $\mu$m. (——), resist bottom; (— · —), resist top (Nomura, 1981). (1) Difference in the stopping power right after beam incidence. (2) Large backward electron spreading. (3) Difference in the background level to the signal. (4) Difference in the background intensity between the top and bottom layers of the resist.

by the forward electrons to that by the backward electrons is proportional to $E_0^{1.7}$. An energy increase from 20 to 40 keV yields an improvement of the ratio by a factor of $(40/20)^{1.7} = 3.2$. (4) The EID curves at the top and the bottom come closer with increasing energy. This means that a vertical side-wall in developed profiles can be obtained more easily.

According to Kyser and Ting (1979), let us describe the energy dependence of intraproximity effects for line exposures with various widths (called intraline proximity effects). Their MC results of the energy density absorbed in the bottom layer of a 1 $\mu$m PMMA resist film are shown in Fig. 31 at 10 and 30 keV. The edge slope of the incident beam is 0.125 $\mu$m for each line. When the threshold energy for successful development is set to a high energy level appropriate for wider lines, then finer lines fail to be developed out. When the

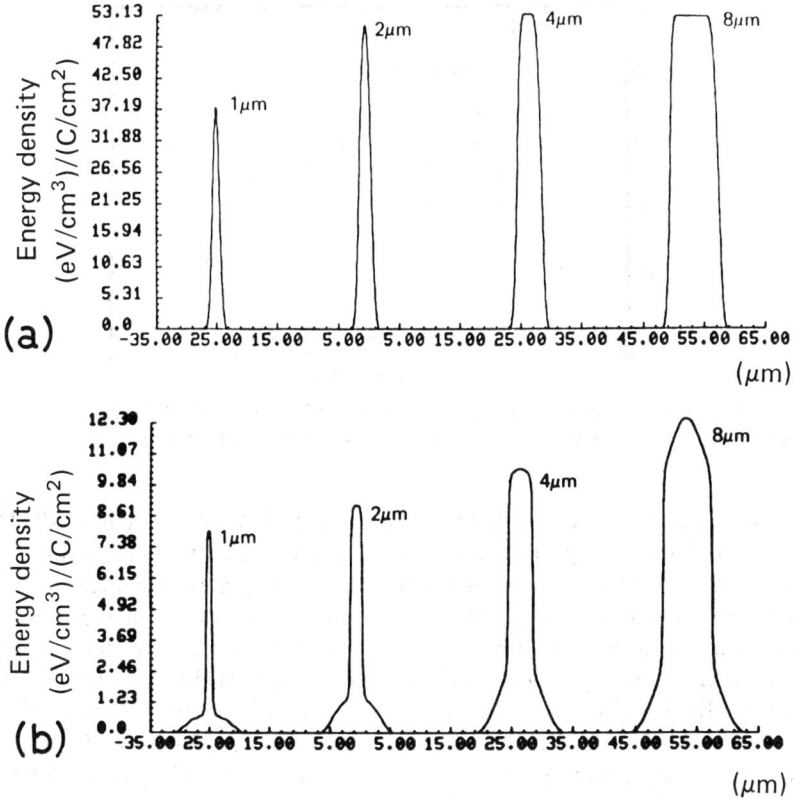

Fig. 31.    Simulated energy density absorbed in a 1.0 μm PMMA film for linewidths of 1, 2, 4, and 8 μm written at (a) 10 and (b) 30 keV (Kyser and Ting, 1979). Scale factor $= 10^{25}$.

threshold energy is set to a lower energy level, wider lines are overetched although all the lines are developed. Note that at 10 keV the top of the widest curve is flat because of the short electron range, while at higher energies the top does not reach saturation; moreover the curves have tails due to the collection of the background [characteristics (2) in Fig. 30]. To see the variation of the saturation value with line width, they calculated the maximum absorbed energy for an arbitrary line normalized by that for an infinitely wide line, $E' = E_{max}/E$. This is shown in Fig. 32 for a 1.0 μm thin resist film on Si with a parameter of beam energy. At 10 keV the $E'$ curve saturates for line widths greater than about 3 μm. As energy increases, the saturation is slow. The complete saturation requires a width more than twofold as wide as the electron range. For lines with larger widths than 2–3 μm a dose modulation is not necessary at 10 keV. At medium energies the saturation occurs at larger line widths.

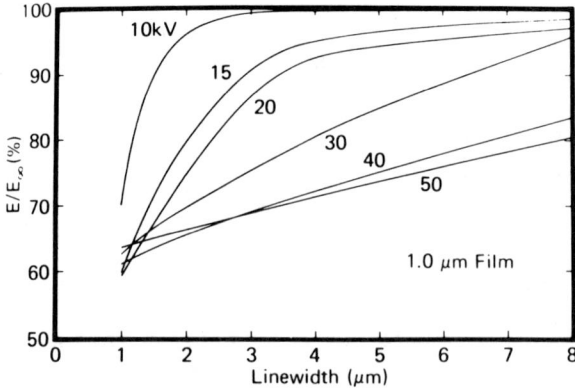

FIG. 32.   Energy density absorbed at 1.0 $\mu$m depth for a 1.0 $\mu$m PMMA film on Si versus linewidth and beam voltage, relative to a large linewidth at the same voltage (Kyser and Ting, 1979).

On the contrary, at very high energies, say 50 keV, the slope is small. This means that the background collection is gradual because the background scattered into a large area, as mentioned in the third characteristic of Fig. 30. Therefore, no exposure dose correction may be necessary over a wide range of line widths so far as single lines are concerned. This decrease of the proximity effect with increasing energy has been reported by Parikh (1980). For dense patterns the interline proximity effect has to be considered. Nomura (1981) studied this effect by introducing the resist contrast function (RCF) into exposure array patterns of unequal lines (width $w$) and spaces (width $s$). According to Broers (1981), the RCF is defined as follows:

$$RCF = (E_{max} - E_{min})/(E_{max} + E_{min}) \qquad (74)$$

where $E_{max}$ and $E_{min}$ are the maximum absorbed energy in the exposed region and the minimum absorbed energy in the unexposed region, respectively. The results for line widths of 0.5, 1.0, and 3.0 $\mu$m are plotted in Fig. 33 as a function of $w/(w + s)$, which corresponds to the ratio of the exposed area to the total area. As the value of $w/(w + s)$ increases, the line pattern becomes more dense. The results are shown at the bottom ($z = 0.75-0.80$ $\mu$m) layers of the 0.80 $\mu$m resist film. At the 3 $\mu$m width the value of RCF reversed. Namely, the RCF at 40 kV is greater than that at 20 kV in the region of $w/(w + s)$ above about 0.7. The range where the high voltage dominates expands with a decreasing width of lines. This fact shows that high voltage is more favorable for dense patterns due to characteristics (2) in Fig. 30. The reason why the RCF has larger values for smaller values of $w/(w + s)$ is that the spacing becomes large compared to the electron range and the value of $E_{min}$ decreases in the center of the spacing. In addition to the above superiority of the high voltage, the forward electron

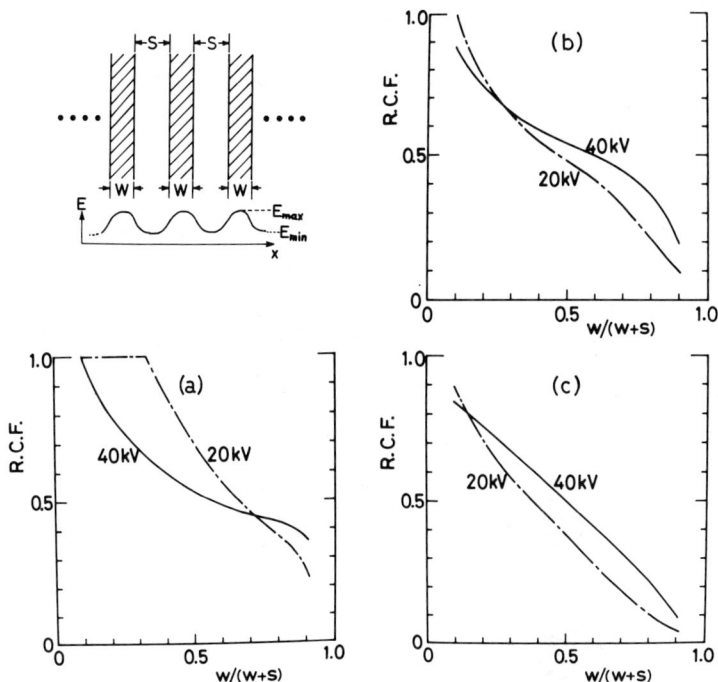

FIG. 33.   The RCF values at $w$ values of (a) 3.0, (b) 1.0, and (c) 0.5 $\mu$m are compared at 20 and 40 kV as a function of pattern occupancy $w/(w + s)$ (Nomura, 1981). $z = 0.8\ \mu$m.

spreading decreases as the beam energy increases (Nomura *et al.*, 1981). The advantage of the use of high-energy beams has been reported experimentally (Neill and Bull, 1980; Yoshimi *et al.*, 1982). For general applications, however, the pattern density may have to be taken into consideration.

## D. *Comparisons of Developed Line Profiles between Monte Carlo Calculations and Experiments*

There are many publications about etched profiles, for example, on their development time dependence and their dose dependence, etc. It will be very important for applications to lithography as to how accurate the predictions of etching profiles can be made. The reader can refer to a paper by Neureuther (1983) to learn the contribution of IC process modeling and simulation in technology design. As described in Section IV, resist modeling is established based on both the spatial distribution of the absorbed energy density and the solubility rate for resist etching. Usually the calculations are made for line electron source exposures, where the proximity effect is included because of

easy comparisons with experiments. Experimental work is performed by
development with an appropriate solvent following a single or multiple line
exposure on a resist film. After profile development, the sample is cleaved
normal to the exposed line, and its cross section is observed in a scanning
electron microscope. (See the work by Phang and Ahmed, 1979.) A typical
result is shown in Fig. 34, comparing experiment and theories. The experimental conditions are shown at the bottom of the figure. The developer is a
mixture of 1 part MIBK and 3 parts IPA. The development time is fixed to
1 min. The dose is varied from $1.0 \times 10^{-4}$ to $1.0 \times 10^{-3}$ $C/cm^2$. The energy
dissipation is obtained by MC calculations based on the single scattering
model. Since the developer is a weak solvent, the threshold model is usually
applicable. However, the development process may be important for a thick
resist film. Thus, they investigated developed profiles with both threshold and

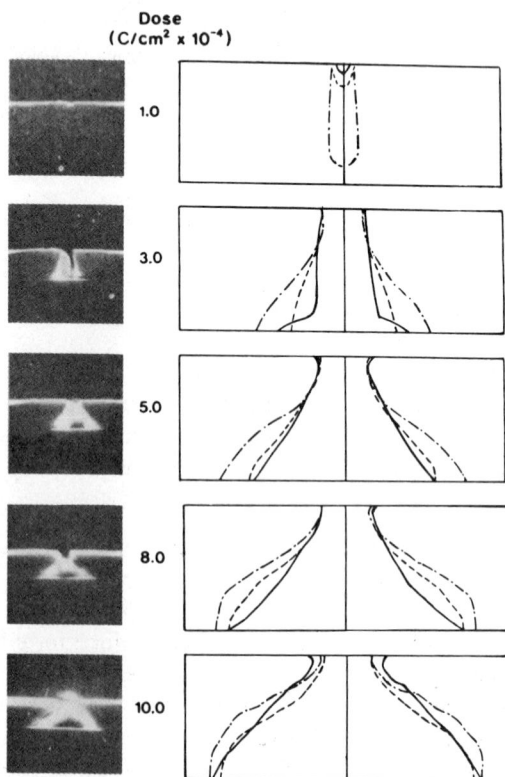

FIG. 34.   Developed profiles of isolated line exposures for a PMMA resist film of 2.05 $\mu$m
thickness on Si. Experimental profiles (——) are compared with the two theoretical profiles, the
threshold (— · —) and the development (– – –) models (Phang and Ahmed, 1979). Beam energy,
15 keV; beam diameter, 0.5 $\mu$m.

development models. The experimental profiles shown by a solid line show better agreement with those obtained with the development model rather than with the threshold model. Down to one-third of the film thickness, both theories predict the experimental profiles well. They have done experiments on two adjacent line exposures and obtained a similar result, confirming that the threshold model can give good results for a thin resist film of 0.7 $\mu$m.

Next, let us see the work with a 1:1 solution of MIBK and IPA which has a medium etching rate between a strong developer MIBK and a weak developer MIBK:IPA = 1:3. In Fig. 35 are shown the results of Neureuther *et al.* (1979) at doses of 60, 80, and 100 $\mu$C/cm$^2$. The resist thickness is 1.0 $\mu$m. The line width is a quarter micrometer. The development time is 180 sec and simulated profiles are shown at every 20 sec. The parameters used in the solubility rate equation are determined by large areal exposure experiments. They first tried to compare experiment and theory through the use of an estimated beam size of 0.1 $\mu$m diameter, but were not successful. The beam size was changed to about 0.25 $\mu$m diameter in order to match top and bottom rounding and sidewall slope of the profiles. Then, quite good agreement in the dose

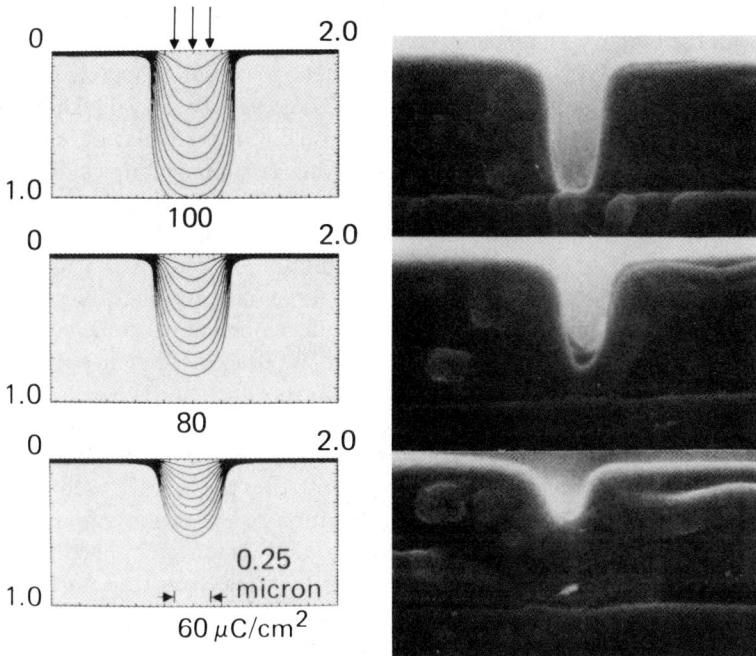

FIG. 35.    Simulated and observed line edge profiles for a 0.25 $\mu$m line in 1.0 $\mu$m film at doses of 60, 80, and 100 $\mu$C/cm$^2$. Simulated contours are shown at 20 sec intervals. The final contour is at 180 sec (Neureuther *et al.*, 1979).

dependence of etching speed is obtained. They have also written a larger line of 0.5 μm width and multilines with a spacing of 0.5 μm. General agreement is obtained in shape and dose dependence. Better agreement is obtained for larger lines. However, discrepancies are seen between both results. For example, the simulated profiles show a rounder etching front, a sharper edge roundness at the top surface, and a somewhat broader shape.

Murata et al. (1979) have also done a similar line exposure experiment by using the same developer as Neureuther et al. (1979), but with varying development times. A typical result from cross sections of developed patterns at 20 keV is shown in Fig. 36 at a dose of $2.4 \times 10^{-3}$ C/cm$^2$. The areal charge density in the scan field is used as the dose of a single line exposure. The profiles are shown at every 15 sec. In the theoretical results the sequential profiles are shown at each development time. In this work the Spencer–Fano formula is incorporated for energy loss calculations in the MC program, which gives smaller energy loss near the surface. This results in a slow development at the initial stage. The beam size is 5500 Å in diameter, which is measured from a line spectrum of the secondary electron signal when the beam is scanned over a fine gold wire. The present beam size is much larger than that used in the experiment by Neureuther et al. (1979). This might be an important difference between the two experiments.

Murata et al. (1979) used the same parameter in the solubility rate equation [Eq. (69)] as those of Neureuther et al. (1979) except for the value of $S_2$. The value is adjusted to fit the experimental profile at a particular development time, then the determined value is used for other development times. The value of $S_2$ newly determined is about 0.5 Å/sec. They have also performed experiments with single, three, and six line exposures at a dose of $1.2 \times 10^{-3}$ C/cm$^2$. The line spacing is 5000 Å for multiline exposures. They found $S_2 = 0.23$, 0.81 and 0.89 Å/sec for single, three, and six lines, respectively. The $S_2$ values should coincide since the development is done under the same experimental conditions for all samples. It seems that it approaches the value of $S_2 = 1.0$ Å/sec for a large areal exposure.

Neureuther et al. (1979) investigated changes in profile shape by varying parameters such as solubility rate constants and beam diameter, but did not find significant improvement in the agreement. After this work, Rosenfield and Neureuther (1981) investigated a generalization of the etching model. They modified the string model by increasing the vector component of the development rate along the direction of primary electron trajectories. This idea is based on the physical mechanism that the exposure of high-energy electrons generates volatile products and micropores, and they promote etching along their trajectories. This preferential etching leads to improvements in the agreement between the simulated and experimental profiles. However, the extent of the preferential etching is probably related to exposure

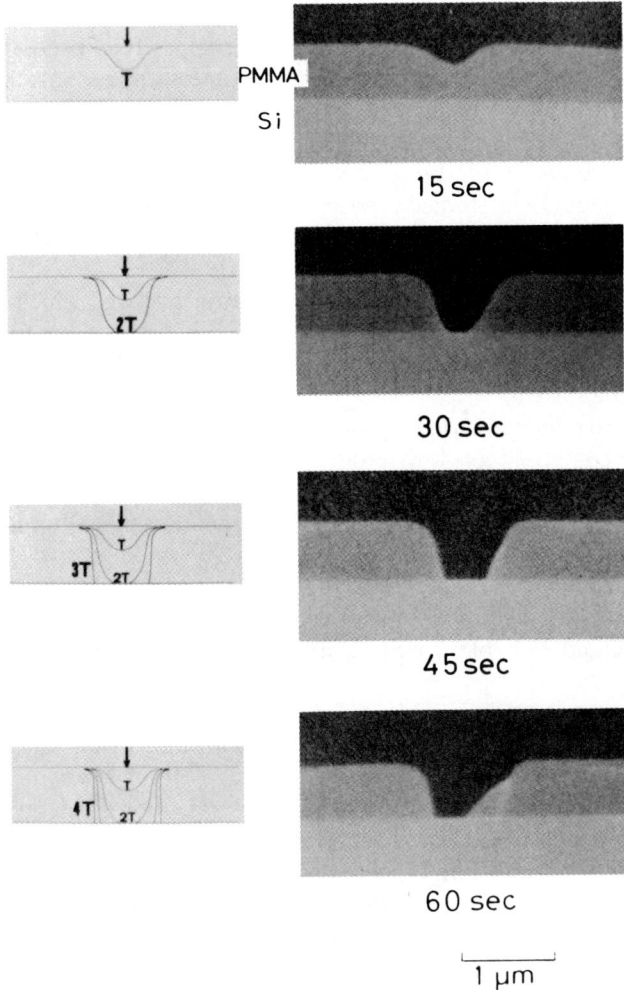

FIG. 36.    The time evolution of cross-sectional profiles for a single line exposure at a dose of $2.4 \times 10^{-3}$ C/cm². $T = 15$ sec, $S_2 = 0.5$ Å/sec (Murata *et al.*, 1979).

dose and resist material, etc. Its dependence is not given for arbitrary patterns. Further studies are required for general applications to process technology.

### E.  Exploratory Studies

As already mentioned, electron scattering influences the developed resist pattern profile through various parameters such as dose, development time, resist thickness, resist material, substrate material, beam size, beam energy, etc.

The appropriate control of pattern profiles is important to implement the optimum fabrication process, e.g., undercut profiles for lift-off techniques (Hatzakis, 1969). MC simulations can be conducted under various process conditions, realistic or imaginary, to understand their effects on developed patterns and to utilize the results for pattern correction. Many papers have been published along this direction.

Nakata et al. (1978) studied exposure adjustment for resist pattern defining a source drain region of an FET device with MC calculations of electron scattering.

Kyser and Pyle (1980) developed a software package which simulates processes of electron scattering with an MC method and subsequent development. They investigated proximity effects from various aspects based on this modeling system.

Todokoro (1980) reported a profile simulation in double-layer resist films which are composed of low-sensitivity resists on high-sensitivity resists (LO/HI) or the reverse combination. The technique is useful in the lift-off process for the submicrometer regime. Experimental verification is given for the simulated results.

Rosenfield and Neureuther (1981) and Rosenfield et al. (1981) utilized MC calculations both to study the use of exposure pattern bias in electron-beam lithography and to explore writing strategies such as local initial thinning techniques.

Chang et al. (1981) made MC calculations to estimate electron scattering effects on the fundamental limitation for both resolution and linewidth control.

Paraszczak et al. (1983) studied the spatial distribution of energy deposited with MC simulations and experiments in a sample of bilayer structure where a thin polysiloxane resist film is coated on top of thick AZ1350J photoresists. This thin negative resist has high etching resistance for an oxygen plasma. Both results have shown good agreement.

Proximity effects in shaped electron-beam lithography have been examined by Okubo and Takamoto (1983), Augur et al. (1985), and Chen et al. (1985) based on MC calculations. Okubo and Takamoto (1983) discuss pattern fidelity by introducing several characteristic parameters such as energy efficiency, which represents how effectively exposed energy is utilized within a designed pattern, energy density slope, roundness at the pattern edge corner, and energy ratio, which measures the extent of the interproximity effect between two adjacent rectangular patterns. They evaluated those parameters as a function of beam energy or beam width and found desirable experimental conditions of beam edge width $< 0.2 \ \mu$m in the regime of 0.5 $\mu$m lithography. Augur et al. (1985) also show the advantage of the use of high beam energy to obtain better pattern fidelity. This paper tries to understand characteristics of energy deposition in a resist and to explore correction schemes for the proximity effect.

Hasegawa and Iida (1985) explored the effects of heavy atoms such as Sn and Zr doped into resists for dry-etch resistance enhancement on lithography resolution. The edge slope in the lateral distribution of the absorbed energy density for a rectangular pattern showed a clear degradation of resolution with increasing atomic number of the dopant. Similar MC applications can be studied to examine the effects of metals doped into x-ray resists to enhance the resist sensitivity in the future.

## F.  Resolution Limit

Many experimental attempts have been made to pursue ultra-high-resolution lithography. The results are summarized in work by Kyser (1982, 1983), including the experimental conditions. As can be seen in Table VI, where some recent works are added, a resolution of about a few hundred angstroms is obtained, except for special cases such as exposures on inorganic resists. At present a probe diameter of 5 Å can be obtained at high electron energies with sufficient beam current to expose a resist. Such high-energy electrons have less forward scattering spreading. In addition, a thin resist film and a thin substrate are used in these experiments. It seems that the primary electron almost penetrates straight, and the spread of energy dissipation is caused by the generated secondary electrons. Therefore, theoretical investigations of resolution have to account for these electrons. A hybrid model of MC simulation developed by Murata et al. (1981) is useful for this subject. The calculated trajectories in a 0.4 $\mu$m PMMA film (without substrate) through use of the knock-on model are shown separately for the primary and secondary electrons in Fig. 37. The incident electron energy is 20 keV. The forward electron scattering spreading seems to be significant due to a relatively large film thickness. The value of $\varepsilon_c$ [Eq. (26)] is 0.001. Then, the energy of secondary electrons generated by 20 keV electrons is distributed between 20 eV and 10 keV, neglecting the energy loss in the thin film. As is seen from the trajectories for secondaries, secondary electrons generated with lower energy move nearly perpendicular to the direction of the primaries. On the other hand, high-energy secondaries tend to move forward upon generation. Although high-energy secondaries have a long electron range, only a few are generated, while low-energy secondaries have a short range although they have a large generation rate. Consequently, medium-energy electrons of a few hundred eV to 1 keV may have an important influence on resolution, having a corresponding range of several tens of Å to 440 Å. These secondary electrons give an additional spreading to the forward spreading. The difference between

## TABLE VI
### Literature References on High-Resolution Electron-Beam Lithography[a]

| Reference | Target structure | Electron-beam parameters | Development parameters | Linewidth resolution | Comments |
|---|---|---|---|---|---|
| Sedgwick et al. (1972) | Thin PMMA on 1500 Å $Si_3N_4$ | 25 kV SEM $1.2 \times 10^{-10}$ C/cm | — | 600 Å single lines | 600 Å Al lift-off |
| Broers et al. (1976) | 100 Å PdAu on 100 Å carbon | 45 kV STEM 5 Å beam $1 \times 10^{-6}$ C/cm | Polymerized contamination | 80 Å | PdAu ion etched |
| Broers et al. (1978) | 1100 Å PMMA/ 225 Å PdAu on 600 Å $Si_3N_4$ | 56 kV STEM 10 Å beam 500 $\mu$C/cm$^2$ | 15, 45 sec 1:3 (MIBK:IPA) | 250 Å lines/gaps | |
| Howard et al. (1980) | 1500 Å PMMA/ 4000 Å P(MMA/MAA) on thick Si | 30 kV SEM $2.4 \times 10^{-9}$ C/cm | 5–10 sec 1:2 (cell:meth) | 400 Å lines/gaps | 300 Å Au lift-off |
| Broers (1981) | 300 Å PMMA on 600 Å $Si_3N_4$ | 50 kV STEM 10 Å beam 300 $\mu$C/cm$^2$ | 50 sec 1:2 (MIBK:IPA) ** = 4 Å/sec | 225 Å lines/gaps | 50 Å AuPd shadow coatings |
| Beaumont et al. (1981) | 600 Å PMMA on 300 Å carbon | 50 kV SEM 80 Å beam | 30–45 sec 1:3 (MIBK:IPA) | 160 Å lines 370 Å gaps | PtPd lift-off |
| Isaacson and Murray (1981) | 300 Å NaCl on 100 Å carbon | 100 kV STEM 5 Å beam | | 15 Å lines 30 Å gaps | In situ develop |
| Lee and Ahmed (1981) | 500–2000 Å x-link/1000– 3000 Å PMMA | 50 kV STEM 10 Å beam 1000 $\mu$C/cm$^2$ | 10 sec pure MIBK + 45 sec in 1:3 on 1000 Å $Si_3N_4$ | 150 Å lines 500 Å gaps | Resist only |
| Howard et al. (1983) | 500 Å PMMA on Si or GaAs | 20–120 kV STEM (0.4–2 nC/cm) 20 Å beam | 3:7 (ethylene glycol monoethyl ether: methanol | 100 Å | Resist or AuPd lift-off |
| Emoto et al. (1985) | 2300 Å PMMA on thick Si | 50 kV EBL system 25 Å beam 1.05 nC/cm | 30 sec 1:3 (MIBK:IPA) | 80 Å 1000 Å period | Resist only |

[a] Kyser, 1982.

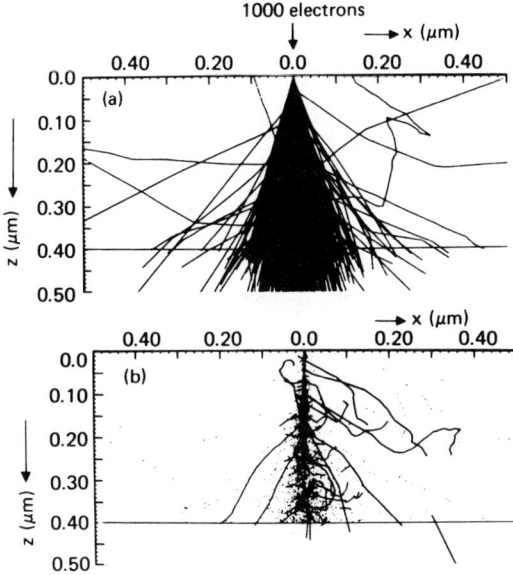

FIG. 37. Trajectories of (a) the primary and (b) the secondary electrons in an isolated 4000 Å thin PMMA film with the hybrid model. There are 1000 electrons of 20 keV incident at the origin (Murata *et al.*, 1981).

the models with and without secondary electron generation is clearly seen by drawing the equienergy contours. The results are shown in Fig. 38. The old model or the conventional single scattering model predicts a conic shape starting with the surface, which represents the depth variation of the forward spreading. The new round contours show a significant discrepancy from the

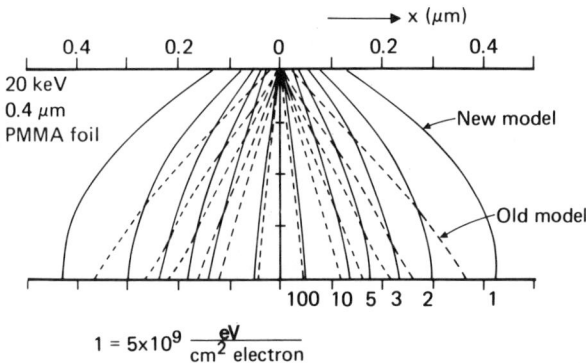

FIG. 38. Equienergy density contours for an isolated 4000 Å thin PMMA film at 20 keV. The results with the new (hybrid) and old (conventional) models are compared (Murata *et al.*, 1981).

old ones, especially near the resist surface. If one of the equienergy contours reaches the threshold energy, the resist develops out along the contours. Kyser (1983) reported the linewidths of latent images with various absorbed energy density for various electron energies. His results for a 500 Å film are shown in Table VII. Since he used a cell size of 20 Å, the minimum full width is 40 Å. This is the reason why the values at the top line in the table have a value of 40 Å for energies larger than 25 keV. The contour with an absorbed energy of $1 \times 10^{30}$ eV/C cm$^2$ will reach the threshold energy of $D_c = 1 \times 10^{22}$ eV/cm$^3$ at a dose of $1 \times 10^{-8}$ C/cm. Then, the linewidths are 300, 200, 100, and 80 Å at 25, 50, 75, and 100 keV, respectively. These values are very similar to those given in Table VI, although the reason to adopt the energy of $1 \times 10^{30}$ eV/C cm$^2$ requires further discussion, including the development process.

The model by Samoto *et al.* (1983) emphasizes low-energy secondaries. They calculated the variation of the fractional exposure as a function of line width and compared it with the experimental data of Broers (1981). The result showed that the calculation underestimated the experiment. The accuracies of both models have to be checked in future work. It should be pointed out here that the finite size of resist molecules must be taken into account in resolution problems, as discussed in Section IV,A.

## G. Negative Resist Pattern Analysis

Negative electron resists are important in practical applications to VLSI fabrication because they have high sensitivity, which leads to high throughput. Only a few papers, however, have been published on the quantitative evaluation of developed resist patterns. This probably comes from the peculiar negative resist characteristics of swelling and shrinkage during development and rinse processes. For further applications more studies have to be

TABLE VII

LINEWIDTHS OF LATENT IMAGE FROM EQUI-ENERGY DENSITY CONTOURS AT VARIOUS ENERGIES FOR A 500 Å FILM[a]

| $E$ (eV/C cm$^2$) | $W$(Å) | | | |
|---|---|---|---|---|
| | 25 keV | 50 keV | 75 keV | 100 keV |
| $1 \times 10^{31}$ | 80 | 40 | 40 | 40 |
| $1 \times 10^{30}$ | 300 | 200 | 100 | 80 |
| $5 \times 10^{29}$ | 500 | 300 | 250 | 200 |
| $2 \times 10^{29}$ | 900 | 600 | 550 | 500 |

[a] Kyser, 1982.

conducted on resist characteristics. Heidenreich *et al.* (1975) and Lin (1975) treated the problem by using a Gaussian dose distribution for a single line scan. Although the lateral electron scattering has been taken into consideration to some extent, it is desirable to include a more accurate distribution of the absorbed energy density in the resist. In this section we discuss an example of work on negative resist patterns which utilize MC calculations (Nakata *et al.*, 1981).

They performed electron exposure experiments on PGMA resist films of 0.6 μm thickness on a Si wafer at 10 keV. The patterns are infinite arrays of equal lines and spaces with 1, 2, 3, and 10 μm widths. The exposure is done with a beam diameter of 0.25 μm and a beam address spacing of 0.25 μm. The relation between the remaining thickness and dose is obtained from the largest areal exposures. These results were already shown in Fig. 13. As described in Section IV, analyses were carried out based on both this contrast curve and the calculated spatial distribution of the absorbed energy density with MC simulations.

The calculated results are shown in Fig. 39 for 1 and 3 μm line patterns. Only the right half is shown due to symmetry. The upper figures are the equienergy contours and the lower ones are developed profiles. The experimental profiles are shown by solid lines. The beam address points are shown by arrows. We can see from comparisons between calculated results that the isocontours for a 3.0 μm line shift to the surface, namely more energy is deposited into the 3.0 μm line due to the intraproximity effect. The equienergy contours have a slope at the pattern edges due to both forward and backward

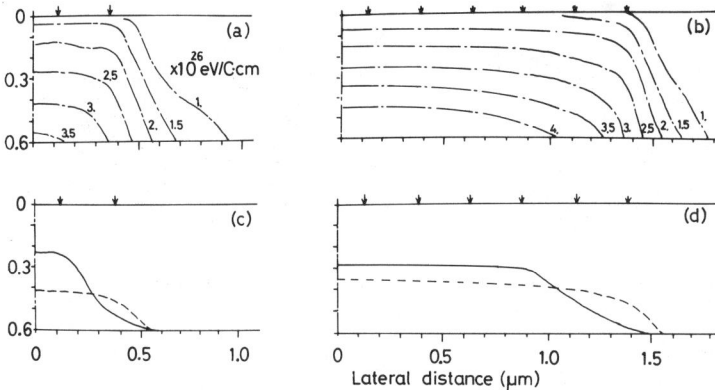

FIG. 39.   (a), (b) The equi-energy density contours and (c), (d) developed cross-sectional profiles for negative resist patterns of lines and spaces. The widths of line and space are: (a), (c) 1 μm and (b), (d) 3 μm. The initial beam energy is 10 keV. (———), experiment, (– – –), MC calculation. Profiles are compared at a dose of $4.0 \times 10^{-7}$ C/cm$^2$ (Nakata *et al.*, 1981; Murata, 1982).

electron scattering. The theoretical developed profile (dashed line) has a slope at the pattern edge, reflecting the slope of the equienergy contours, and the remaining thickness of the 3 $\mu$m line is larger than that of the 1 $\mu$m line. However, the experimental profiles deviate significantly from the calculated ones, as anticipated from previous work (Heidenreich *et al.*, 1975; Lin, 1975). This discrepancy in the shape suggests that the developed pattern shrinks in the horizontal direction, being stuck to the Si surface by stronger adhesion, and swells in the vertical direction. This change keeps the area of the cross section about constant, which is, in other words, the resist volume.

These assumptions can be clearly seen from another experiment which gives three times more dose at the edge as others. The results are shown in Fig. 40. Very large energy absorption occurs at the edge by the extra dose. More energy is deposited with a 1 $\mu$m line than a 3.0 $\mu$m line in the vicinity of the edge because the extra dose at the other side is given in a shorter distance. Both experimental and calculated results have a high peak near the pattern edge. We can easily infer the manner in which the hill moves by the swelling and shrinkage. Another interesting point is that the deformation becomes small as the dose increases. This is because resist material with a larger gel fraction is hard to deform.

Recently, Suzuki *et al.* (1984) reported theoretical and experimental studies on negative resist profiles in x-ray lithography. They modified the negative resist modeling by incorporating the resist deformation under the assumption that the two-dimensional shrinkage occurs isotropically. The agreement is good between experiment and theory. This modeling can be also applied to the present problem.

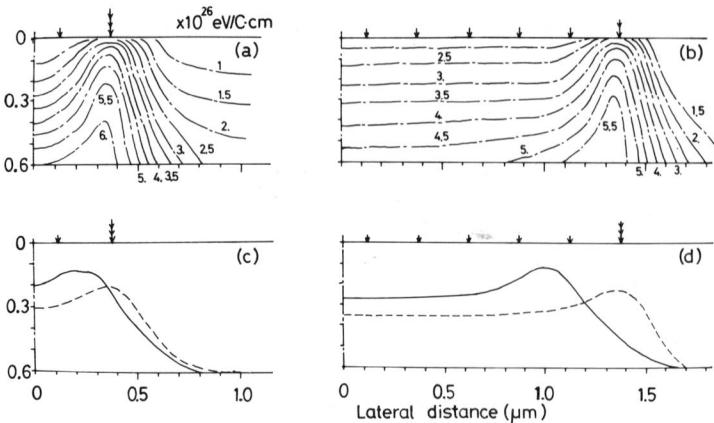

FIG. 40.    (a), (b) The equi-energy density contours and (c), (d) developed cross-sectional profiles for the same patterns as in Fig. 39 except with extra doses at the edges (Murata, 1982).

## VI. Applications to X-Ray Lithography

Conventional photolithography replication techniques cannot respond to future submicrometer lithography requirements because of the diffraction limit. One promising technique is x-ray lithography, although it still has some problems to be solved such as mask alignment, mask material, and an x-ray source with high brightness (for example, Wittels, 1980; Neureuther, 1980; Trotel and Fay, 1980; Watts and Maldonado, 1982; Triplett and Hollman, 1983; and Shimukunas, 1984). Pattern analysis is also one of the important problems. As described in the MC modeling, incident x rays produce photo- and Auger electrons which sensitize a resist. The spreading of these electrons determines the ultimate resolution in x-ray lithography. However, in practical applications the Fresnel diffraction effect and the penumbral effect cannot be neglected with a finite mask–wafer gap and with a finite source size, respectively. Therefore, various compromises are necessary. For example, x rays with shorter wavelengths degrade the resolution, although the Fresnel diffraction effect decreases, and the desired resolution decides the distance between a wafer and the x-ray source based on the penumbral effect, which is associated with the device throughput. It is seen from the above fact that generated electrons play an important role for implementing an efficient x-ray exposure.

### A. Trajectory Plotting of Generated Electrons

In Fig. 41 simulated trajectories of generated electrons by incident x rays are plotted on a two-dimensional chart (Murata *et al.*, 1985). The result is calculated for a synchrotron radiation source with an energy of 1.0 GeV, as shown in Fig. 11. Photons numbering 50,000 in the wavelength range of 0–20 nm are exposed on a 200 nm PMMA film through a 3.0 $\mu$m mask substrate, where small-energy photons are mostly absorbed. As can be seen from the figure, electrons with short traveling paths are concentrated along the incident axis, which are mainly Auger electrons. Electrons with various traveling paths, relatively long, extend to both sides from the axis, which are mainly photoelectrons with various energies. Note that the variation of path length is also caused by the projection on the two-dimensional chart. In this example the photons with energies of 0.6–2 keV are mainly absorbed in the resist. Most of the electrons are within a lateral range of 25 nm. We should point out here that the fast secondary electron emission by high-energy electrons is analogous to the present electron emission phenomenon, as previously stated.

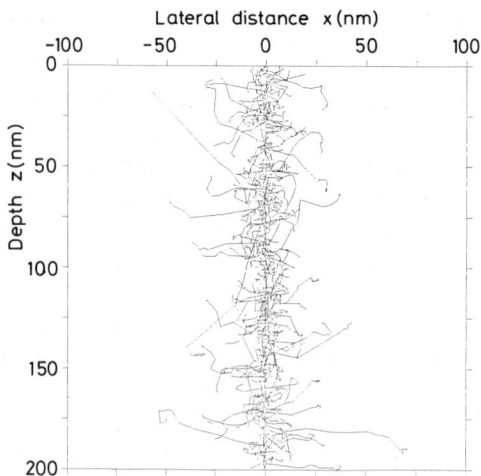

FIG. 41. Simulated trajectories of photo- and Auger electrons generated in a 0.2 μm PMMA film by an irradiation of synchrotron radiation of 1 GeV. 50,000 photons between the wavelength range of 0–20 nm are incident on a Si mask substrate (Murata *et al.*, 1985).

## B. Lateral Distributions of Energy Absorption

One of the most interesting physical quantities is the lateral spreading of electron scattering. The resolution can be evaluated from this scattering. Qualitative observation can be made by trajectory plotting, as has been shown in Fig. 41. Let us examine the lateral distribution of the absorbed energy density for a 1.0 μm PMMA thin film without substrate. The MC results are shown in Fig. 42 for a line source of synchrotron radiation and monochromatic x rays of Cu $L_\alpha$, Al $K_\alpha$, and Mo $L_\alpha$ lines (Murata, 1985b). They are obtained by averaging the energy density over the depth. As seen from the figure, there is a clear difference in the curve shape between SOR and monochromatic x rays. In the curves for three characteristic lines there are two features: a sharp peak in the central region and a Gaussian-like background with low intensity. Auger electrons with smaller energies contribute mostly to the former peak, and photoelectrons with larger energies produce the latter background. The peak distribution does not change very much because the Auger electron energy is fixed, but the background distribution does change with a photoelectron energy which is determined by the characteristic x-ray energies. Namely, the Mo $L_\alpha$ line (2.29 keV) shows the largest spreading.

On the other hand, the SOR yields a monotonically decreasing curve. This seems to be caused by incident photons with continuously distributed energy. We can see a broader distribution at a higher electron orbital energy of 1.3 GeV because more photons are incident with higher energies.

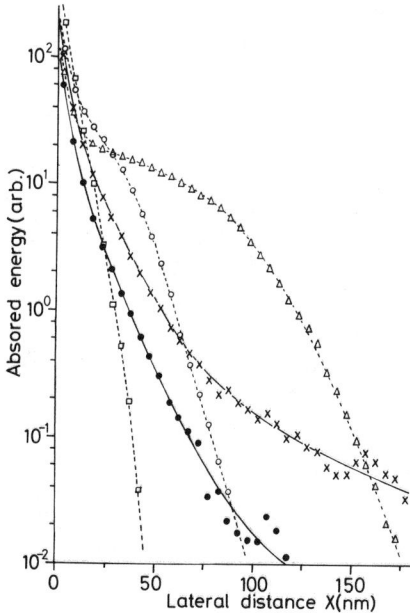

FIG. 42. The lateral distributions of absorbed energy for various x-ray sources (Murata, 1985b). ( ×), 1.3 GeV; (●), 1.0 GeV; (△), Mo $L_\alpha$, 2.3 keV; (○), Al $K_\alpha$, 1.5 keV; (□), Cu $L_\alpha$, 0.93 keV.

The use of such x-ray sources causes a proximity effect within a lateral distance of about 150 nm, depending on the x-ray energy. The resolution of x-ray lithography is not clear at this stage since development processes are concerned. A relative comparison can be made among them. Taking the full width where the energy decreases from the peak in the center to $\frac{1}{10}$, they are 22, 48, 92, 24, and 28 nm for Cu $L_\alpha$, Al $K_\alpha$, Mo $L_\alpha$, and SORs at 1.0 and 1.3 GeV, respectively.

The reliability of these calculations has been investigated by Murata et al. (1985) in comparison with experimental data for the electron range by Rishton et al. (1983). In the present study an accurate description of electron behavior is necessary in the low-energy range of a few hundred eV to about 3 keV. The comparison shows that the accuracy is within about 20%.

## C. Substrate Effect on Energy Absorption

Only a part of the incident x rays are absorbed in a resist, about 10% for Al $K_\alpha$ x rays in a 1.0 μm PMMA film. Most of them are absorbed in the substrate. Photoelectrons and Auger electrons generated in the substrate are scattered back and deposit a large amount of energy in the resist near the interface. When the photon energy is large, they can spread into a wide region. This phenomenon corresponds to the proximity effect caused by electrons backscattered from a substrate. MC calculations can reveal this effect easily.

A typical result for a 1.0 μm PMMA resist is shown in Fig. 43 with SOR at 1.3 GeV (Murata *et al.*, 1985). The equienergy contours with Si and Au substrates (solid lines) are compared with those for an isolated resist film (dashed lines). Only the region up to 0.15 μm from the bottom and the right half are shown in the figure. In the upper region the effect is small. In this case many high-energy photons are incident on the substrate, which can hardly be absorbed in the resist. The contours without substrates show a large reduction near the interface due to escaped electrons. With the Si substrate this reduction is almost compensated, although it is a little excessive. With the Au substrate, however, the backscattered electron effect is too excessive.

Developed profiles are shown in Fig. 44 for the above cases with a solvent of MIBK (Murata *et al.*, 1985). With the Au substrate the profile has a relatively strong undercut. This excessive dose near the interface can be seen in a different way by plotting the energy as a function of depth, as shown in Fig. 45 (Murata *et al.*, 1985). The results are shown for three different cases of PMMA films without and with Si and Au substrates. The curves have a gradual slope with depth because of the absorption of especially low-energy photons. The decrease at the top surface is due to escaped electrons. The peaking near the interface is due to the backscattering effect. We can see that

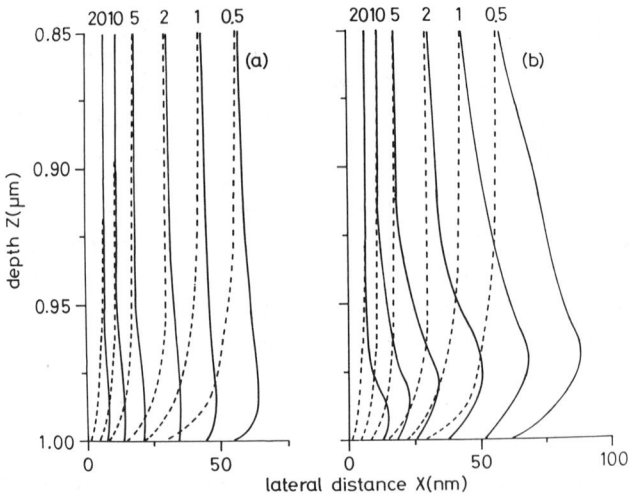

FIG. 43. Substrate effect on absorbed energy contours in a 1.0 μm PMMA resist film for a SOR x-ray source. (———), the contours with (a) Si and (b) Au substrates. (– – –), the contours without substrates. Only the contours are illustrated near the interface (Murata *et al.*, 1985).

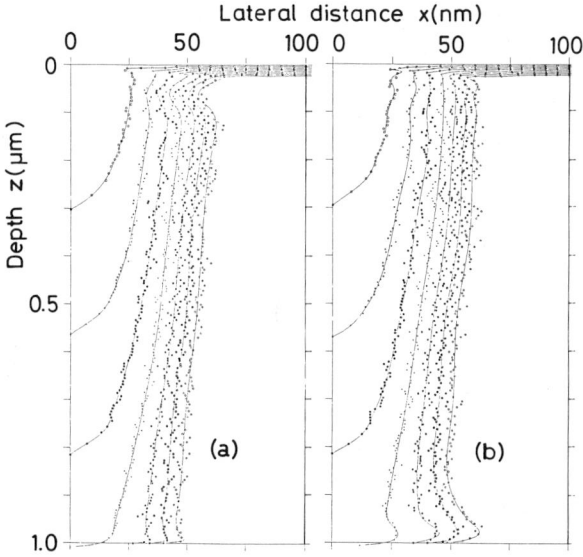

FIG. 44.   Simulation of developed profiles for the same conditions as in Fig. 43, 1.3 GeV, with substrates of (a) Si and (b) Au (Murata *et al.*, 1985).

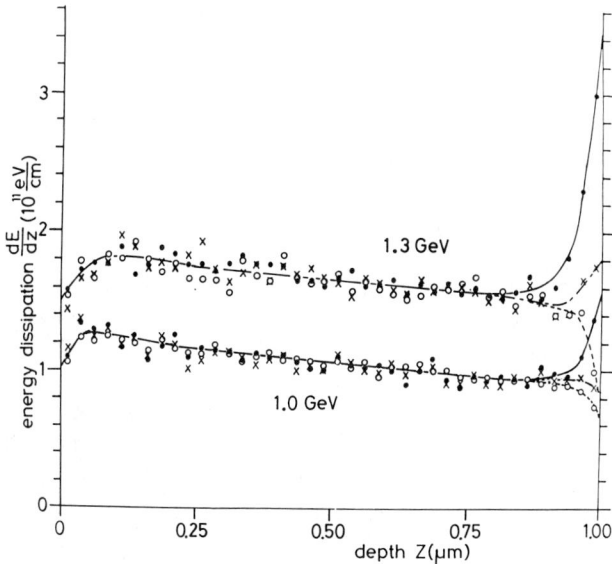

FIG. 45.   Energy dissipation in a 1.0 μm PMMA resist film as a function of depth with an SOR source of 1.0 and 1.3 GeV. Note large differences in energy dissipation near the interface among various substrates (Murata *et al.*, 1985). (○), (− − −), Isolated film; (×), (— · —), Si substrate; (●), (——), Au substrate.

about twice the energy deposited by the primary x rays is deposited at the interface with the Au substrate. This effect is demonstrated by Semenzato *et al.* (1985) for the tungsten $M_\alpha$ line with bremsstrahlung background. The effect might be a serious problem in the practice of pattern replication because a relatively large dose is deposited near the interface even in the masked area, owing to a finite mask contrast.

### D. *Other Effects on Developed Profiles*

Studies of the ultimate resolution can be done with MC simulations. The results with solvents of MIBK and MIBK:IPA = 1:3 have been reported by Murata (1985a) for Al $K_\alpha$, including the mask contrast effect. Semenzato *et al.* (1985) have performed MC simulations to investigate etching profiles under various experimental conditions such as x-ray source, source size (penumbral effect), and mask pattern shape. In order to establish an efficient exposure system, further studies are required.

### VII. APPLICATIONS TO REGISTRATION MARK DETECTION IN ELECTRON-BEAM LITHOGRAPHY AND MASK INSPECTION TECHNOLOGY

Accurate detection of pattern registration marks plays an important role in a direct writing electron-beam machine. If patterns are desired to be written with a line resolution of 0.1 $\mu$m, an accuracy of 0.02 $\mu$m seems to be required for alignment (Lin *et al.*, 1981). For this purpose a scanning fine electron beam can be used in the same manner as in scanning electron microscopy, where backscattered electrons are detected as a signal. This high accuracy to find a position can also be used for x-ray mask inspection. The inspection of a mask with a high pattern density of submicrometer features cannot be achieved by visible light because of low resolution. A general scheme to detect signals from the registration mark or the mask pattern is shown in Fig. 46, where the quartered annular semiconductor detector is demonstrated. The signal spectra are processed to find an accurate position. The advantage of the backscattered electron signal is the following. Backscattered electrons are energetic enough to transmit signals through a coated resist film when the primary electrons produce them by hitting the registration mark. For mask inspection, backscattered electrons can give a large signal difference between the opaque region (usually heavy material is used to gain high x-ray absorption) and the transparent region (light material for a mask substrate is used for low x-ray absorption).

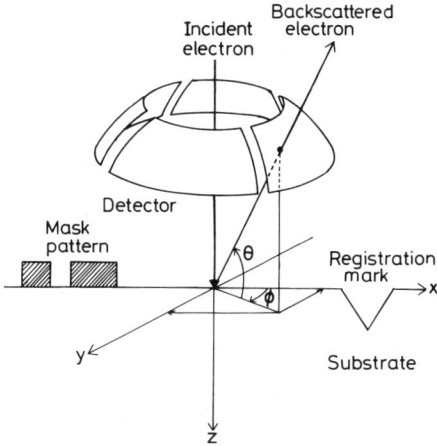

FIG. 46.    Schematic diagram of both registration mark detection and mask inspection.

The accuracy with which we can identify the position is determined by the signal dependence on the incident beam energy, the resist thickness, the shape of the registration mark, and the atomic number of the mark. It also depends on the azimuthal and zenith angles of detection. Investigations of these problems can be done by MC calculations including a special geometry for registration marks or mask patterns. The single scattering model based on the continuous slowing-down approximation has been applied to the following subject.

### A.  Registration Mark Detection

Aizaki (1978) studied the total signals of backscattered electrons from steps, and grooves with various depths. Lin *et al.* (1980, 1982) reported the signal formation from 54.6° tapered silicon steps. They checked the quality of the signal by varying the incident electron energy and found that the contrast increased with increasing step depth and saturated at a step depth of 40% of the Bethe range, which is associated with the maximum penetration depth of the backscattered electrons. Here the contrast is defined as the maximum mark signal difference normalized by the background signal. They also studied the effect of the energy distribution of backscattered electrons. The semiconductor detector has an energy dependence of the sensitivity because the carrier generation rate increases with incident energy. Namely, high-energy electrons have a higher sensitivity. Their conclusion is that the energy-weighted signals of backscattered electrons give a better quality than the number-weighted signals.

Their research was extended to resist-coated silicon steps. Examples of trajectory plotting are shown in Fig. 47, calculated by Lin *et al.* (1981). The incident energy is 20 keV. Figure 47a is a case in which the resist coating keeps the same topography as the silicon tapered step. Figure 47b is a case in which the step is made planar by the resist coating. A general resist-coated configuration is between these two extreme cases. The calculated backscattering yield for various conditions is shown in Fig. 48 (Lin *et al.*, 1981) when the

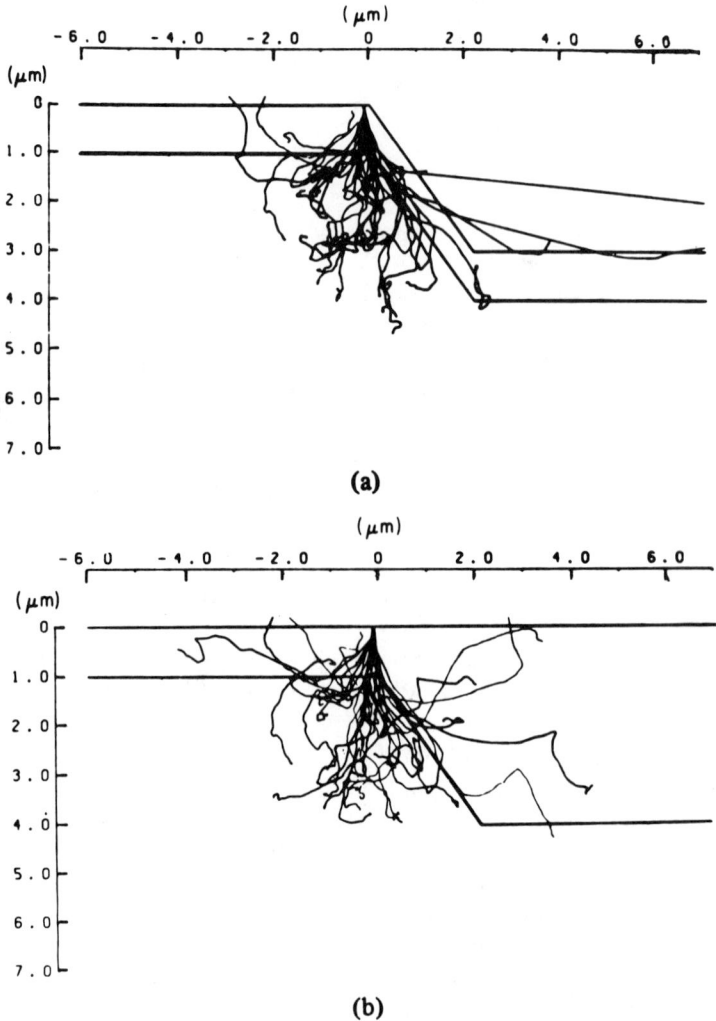

(a)

(b)

FIG. 47.   Simulated electron trajectories at 20 keV for a resist-coated Si step. (a) Resist profile in a step which ideally follows the Si step. (b) Resist with a planar surface (Lin *et al.*, 1981).

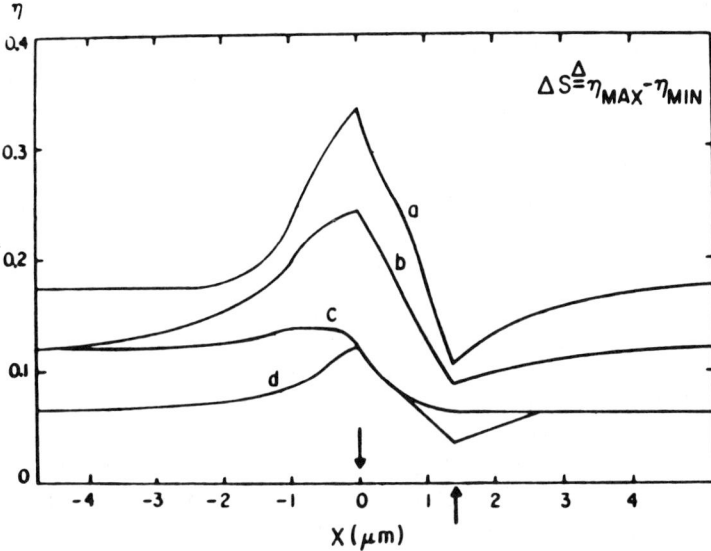

FIG. 48.   Simulated total backscattered electron signals for a 20 keV beam scanning across a 2 $\mu$m step of inclination angle 54.7°. (a) A pure Si step. (b) A Si step covered with a 1 $\mu$m PMMA whose profile ideally follows the Si step. (c) A Si step covered with a 1 $\mu$m PMMA whose profile is a flat surface. (d) A pure PMMA step. The arrows indicate the top and bottom corners of the tapered Si step, respectively (Lin *et al.*, 1981).

beam crosses the step, corresponding to Fig. 47. As seen from Fig. 48, a characteristic feature of the step is expressed by the increasing number of backscattered electrons from the step side as the beam approaches the step. The yield is maximum at the top edge of the step in all cases except case c. When the beam goes down the slope, the yield decreases because the possibility of electrons to be backscattered becomes small on the left side of the beam, and the area decreases from the step on the right side of the beam. The yield is minimum at the bottom edge when the backscattering has the maximum barrier, i.e., the step on the left. When the beam goes away from the step, the yield recovers to normal. With a sample made planar, the step variation of the yield is decreased (curve c). As seen in Fig. 47b, several electrons do not go out of the sample as backscattered electrons with a flat surface.

Stephani (1979) investigated backscattered electron signals from vertical steps, grooves, and pedestals of 1 $\mu$m depth, separating the takeoff angles into three regions ($0° \leqq \theta \leqq 30°$, $30° \leqq \theta \leqq 60°$, $60° \leqq \theta \leqq 90°$) and azimuthal angles into four parts, as shown in Fig. 46. One of their conclusions is that the contrast (the maximum signal difference) in the backscattered electron image decreases with an increasing takeoff angle. This was also verified by Lin *et al.*

(1982). This is a natural conclusion because those electrons which back-scattered at high exit angles must have penetrated deep into the sample, have suffered multiple scattering there, and have lost information of the sample geometry. Those electrons which exit at low takeoff angles are mostly single and large-angle backscattered electrons which are sensitive to the sample geometry. In this way MC calculations can reveal various characteristics of backscattered electrons in the detection technique for registration marks.

### B. Mask Inspection

Recent work on mask inspection by Rosenfield (1984) and Rosenfield *et al.* (1983, 1985) which has been done with MC simulations, is discussed here. A mask is fabricated by forming mask patterns with heavy material (typically gold) on a light material substrate (typically silicon membrane). Different from the regular registration mark detection, the backscattered electron image has atomic number contrast as well as topographical contrast. We must distinguish the signal for the correct mask configuration from the signal for mask defects, typically the presence or absence of masking material. Therefore, it is important to know what signals are obtained from the backscattered electrons. In Fig. 49 simulated trajectories are projected onto the $x-z$ plane (Rosenfield *et al.*, 1983). An example of mask patterns is composed of two gold steps with widths of 0.25 $\mu$m and a 0.85 $\mu$m wide gold step spaced by 0.3 $\mu$m. The program was developed to follow scattered electrons in the sample with this geometry. Some incident electrons are tangled within the wider step due to the large stopping power of heavy materials. We can see clearly four electrons which travel through the steps. In actual inspection of

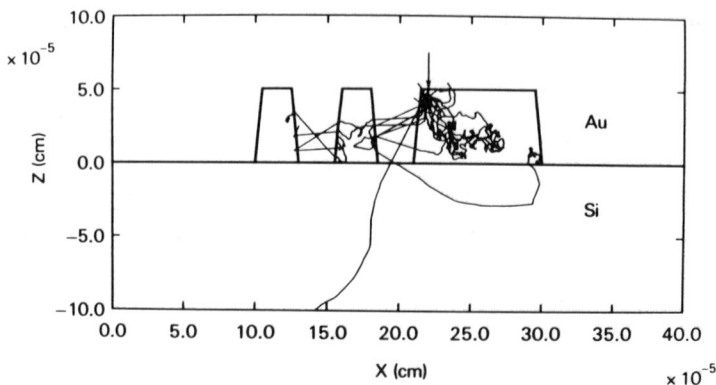

Fig. 49. Simulated electron trajectories in a typical mask pattern where mask inspection is performed by a scanning electron beam (Rosenfield *et al.*, 1983). Au thickness = 0.50 $\mu$m; beam energy = 25 kV; 15 incident electrons at $x = 2.20$ $\mu$m.

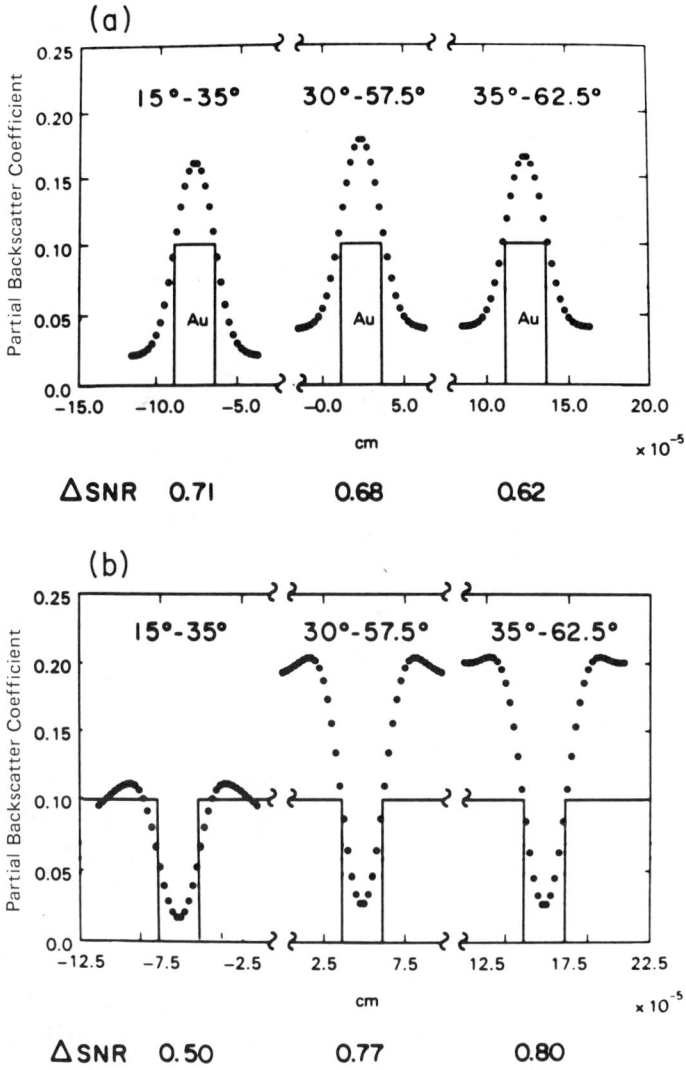

FIG. 50.   The simulated energy signals for (a) a 0.25 $\mu$m gold step and (b) a 0.25 $\mu$m hole in a gold film on Si for three different takeoff angle ranges (Rosenfield *et al.*, 1983). The values of $\Delta$SNR are shown at the bottom of the figure.

practical mask patterns, complicated electron scattering properties have to be understood sufficiently.

The results calculated by Rosenfield *et al.* (1983) are shown in Fig. 50 for two patterns of a single Au step and hole, separating the takeoff angles into three regions. The ordinate shows the backscattered electron intensity which is weighted by each electron energy for direct comparison with experimental data, detected by the solid detector. They reported that the agreement is good between theory and experiment. A more reasonable comparison with experiments should be made by taking account of the signal-to-noise ratio (SNR). The signal is given by the difference of the maximum and the minimum intensities, $I_{max} - I_{min}$. Assuming that the shot noise is given by the square root of the maximum intensity, they introduced the following normalized SNR between bulk silicon (substrate) and 0.5 $\mu$m thick gold film on silicon as follows:

$$\Delta SNR = \frac{(I_{max} - I_{min})/\sqrt{I_{max}}}{(I_{max\,Au} - I_{min\,Si})/\sqrt{I_{max\,Au}}} \tag{75}$$

This value is represented below the figure. In the step case high backscattering coefficients are obtained in all angle regions, where many electrons are backscattered from the step side. However, the $\Delta SNR$ is the highest at low exit angles because the background intensity is low from the silicon surface. In each angle region the signal shape is good.

On the contrary, for the 0.25 $\mu$m hole a value of $\Delta SNR$ is very small in the low-angle region because the backscattering coefficients outside the hole are small at low exit angles. Therefore, a compromise is necessary for mask inspection of patterns with small steps and holes contained. As suggested by Rosenfield *et al.* (1983), takeoff angles of $30°-57.5°$ are appropriate in this case. In mask inspection a medium takeoff angle is preferable, different from the registration mark detection.

They also investigated values of $\Delta SNR$ for various sizes of steps or holes at the optimum takeoff angle, and found that a small gold step had the lowest value. Their other studies are concerned with various effects on the back scattered electron signals, such as electron-beam size, electron-beam energy, etc. In practice, however, as mentioned in Fig. 49, the effect of adjacent patterns has to be taken into account.

### ACKNOWLEDGMENTS

The authors would like to express their hearty appreciation to G. Shinoda (Emeritus Professor of Osaka University) for his encouragement and for supporting the opportunity of their first collaboration at the IBM Research Laboratory (San Jose) in 1972.

One of the authors (K. M.) is grateful to K. Nagami (University of Osaka Prefecture) for his constant encouragement and to R. Shimizu (Osaka University) for his useful discussions.

## REFERENCES

Adesida, I. and Everhart, T. E. (1980). *Proc. Int. Conf. Electron Ion Beam Sci. Technol., 9th*, p. 189.

Adesida, I., Everhart, T. E., and Shimizu, R. (1979). *J. Vac. Sci. Technol.* **16**, 1743.

Agarwal, B. K. (1979). "X-ray Spectroscopy," p. 181. Springer-Verlag, Berlin.

Aizaki, N. (1978). *Jpn. J. Appl. Phys.* **18** (Suppl. 18), 319.

Aizaki, N. (1979). *J. Vac. Sci. Technol.* **16**, 1726.

Atoda, N., and Kawakatsu, H. (1976). *J. Electrochem. Soc.* **123**, 1519.

Augur, R. A., Jones, G. A. C., and Ahmed, H. (1985). *J. Vac. Sci. Technol.* **B3**, 429.

Beaumont, S. P., Bower, P., Tamura T., and Wilkinson, C. (1981). *Appl. Phys. Lett.* **38**, 436.

Berger, M. J. (1963). *In* "Methods in Computational Physics" (B. Alder, S. Fernback, and M. Rotenberg, eds.), Vol. 1, p. 135. Academic Press, New York.

Berger, M. J. (1971). *Proc. Symp. Microdosimetry, 3rd*, p. 157.

Berger, M. J., and Seltzer, S. M. (1964). *Natl. Acad. Sci. Natl. Res. Council* (1133), p. 205.

Bethe, H. A. (1933). *Handb. Phys.* **24**, 273.

Bethe, H. A. (1953). *Phys. Rev.* **89**, 1256.

Betz, H., Fey, F. K., Heuberger, A., and Tischer, P. (1979). *IEEE Trans. Electron Devices* **ED-26**, 693.

Bishop, H. E. (1966). *In* "X-ray Optics and Microanalysis" (R. Castaing, P. Descamps, and J. Philibert, eds.), p. 153. Hermann, Paris.

Broers, A. (1981). *J. Electrochem. Soc.* **128**, 166.

Broers, A., Molzen, W., Cuomo, J., and Wittels, N. (1976). *J. Appl. Phys.* **29**, 596.

Broers, A., Harper, J., and Molzen, W. (1978). *Appl. Phys. Lett.* **33**, 392.

Brown, D. B., Wittry, D. B., and Kyser, D. F. (1969). *J. Appl. Phys.* **40**, 1627.

Chang, T. H. P. (1975). *J. Vac. Sci. Technol.* **12**, 1271.

Chang, T. S., Kyser, D. F., and Ting, C. H. (1981). *IEEE Trans. Electron Devices* **ED-28**, 1295.

Chapman, F. M., Jr., and Lohr, L. L., Jr. (1974). *J. Am. Chem. Soc.* **96**, 4731.

Charlesby, A. (1954). *Proc. R. Soc. London Ser. A* **222**, 542.

Chen, A. S., Neureuther, A. R., and Pavkovich, J. M. (1985). *J. Vac. Sci. Technol.* **B3**, 148.

Chung, M. S. C., and Tai, K. L. (1978). *Proc. Int. Conf. Electron Ion Beam Sci. Technol., 8th*, p. 242.

Colbert, H. M. (1974). Sandia Laboratories Report No. SLL-74-0012.

Cosslett, V. E., and Thomas, R. N. (1964). *Br. J. Appl. Phys.* **15**, 235, 883, 1283.

Curgenven, L., and Duncumb, P. (1971). Tube Investments Research Report No. 303, Saffron Walden, Wssex, England.

Duncumb, P., and Reed, S. J. B. (1968). *In* "Quantitative Electron Probe Micro Analysis" (K. F. J. Heinrich, ed.), Special Publ. 298, p. 133, NBS, Washington, D.C.

Eggarter, E. (1975). *J. Chem. Phys.* **62**, 833.

Ehrenberg, W., and King, D. E. N. (1963). *Proc. Phys. Soc. London* **81**, 751.

Einspruch, N. G. (1981). "VLSI Electronics Microstructure Science," Vol. 1, p. 7. Academic Press, New York.

Emoto, F., Gamo, K., Namba, S., Samoto, N., and Shimizu, R. (1985). *Jpn. J. Appl. Phys.* **24**, L809.

Everhart, T. E., and Hoff, P. H. (1971). *J. Appl. Phys.* **42**, 5837.

Everhart, T. E., Herzog, R. F., Chung, M. S., and DeVore, W. J. (1971). *Proc. Int. Conf. X-ray Opt. Microanal. 6th* p. 81.

Feder, R., Spiller, E., and Topalian, J. (1975). *J. Vac. Sci. Technol.* **12**, 1332.

Goldstein, J. I., Newbury, D. E., Echlin, P., Joy, D. C., Fiori, C., and Lifshin, E. (1981). *In* "Scanning Electron Microscopy and X-ray Microanalysis." Plenum, New York.

Green, M. (1963). *Proc. Phys. Soc.* **82**, 204.

Greeneich, J. S. (1973). PnD. thesis, university of California, Berkeley.

Greeneich, J. S. (1974). *J. Electrochem. Soc.* **121**, 1669.

Greeneich, J. S. (1975). *J. Electrochem. Soc.* **122**, 970.

Greeneich, J. S. (1979). *J. Vac. Sci. Technol.* **16**, 1749.

Greeneich, J. S., and Van Duzer, T. (1974). *IEEE Trans. Electron Devices* **ED-21**, 286.

Grobman, W. D., and Speth, A. J. (1978). *Proc. Int. Conf. Electron Ion Beam Sci. Technol., 8th* p. 276.

Gryzinski, M. (1965). *Phys. Rev.* **138**, A336.

Hasegawa, S., and Iida, Y. (1985). *IEEE Trans. Electron Devices* **ED-32**, 95.

Hatzakis, M. (1969). *J. Electrochem. Soc.* **116**, 1033.

Hatzakis, M. (1971). *Appl. Phys. Lett.* **18**, 7.

Hatzakis, M., and Broers, A. N. (1971). *Rec. Symp. Electron, Ion Laser Beam Technol., 11th* p. 337.

Hatzakis, M., Ting, C. H., and Viswanathan, N. S. (1974). *Proc. Int. Conf. Electron Ion Beam Sci. Technol., 6th* p. 542.

Hawryluk, R. J. (1981). *J. Vac. Sci. Technol.* **19**, 1.

Hawryluk, R. J., and Smith, H. I. (1972). *Proc. Int. Conf. Electron Ion Beam Sci. Technol., 5th*, p. 51.

Hawryluk, R. J., Soares, A., Smith, H. I., and Hawryluk, A. M. (1974a). *Proc. Int. Conf. Electron Ion Beam Sci. Technol., 6th*, p. 87.

Hawryluk, R. J., Hawryluk, A. M., and Smith, H. I. (1974b). *J. Appl. Phys.* **45**, 2551.

Hawryluk, R. J., Smith, H. I., Soares, A., and Hawryluk, A. M. (1975). *J. Appl. Phys.* **46**, 2528.

Hawryluk, R. J., Hawryluk, A. M., and Smith H. I. (1982). *J. Appl. Phys.* **53**, 5985.

Heidenreich, R. D. (1977). *J. Appl. Phys.* **48**, 1418.

Heidenreich, R. D., Thompson, L. F., Feit, E. D., and Melliar-Smith, C. M. (1973). *J. Appl. Phys.* **44**, 4039.

Heidenreich, R. D., Ballantyne, J. D., and Thompson, L. F. (1975). *J. Vac. Sci. Technol.* **12**, 1284.

Heinrich, K., Betz, H., Heuberger, A., and Pongratz, S. (1981). *J. Vac. Sci. Technol.* **19**, 1254.

Herzog, R. F., Greeneich, J. S., Everhart, T. E., and Van Duzer, T. (1972). *IEEE Trans. Electron Devices* **ED-19**, 635.

Horiguchi, S., Suzuki, M., Kobayashi, T., Yoshino, H., and Sakakibara, Y. (1981). *Appl. Phys. Lett.* **39**, 512.

Howard, R. E., Hu, E., Jackel, L. D., Grabbe, P., and Tennant, D. (1980). *Appl. Phys. Lett.* **36**, 592.

Howard, R. E., Craighead, H. G., Jackel, L. D., and Mankiewich, P. M. (1983). *J. Vac. Sci. Technol.* **B1**, 1101.

Hubbell, J. H. (1977). *Radiat. Res.* **70**, 58.

Hundt, E., and Tischer, P. (1978). *J. Vac. Sci. Technol.* **15**, 1009.

Ichimura, S., Aratama, M., and Shimizu, R. (1980). *J. Appl. Phys.* **51**, 2853.

Issacson, M., and Murray, A. (1981). *J. Vac. Sci. Technol.* **19**, 1117.

Jackel, L. D., Howard, R. E., Mankiewich, P. M., Craighead, H. G., and Epworth, R. W. (1984). *Appl. Phys. Lett.* **45**, 698.

James, F. (1980). *Rep. Prog. Phys.* **43**, 73.

Jewett, R. E., Hagouel, P. I., Neureuther, A. R., and Van Duzer, T. (1977). *Polymer Eng. Sci.* **17**, 381.

Jones, F., and Hatzakis, M. (1978). *Proc. Int. Conf. Electron Ion Beam Sci. Technol., 8th*, p. 256.

Jones, F., and Paraszezak, J. (1981). *IEEE Trans. Electron Devices* **ED-28**, 1544.

Karapiperis, L., Adesida, I., Lee, C. A., and Wolf, E. D. (1981). *J. Vac. Sci. Technol.* **19**, 1259.

Kato, T., Yahara, Y., Nakata, H., Murata, K., and Nagami, K. (1978). *J. Vac. Sci. Technol.* **15**, 934.

Kotera, M., Murata, K., and Nagami, K. (1981). *J. Appl. Phys.* **52**, 997, 7403.

Kratschmer, E. (1981). *J. Vac. Sci. Technol.* **19**, 1264.

Krefting, E. R., and Reimer L. (1973). In "Quantitative Analysis with Electron Microprobes and Secondary Ion Mass Spectroscopy" (E. Preuss, ed.), p. 114. Zentralbibliothek der KFA, Julich.

Ku, H. Y., and Scala, L. C. (1969). J. Electrochem. Soc. 116, 980.

Kulchitsky, L. A., and Latyshev, G. D. (1942). Phys. Rev. 61, 254.

Kyser, D. F. (1981). Scanning Electron Microsc. p. 47.

Kyser, D. F. (1982). Proc. Asilomer Conf. Electron Beam Interact. Solids p. 331; (1983). J. Vac. Sci. Technol. B1, 1391.

Kyser, D. F., and Murata, K. (1974a). IBM J. Res. Dev. 18, 352.

Kyser, D. F., and Murata, K. (1974b). Proc. Int. Conf. Electron Ion Beam Sci. Technol., 6th p. 205.

Kyser, D. F., and Pyle, R. (1980). IBM J. Res. Dev. 24, 426.

Kyser, D. F., and Ting, C. H. (1979). J. Vac. Sci. Technol. 16, 1759.

Kyser, D. F., and Viswanathan N. S. (1975). J. Vac. Sci. Technol. 12, 1305.

Kyser, D. F., Schreiber, D. E., and Ting, C. H. (1980). Proc. Int. Conf. Electron Ion Beam Sci. Technol., 9th, p. 255.

Lee, K., and Ahmed, H. (1981). J. Vac. Sci. Technol. 19, 946.

Lenz, V. F. (1954). Z. Naturforsch. A9, 185.

Lewis, H. W. (1950). Phys. Rev. 78, 526.

Lin, L. H. (1975). J. Vac. Sci. Technol. 12, 1289.

Lin, Yi-Ching and Neureuther, A. R. (1981). IEEE Trans. Electron Devices ED-28, 1397.

Lin, Yi-Ching, Adesida, I., and Neureuther, A. R. (1980). Appl. Phys. Lett. 36, 672.

Lin, Yi-Ching, Neureuther, A. R., and Adesida, I. (1982). J. Appl. Phys. 53, 899.

Love, G., Cox, M. C. G., and Scott, V. O. (1978). J. Phys. D 11, 7.

McDonald, I. R., Lamki, A. M., and Delancy, C. F. G. (1971). J. Phys. D4, 1210.

Maldonado, J. R., Coquin, G. A., Maydan, D., and Somekh, S. (1975). J. Vac. Sci. Technol. 12, 1329.

Mladenov, G. M., Braun, M., Emmoth, B., and Biersack, J. P. (1985). J. Appl. Phys. 58, 2534.

Moller, C. (1931). Z. Phys. 70, 786.

Mott, N. F., and Massey, H. S. W. (1965). In "The Theory of Atomic Collisions," 3rd Ed. Clarendon, Oxford.

Motz, J. W., Olsen, H., and Koch, H. W. (1964). Rev. Mod. Phys. 36, 881.

Munro, E. (1980). Adv. Electron. Electron Phys. 13B, 73.

Murata, K. (1982). Proc. Asilomer Conf. Electron Beam Interact. Solids p. 311.

Murata, K. (1985a). J. Appl. Phys. 57, 575.

Murata, K. (1985b). To be published.

Murata, K., Matsukawa, T., and Shimizu, R. (1971). Jpn. J. Appl. Phys. 10, 678; Proc. Int. Conf. X-ray Opt. Microanaly., 6th p. 105.

Murata, K., Kotera, M., and Nagami, K. (1978a). Jpn. J. Appl. Phys. 17, 1671.

Murata, K., Nomura, E., Nagami, K., Kato, T., and Nakata, H. (1978b). Jpn. J. Appl. Phys. 17, 1851.

Murata, K., Nomura, E., Nagami, K., Kato, T., and Nakata, H. (1979). J. Vac. Sci. Technol. 16, 1734.

Murata, K., Kyser, D. F., and Ting, C. H. (1981). J. Appl. Phys. 52, 4396.

Murata, K., Kotera, M., Nagami, K., and Namba, S. (1985). IEEE Trans. Electron Devices ED-32, 1694.

Nakase, M., and Yoshimi, M. (1980). IEEE Trans. Electron Devices ED-27, 1460.

Nakata, H., Kato, T., Murata, K., and Nagami, K. (1978). Proc. Int. Conf. Electron Ion Beam Sci. Technol., 8th p. 393.

Nakata, H., Kato, T., Murata, K., Hirai, Y., and Nagami, K. (1981). J. Vac. Sci. Technol. 19, 1248.

Neill, T. R., and Bull, C. J. (1980). Electron. Lett. 16, 621.

Neureuther, A. R. (1983). *Proc. IEEE* **71,** 121.

Neureuther, A. R. (1980). *In* "Synchrotron Radiation Research" (H. Winick and S. Doniach, eds.), p. 223. Plenum, New York.

Neureuther, A. R., Kyser, D. F., and Ting, C. H. (1979). *IEEE Trans. Electron Devices* **ED-26,** 686.

Nigam, B. P., Sundaresan, M. K., and Wu Ta-You (1954). *Phys. Rev.* **115,** 491.

Nomura, E. (1981). PhD. thesis, University of Osaka Prefecture, Osaka.

Nomura, E., Murata, K., and Nagami, K. (1979). *Jpn. J. Appl. Phys.* **18,** 1353.

Nomura, E., Murata, K., and Nagami, K. (1981). *Phys. Stat. Solidi (a)* **63,** 281.

Nosker, R. W. (1969). *J. Appl. Phys.* **40,** 1872.

Okada, K., and Matsui, J. (1983). *Jpn. J. Appl. Phys.* **22,** L810.

Okubo, T., and Takamoto, K. (1983). *Jpn. J. Appl. Phys.* **22,** 1335.

Onga, S., and Taniguchi, K. (1985). *Symp. VLSI Technol. Dig.* p. 68.

Owen, G., and Rissman, P. (1983). *J. Appl. Phys.* **54,** 3573.

Paraszczak, J., Kern, D., Hatzakis, M., Bucchignano, J., Arthur, E., and Rosenfield, M. (1983). *J. Vac. Sci. Technol.* **1,** 1372.

Parikh, M. (1979). *J. Appl. Phys.* **50,** 4371, 4378, 4383.

Parikh, M. (1980). *J. Appl. Phys.* **51,** 700.

Parikh, M., and Kyser, D. F. (1979). *J. Appl. Phys.* **50,** 1104.

Phang, J. C. H., and Ahmed, H. (1979). *J. Vac. Sci. Technol.* **16,** 1754.

Rao-Sahib, T., and Wittry, D. B. (1974). *J. Appl. Phys.* **45,** 5060.

Reimer, L. (1968). *Optik* **27,** 86.

Reimer, L., Gilde, H., and Sommer, K. H. (1970). *Optik* **30,** 590.

Rishton, S. A., Beaumont, S. P., and Wilkinson, C. D. W. (1983). *Proc. Int. Conf. Electron Ion Beam Sci. Technol., 10th* p. 211.

Ritsko, J. J., Brillson, L. J., Bigelow, R. W., and Fabish, T. J. (1978). *J. Chem. Phys.* **69,** 3931.

Rosenfield, M. G. (1984). PhD. thesis, University of California, Berkeley.

Rosenfield, M. G., and Neureuther, A. R. (1981). *IEEE Trans. Electron Devices* **ED-28,** 1289.

Rosenfield, M. G., Neureuther, A. R., and Ting, C. H. (1981). *J. Vac. Sci. Technol.* **19,** 1242.

Rosenfield, M. G., Neureuther, A. R., and Viswanathan, R. (1983). *J. Vac. Sci. Technol.* **B1,** 1358.

Rosenfield, M. G., Goodman, D. S., Neureuther, A. R., and Prouty, M. D. (1985). *J. Vac. Sci. Technol.* **B3,** 377.

Saitou, N. (1973). *Jpn. J. Appl. Phys.* **12,** 941.

Saitou, N., Munakata, C., and Honda, Y. (1972). *Jpn. J. Appl. Phys.* **11,** 1061.

Saito, Y., Yoshihara, H., and Watanabe, I. (1982). *Jpn. J. Appl. Phys.* **21,** L52.

Samoto, N., Shimizu, R., Hashimoto, H., Adesida, I., Wolf, E., and Namba, S. (1983). *J. Vac. Sci. Technol.* **B1,** 1367.

Schneider, D. O., and Cormack, D. V. (1959). *Radiat. Res.* **11,** 418.

Schwinger, J. (1949). *Phys. Rev.* **75,** 1912.

Seah, M. P., and Dench, W. A. (1979). *Surf. Interface Anal.* **1,** 2.

Sedgwick, T. O., Broers, A. N., and Agule, B. J. (1972). *J. Electrochem. Soc.* **119,** 1796.

Seltzer, S. M. (1974). Natl. Bur. Stand. Rept. NBSIR 74, Washington, D.C.

Semenzato, L., Eaton, S., Neukermans, A., and Jaeger, R. P. (1985). *J. Vac. Sci. Technol.* **B3,** 245.

Shaw, C. H. (1981). *J. Vac. Sci. Technol.* **19,** 1286.

Shimizu, R. (1974). *J. Appl. Phys.* **45,** 2107.

Shimizu, R., and Everhart, T. E. (1972). *Optik* **36,** 59.

Shimizu, R., and Everhart, T. E. (1978). *Appl. Phys. Lett.* **33,** 784.

Shimizu, R., Murata, K., and Shinoda, G. (1966). *In* "X-ray Optics and Microanalysis, (R. Castaing, P. Descamps, and J. Philibert, eds.), p. 127. Hermann, Paris.

Shimizu, R., Ikuta, T., Everhart, T. E., and DeVore, W. J. (1975). *J. Appl. Phys,* **46,** 1581.

Shimukunas, A. R. (1984). *Solid State Technol.,* September, p. 192.

Shinoda, G., Murata, K., and Shimizu, R. (1968). Special Publ. 298, p. 155. Natl. Bur. Stand., Washington, D.C.

Spencer, L. V., and Fano, U. (1954). *Phys. Rev.* **93,** 1172.

Stephani, D. (1979). *J. Vac. Sci. Technol.* **16,** 1739.

Stephani, D., and Kratschmer, E. (1981). *Proc. Microcircuit Eng.* p. 228.

Suzuki, Y., Yoshioka, N., and Yamazaki, T. (1984). *J. Vac. Sci. Technol.* **B2,** 301.

Thornton, P. R. (1979). *Adv. Electron. Electron Phys.* **48,** p. 271; (1980). **54,** p. 69.

Tischer, P., and Hundt, E. (1978). *Proc. Int. Conf. Electron Ion Beam Sci. Technol., 8th* p. 444.

Todokoro, Y. (1980). *IEEE Trans. Electron Devices* **ED-27,** 1443.

Todokoro, Y. (1981). *Trans. IECE* **J-64-C,** 353 (in Japanese).

Triplett, B. B., and Hollman, R. F. (1983). *Proc. IEEE* **71,** 585.

Trotel, J., and Fay, B. (1980). *In* "Electron Beam Technology in Microelectronic Fabrication" (G. R. Brewer, ed.), p. 309. Academic Press, New York.

Ueberreiter, K. (1968). *In* "Diffusion in Polymers" (J. Crank and G. S. Park, eds.), p. 219. Academic Press, London.

Veigele W. J. (1973). *At. Data* **5,** 51.

Viswanathan, N. S., Pyle R., and Kyser, D. F. (1976). *Proc. Int. Conf. Electron Ion Beam Sci. Technol., 7th* p. 218.

Watts, R. K., and Maldonado, J. R. (1982). *In* "VLSI Electronics Microstructure Science" (N. G. Einspruch, ed.), Vol. 4, p. 55. Academic Press, New York.

Wentzel, G. (1927). *Z. Phys.* **40,** 590.

Wilson, R. (1964). *Phys. Rev.* **60,** 749.

Wittels, N. D. (1980). *In* "Fine Line Lithography" (R. Newman, ed.), Vol. 1, p. 1. North-Holland Publ., Amsterdam.

Wolf, E. O., Ozdemir, F. S., Perkins, W. E., and Coane, P. J. (1971). *Rec. Symp. Electron, Ion, Laser Beam Technol., 11th* p. 331.

Yoshimi, M., Takahashi, M., Kawabuchi, K., Kato, K., and Takigawa, T. (1982). *Proc. Int. Conf. Solid State Devices, 14th,* Tokyo; (1983). *Jpn. J. Appl. Phys. (Suppl. 22-1),* 179.

# Invariance in Pattern Recognition

HARRY WECHSLER

*Department of Electrical Engineering*
*University of Minnesota*
*Minneapolis, Minnesota 55455*

## I. Introduction

Our concern in this survey is about invariance in pattern recognition. The goal is to describe computational techniques for recognizing patterns, invariant to the distortions they might have been subject to. The distortions, to be described and dealt with in the next sections, can be geometric in nature, or they can be caused by changes in illumination and/or motion. Our treatment cannot be exhaustive; still, we cover enough ground to give the reader the right flavor of the problem.

Pattern recognition (PR) is a subtopic within computer vision (CV), one of the application areas of artificial intelligence (AI). To motivate the methodology and approach used in the context of invariance in pattern recognition we consider that it is more than appropriate to discuss first AI in general, and CV in particular.

AI deals with the types of problem solving and decision making that humans continuously face in dealing with the world. Such activity involves by its very nature complexity, uncertainty, and ambiguity which can "distort" the phenomena (e.g., imagery) under observation, Still, following the human example, any artificial (vision) system should process information such that the result is invariant to the vagaries of the data acquisition process.

Following Gevarter (1985) AI issues belong to two classes:

(1) Basic research
  (a) Modeling and representation of knowledge (K)
  (b) Common sense and reasoning
  (c) Heuristic search
  (d) AI languages (software) and tools (hardware)
(2) Application areas
  (a) Expert systems
  (b) Computer vision
  (c) Natural language proccessing
  (d) Problem solving and planning

We consider next only those aspects relevant to our discussion. Any computational process, as will be seen later, is characterized by a particular representation used to encode the knowledge as related to the task at hand. The knowledge, which then becomes part of the system memory, represents an implicit encoding of the world surrounding us. For both storage and computational reasons, not everything in the world can be stored, but it might yet require processing and recognition. To that end, the implicit knowledge has to be algorithmically processed and made explicit. This last step is known as inference or reasoning. Computer vision represents the computational study of processes which produce useful descriptions from visual input. Computational models, central to the study of all of AI, then describe in a precise (algorithmic) way how a solution can be found. Expert systems (ES) are intelligent computer programs that use knowledge and inference procedures to solve (recognition) problems that are difficult enough to require significant human expertise for their solution. The knowledge necessary to perform at such a level, plus the inference procedures used, can be thought of as a model of the expertise of the best practitioner in the field (e.g., the human visual system). It is the task of the knowledge engineer (KE) to bring expert knowledge to bear on problem solving.

AI is highly relevant to the latest (technological) information revolution we are witnessing. In particular, computer vision is useful, among other things, to robotic applications by making available a closed-loop control, to inspection by allowing 100% automatic flaw detection, and to autonomous vehicle navigation in hostile environments.

There is a lot of terminology concerning CV and its subareas, and it is proper to label those subtopics according to their meaning. Pattern recognition (PR) usually means classification/discrimination among different object classes as described by a measurement vector **x**. Statistical techniques are then used on knowledge represented as **x** toward classification. For the goal of our presentation knowledge is not restricted to representation of the **x** type but it can assume any generic form as long as it is appropriate for our task. Image processing (IP) provides for different transformations, task related, to be applied on a particular image representation. Image understanding (IU) aims at a constrained search whose goal is to parse and interpret the data. Data compression is concerned with reducing the load on a communication line, and PR techniques can facilitate the task. There are many textbooks (Fukunaga, 1972; Duda and Hart, 1973; Castleman, 1978; Rosenfeld and Kak, 1982; Ballard and Brown, 1982) on each of the topics, and the reader is encouraged to consult them if he/she desires so. There are a number of generic models of processing involving the above topics, and to emphasize their relationships we conclude the introductory section by presenting next just two of the available models.

*Model (A)*: Preprocessing → feature extraction → classification (PR) → postprocessing. Here, the preprocessing stage, using image processing techniques, attempts to provide some standard image by eliminating distortions caused by uneven illumination or motion. Then, following feature extraction (numerical or symbolic), classification is performed and the image can be assigned to one among a prespecified set of classes. Postprocessing might "clean" the results based on *a priori* knowledge (modeling).

*Model (B)*: There is a loop of both bottom-up (IP, feature extraction), hypothesis (model) formation and top-down (model instantiation, prediction, and verification). The goal is that of image understanding. In terms of our stated goal of invariance in PR, the bottom-up part provides an image representation (REP), the models are stored in memory, and there is a match between REP and the memory. Based on the result, identity of an object can be assessed or the loop is again traversed, until enough information and confidence in one particular model is gained.

## II. Computational Vision

Most vision research, whether studying biological or computer vision, has a common goal of understanding the general computational nature of visual perception. It is widely acknowledged that synergism of various fields like neurophysiology, psychophysics, the physics related to image formation, and signal analysis can further the progress toward the above goal. And indeed, it is the very availability of such a synergetic approach that made computer vision the premier area of both development and success within AI. Furthermore, computational vision research is most beneficial when it leads to the development of general theories/paradigms and/or hypotheses that can be subjected to widespread empirical scrutiny. In terms familiar to AI research, a unified treatment of visual processing tasks is characteristic of the so-called "deep systems," that is, systems which derive their utility from a compact collection of fundamental principles.

The seminal work of Marr (1982) considered computational vision as an information-processing task, and it defined the three levels at which any machine vision system carrying out a visual task must be understood. There is a first, basic computational theory level which specifies what is the goal of the computation, why it is appropriate, and what is the strategy by which it can be carried out. Next levels deal with representation and algorithm (i.e., how is the computational theory implemented in terms of input/output representations and transformation algorithms), and hardware implementation, respectively.

The challenge of the visual recognition problem stems from the fact that the interpretation of a three-dimensional (3D) scene from a single two-dimensional (2D) image is confounded by several dimensions of variability. Such dimensions include uncertain perspective, position, orientation, and size (pure geometric variability) along with sensor noise, object occlusion, and nonuniform illumination. Vision systems must not only be able to sense the identity of objects despite this variability, but they must also be able to explicitly characterize such variability. This is so because the variability in the image formation process (particularly that due to geometric distortion and varying angle of incident illumination) inherently carries much of the valuable information about the image being analyzed.

Most of the motivation for the state-of-the-art research in computational vision comes from psychophysics and neurophysiology. We present below relevant knowledge coming out of these two scientific fields and the hints/constraints they provide for machine vision systems. Assuming a generic process like

Image representation → Matcher ← Visual memory

aimed at invariant PR, we discuss different options regarding both REP (representation) and ALG (algorithm) and make some suggestions.

It is Gibson (1950) who showed the importance of surface detection as a first step toward recovering the 3D structure, and who motivated so much research toward recovering shape from different available cues such as texture, shade, or motion. Furthermore, Gibson introduced the concept of ecological optics, related to the spatial layout of the environment. Determining spatial-layout organization is central to most visual tasks and it is often crucial to the understanding of an image. Spatial organization can be described in terms of the position and orientation of object surfaces within some system of coordinates. (The idea of cognitive maps and systems of coordinates has been discussed most recently by Haber, 1985.) Basically, the hints and/or constraints which follow from the above discussion mean that

(1)  Surface is primary to perception;
(2)  Humans have an innate capability for making surface perception invariant to linear transformations (LT) and perspective distortions in terms of slant and tilt $(\sigma, \tau)$; and
(3)  Perception is not performed in some void/empty space but rather in some environment surrounding us.

One can identify several major ways for handling the issue of image variability. The approaches can be distinguished according to the type of (memory) prototype pattern(s) that are matched with the computed image representations at the interface with visual memory. Since a vision system may interface with visual memory at several different levels of representation, one can classify each of these interfaces according to the criteria set out below. First, one can consider *viewer-centered versus object-centered* representations. A viewer-centered, as the name implies, is viewpoint dependent. An object-centered representation is a description given in terms of a coordinate system that is attached to the object in 3D space. Examples of such representations include the generalized cylinders (Bindford, 1971). Recent research by Sedgwick and Levy (1985) shows that environment-centered (i.e., object-centered in 3D) representations lead to less variability in match than the corresponding viewer-centered representations. So, the next hint/constraint is that of

(4)   Environment (object-centered in 3D environment) representations.

The next two dimensions along which memory (prototype) representations are considered can be thought of as *multiprototype versus canonical*,
where a canonical representation can characterize the pattern with a single
prototype, and *complete versus incomplete*, where a complete pattern representation encodes sufficient information to allow the appearance of the pattern to
be derived under any viewing condition. Those competing methods together
with psychophysical data regarding human memory suggest that

(5)   Memory is distributed and allows for (partial) key indexing for
recognition. Furthermore, memory is "reconstructive," i.e., it yields the entire
output vector (or a closed approximation of it) even if the input is noisy or only
partially present (i.e., it is occluded).

It is interesting to note that a capability like the above is provided by
distributed associative memories (DAM) (Kohonen, 1984) and that computer
vision research definitely favors canonical and complete representations for
the obvious reasons of parsimony and efficiency. The ACRONYM system
(Bindford, 1982) is characteristic of such an approach, and it can derive any
required representation (viewer dependent) via "deprojection."

How about neurophysiology research? Most of the work (see, for example,
Marcelja, 1980) suggests that the first stages of visual processing lead to the
generation of

(6)   Simultaneous (conjoint) space/spatial-frequency representations characterized by high resolution.

Finally, it is worth remembering that low-level vision (i.e., no *a priori* or
high-level knowledge) does not necessarily involve detection or recognition of
abstract/symbolic features but rather it involves the transformation of the
image into a new functional form which is compatible with direct recognition
of complex patterns treated as holistic gestalts. The same low-level (early)
vision model also implies

(7)   Data independent, extensive parallel processing, rather than serial
and iterative proccessing.

Another issue to be considered is that of *analogical versus nonanalogical*
representations. To qualify as an analog process or representation, it must be
shown that the internal process or representation goes through an ordered
series of intermediate states that have a one-to-one relationship to the intermediate stages in the corresponding external process or object (Cooper,

1983). The experiments performed by Shepard and Metzerl (1981) and the concept of retinotopic mappings seem to suggest such a representation, where the relational structure of external information is preserved in the corresponding internal representation.

We can conclude the section on representations by stating our preference for REPs which are canonical, environment-centered, complete, and analogical.

Regarding algorithms (strategies) in using representations as suggested above, the next two models, shown below, are compatible with both the philosophy discussed above and with the main line of present research. The structure of an ES (expert system) for CV could look like

(1)  Feature extraction (data-driven/bottom-up)
(2)  Hypothesis generation
        feature interpretation
        high-level association
(3)  Prediction (model-driven/top-down)
(4)  Verify (low-level association)
        IF credibility < threshold THEN goto 1
(5)  Final interpretation

Alternatively, the following functional model would include metric information and preserve spatial (layout) properties. Its corresponding long-term memory (LTM) is content-addressable (CA) and no sequential search is required for LTM structures. The model/algorithm looks like:

(1)  Short-term memory (STM) encoding used to look up the LTM
(2)  Secondary checks (predict and verify) follow the lookup
(3)  IF confidence is high THEN halt ELSE goto 1

Such a model will also allow for modality-specific constructive interference between imagery and concurrent preception (multisensory data fusion).

Computational vision in general, and invariant pattern recognition in particular, are not easy to implement. The difficulty comes from both the novelty of the problem and the lack of sufficient knowledge regarding its solution, *and* the harsh computational requirements (storage and speed) due to the huge amounts of data to be processed. However, the CV problem is challenging enough both technologically and intellectually as to channel a lot of energy into finding new solutions and also in developing new AI software and hardware environments.

The next sections survey different computational methods aimed at performing invariant pattern recognition. The criteria set up in this section regarding REP and ALG should be kept in mind when assessing the different solutions being suggested.

## III. REVIEW OF BASIC CONCEPTS

We review here basic definitions related to image definition and processing as well as formal tools like Markov chains, graph theory, signal analysis, statistical pattern recognition, and formal languages. The concepts to be described are then used in later sections to develop methods for invariant pattern recognition.

### A. Digital Image Processing

A formal image is a 2D array $A$, where each $(x, y)$ coordinate is called a pixel. It is easy to extend the concept to a stack of such arrays. First one can think about 3D processing, as for CT (computer tomography), or (temporal) data changing over time. The corresponding arrays will be $A(x, y, z)$ and $A(x, y, t)$, respectively. The image is sensed through a particular portion of the spectrum, using one or more spectral bands, and a mapping $f: A \to N$ is defined for each band, where $N$ is given as the range of values, or equivalently the number of bits $n$ used to encode the function value, i.e., 0 to $2^n - 1$. Therefore, one can assume for the general case $\mathbf{f}(x, y, \mathbf{z})$, where $\mathbf{f}$ is a vector of measurements and $\mathbf{z}$ is the axis along which successive images can be stacked.

Much processing in computer vision is involved with detection and identification of objects imaged as 2D patterns. The task is usually accomplished by outlining the boundary (shape) of the object and performing classification based on boundary derived features. Following or in parallel with boundary detection there is the need to encode (economically) and store the contour. An efficient way of encoding boundaries is the chain code representation suggested by Freeman (1974). It approximates a digital curve by a chain which is an ordered and finite sequence of links. A link $l_i$ is a directed straight-line segment of length $t(\sqrt{2})^s$ and of angle $l_i \times 45°$ referenced to the $x$ axis of a right-handed Cartesian coordinate system. The link $l_i$ is any integer 0 through 7 and $s$ is the modulo-2 value of $l_i$. The grid spacing $t$ is normally set to unity. Different properties related to the boundary being analyzed can be found once the chain is determined. Those properties include the length of the curve, the area enclosed by the curve (in the case of a closed curve), and the curvature. As an example, the perimeter is given as $t(n_e + n_o\sqrt{2})$ where $n_e$ and $n_o$ are the number of even and odd links $l_i$, respectively. Such a measure is sensitive to digitization effects (try to see a digitized circle) and to scale effects. Calibration or polar transformations to be described later are needed before such a feature can be used for discrimination purposes. If $l_j = (a_{j_x}, a_{j_y})$ are the components of the link $l_j$ (taken as a vector in 2D space) the surface of the object $S$ is given as $\Sigma_i a_{i_x}(y_{i-1} + \frac{1}{2}a_{i_y})$. Arguments regarding digitization effects apply to the surface area $S$ as well, even that area measurements are less sensitive due to averaging (Wechsler, 1981). The first generation of commerical vision systems, that for binary vision, is based on deriving and measuring features like area, perimeter, center of gravity, orientation, minimum and maximum radii from the center of the object to its boundary and their orientation, the number of holes, and so on. The object ("blob") is defined using thresholding of the original image and implementation is achieved via connectivity analysis. We will refer back to some of the features mentioned later on, and see what kind of invariance can be achieved by using them.

## B. Markov Chains

Sometimes, one way of performing invariant pattern recognition is to look for some internal structure. To capture such (internal) relationships between object components one can use tools like Markov chains, described in this section, and the MST (minimum spanning tree) or formal languages (grammars), to be described in subsequent sections.

Markov chains are characterized by a sequence of trials whose outcomes, say $o_1, o_2, \ldots$ satisfy two properties:

(1)  Each outcome belongs to a finite set $\{O_1, O_2, \ldots, O_m\}$ called the state space; if the outcome of the $n$th trial is $O_i$, then we say that the system is in state $O_i$ at time $n$.

(2)  The outcome of any trial depends at most on the outcome(s) of the ($n$) immediately preceding trials (finite memory).

There is a transition matrix, $P$, characterizing the Markov chain as a stochastic matrix ($n = 1$), and its entries are given by $p_{ij}$, corresponding to the transitions $O_i \rightarrow O_j$. Correspondingly, one can define $p_{ij}^{(m)}$ as $O_i \rightarrow O_{k_1} \rightarrow \cdots \rightarrow O_{k_{m-1}} \rightarrow O_j$, i.e., going from state $O_i$ into state $O_j$ in exactly $m$ steps. [A square matrix $p = (p_{ij})$ is called a stochastic matrix if each of its rows is a probability vector, i.e., $V_{i,j}\, p_{ij} \geq 0$, and $V_i, \Sigma_j p_{ij} = 1$.] It can be shown that if $A$ and $B$ are stochastic, so is $AB$. A stochastic matrix $A$ is said to be regular if all entries of some power $p^m$ are positive, i.e., $p_{ij}^m > 0$. If $P$ is the transition matrix, then the $n$-step transition matrix $P^{(n)} = P^n$. Assuming that $\mathbf{p}^{(o)} = (p_1^{(o)}, \ldots, p_m^{(o)})$ denotes the initial probability distribution, i.e., the likelihood of starting in one of the states $O_1, O_2, \ldots, O_m$, it can be shown that $P^{(n)}$, i.e., the transition matrix after "enough" $n$ steps, approaches a limit matrix $T$. Informally, this means that the effect of the original state wears off as the number of steps increases, and therefore one can capture the internal structure of the image; such a characteristic found applications within texture analysis and OCR (optical character readers). Formally, if $P$ is a regular stochastic matrix, then

(1)   $P$ has a unique fixed probability vector $\mathbf{t}$ (i.e., $\mathbf{t}P = \mathbf{t}$), and the components of $\mathbf{t}$ are all positive;

(2)   $\lim_{n \to \infty} P^n = T$, where $T = \{\mathbf{t}, \mathbf{t}, \ldots, \mathbf{t}\}$;

(3)   if $\mathbf{p}$ is any probability vector, then the sequence of vectors

$$\{\mathbf{p}P, \mathbf{p}P^2, \ldots\} \to \mathbf{t}.$$

The vector $\mathbf{t}$ stands for the steady (final)-state distribution and it is independent of the initial-state distribution. One should note that if $P$ has 1 as one of the diagonal entries, then $P$ is not regular; i.e., there are forbidden transitions.

### C. Graph Theory

Graph theoretic approaches are used for clustering and they attempt to capture the relationship among object components. Examples are the MST (minimum spanning tree) and the clique. The MST for a graph $G = (V, E)$ of vertices $V$ and links $E$ is a subgraph of $G$ which is a tree and it contains all vertices $V$ of $G$. MST for a weighted graph is a spanning tree of minimal weight (think of connecting a number of cities by a minimum length network of roads). MSTs are used for taxonomy and clustering, and its characteristic graph stands for the internal structure characterizing a particular class of objects. MSTs have been used, among other things, for classifying the traces left by nuclear particles in bubble-chamber analysis by matching (graph-isomorphism) MSTs. Cliques are completely connected sets which are not subsets of larger completely connected sets. Such a characteristic, once detected, can be useful for locating subparts, as would be needed in assembly of mechanical parts.

### D. Signal Analysis

#### 1. Fourier Transforms

A function $f(t)$ is said to be periodic with period $T$ if and only if $f(t) = f(t + nT)$, where $n$ is an integer and $t$ is a continuous time variable. For discrete functions, let the sampling interval be $T_t$, and the observation period $T_0 = NT_t$, where $N$ is the number of samples. Space variables $(x, y)$ can be substituted for time variables. The corresponding terms for the frequency domain are $T_s$ (frequency sampling interval), $f_s = 1/T_t = NT_s$. $\omega$ and $s$ ($\omega = 2\pi s$) are the radial and cyclic frequency variables, respectively, and the corresponding Fourier transforms (FT) are $F(\omega)$ and $F(s)$. The Fourier transformations are defined as follows (Bracewell, 1978):

(1) Radial FT:

$$F(\omega) = \int_{-\infty}^{\infty} f(t) \exp(-j\omega t)\, dt$$

Radial IFT (inverse FT):

$$f(t) = \frac{1}{2}\pi \int_{-\infty}^{\infty} F(\omega) \exp(j\omega t)\, dt$$

(2) Cyclic Fourier transformations:

(a) If the signal is nonperiodic and continuous, then the resulting FT is given by

$$F(s) = \int_{-\infty}^{\infty} f(t) \exp(-j2\pi s t)\, dt = \mathscr{F}\{f(t)\}$$

The (inverse) FT is given by

$$f(t) = \int_{-\infty}^{\infty} F(s) \exp(j2\pi s t)\, ds = \mathscr{F}^{-1}\{\mathscr{F}\{f(t)\}\}$$

where $\mathscr{F}$ is the Fourier operator, and the resulting FT is nonperiodic and continuous.

(b) If the signal is periodic and continuous, then by using the Fourier series (FS) method, one obtains a nonperiodic, discrete spectrum given by the FS components. Specifically

$$f(t) = \sum_{-\infty}^{\infty} c(mT_s) \exp(j2\pi m T_t)$$

where

$$c(mT) = \frac{1}{T} \int_{-\infty}^{\infty} f(t) \exp(-j2\pi m T_t)\, dt$$

(c) If the signal is periodic and discrete, then by using the DFT (discrete Fourier transform) the resulting frequency spectrum which is periodic and discrete is given as

(DFT) $$F(mT_s) = \sum_{n=0}^{N-1} f(mT_t) \exp(-j2\pi mn T_s T_t)$$

(IDFT) $$f(nT_t) = \frac{1}{N} \sum_{m=0}^{N-1} (mT_s) \exp(j2\pi mn T_s T_t), \quad \text{for } 0 \le m, n \le N-1$$

Note that a nonperiodic signal yields a continuous representation and that a periodic signal yields a discrete representation, and vice versa, i.e., a continuous signal yields a nonperiodic representation and a discrete signal yields a periodic representation. General transformations for the discrete case are defined next. Assuming a two-dimensional square array, the forward and inverse transforms are given by

$$F(u,v) = \sum_{x=0}^{N-1} \sum_{y=0}^{N-1} f(x,y)g(x,y,u,v)$$

and

$$f(x,y) = \sum_{u=0}^{N-1} \sum_{v=0}^{N-1} F(u,v)h(x,y,u,v)$$

where $g(x,y,u,v)$ and $h(x,y,u,v)$ are called the forward and inverse transformation kernels, respectively. The forward kernel is said to be separable if

$$g(x,y,u,v) = g_1(x,u)g_2(y,v)$$

and it is said to be symmetric if

$$g(x,y,u,v) = g_1(x,u)g_1(y,v)$$

## 2. Linear Systems

A system is said to be linear if, for any inputs $x_1$, $x_2$ and their corresponding system outputs $y_1$, $y_2$, the output of $ax_1 + bx_2$ is $ay_1 + by_2$, where $a$ and $b$ are constants.

Assume that we deal with a linear system and that $f(t)$ is the input signal; $F(s)$, the spectrum of input signal; $g(t)$, the impulse response; $G(s)$, the transfer function; $h(t)$, the output signal; and $H(s)$ the spectrum of the output signal. For linear systems $H(s) = F(s)G(s)$, and then $G(s) = H(s)/F(s)$, $F(s) \neq 0$, and therefore $g(t) = \mathscr{F}^{-1}\{\mathscr{F}\{h(t)\}/\mathscr{F}\{f(t)\}\}$. Finding $G(s)$ or conversely $g(t)$ is equivalent to system indentification.

### E. Statistical Pattern Recognition

As preparation for our discussion on statistical invariance, we review some prerequisites. Assuming that $X' = [x_1, x_2, \ldots, x_n]$ is a random vector, then the covariance matrix of $X$ is defined as $\Sigma_x = E\{(X - \bar{X})(X - \bar{X})'\}$, where $E$ and $\bar{X}$ stand for the expected value operator and the mean of vector $X$, respectively. Specifically,

$$\Sigma_x = \{\sigma_{ij}^2 \,|\, \sigma_{ij}^2 = E[(x_i - \bar{x}_i)(x_j - \bar{x}_j)], \, i,j = 1,2,\ldots,n\}.$$

It follows that the diagonal terms of the covariance matrix are the variances of $\{x_i\}$ and that the off-diagonal terms are the covariance of $x_i$ and $x_j$. It is obvious that $\Sigma_x$ is symmetric. $\Sigma_x$ can be expressed alternatively as $\Sigma_x = S - \bar{X}\bar{X}'$, where $S$ is called the scatter or autocorrelation matrix of $X$ and is given by $S = E\{XX'\} = [E\{x_i x_j\}]$. Furthermore, if we define the correlation coefficient $r_{ij}$ as $r_{ij} = \sigma_{ij}^2/(\sigma_{ii}\sigma_{jj})$, then $\Sigma_x = \Gamma R \Gamma$, where $\Gamma = $ diagonal$(\sigma_{ii})$ and $R$ is the correlation matrix given by $R = \{r_{ij} | |r_{ij}| < 1, r_{ii} = 1\}$ (Ahmed and Rao, 1975).

## 1. Karhunen–Loève Transform

If $\lambda_i$ and $\phi_i$, $i = 1, 2, \ldots, n$, are the eigenvalues and $N$ eigenvectors of a real symmetric matrix $\Gamma$, then $\Phi\Gamma\Phi' = \Lambda$, where $\Phi = [\phi_1, \phi_2, \ldots, \phi_n]$ is the $(N \times N)$ eigenvector matrix such that $\Phi'\Phi = I$, and $\Lambda = \text{diag}(\lambda_1, \lambda_2, \ldots, \lambda_n)$ is the eigenvalue matrix. The generalized Wiener filter considers a signal $X$ embedded in (uncorrelated) noise $W$, $\bar{X} = \bar{W} = 0$, and finds the filter $A$ such that the expected value of the MSE (mean-square error) between $X$ and $\hat{X} = T^{-1}AT(X + W)$ is minimized, where $T$ is any orthogonal transform. For computational reasons one would like to have a diagonal filter, $A_d$. From the eigenvector theorem, if $T$ is chosen to be the eigenvector matrix of $A_r$, then $A_d = TA_rT'$, where $A = TA_rT'$, $A_r = \Sigma_X(\Sigma_X + \Sigma_W)^{-1}$, $A_r$ is symmetric, the filter $A$ is indeed a diagonal matrix, and the transform is called the KLT (Karhunen–Loève transform). Both processes of feature selection and data compression can be considered within the context of KLT. Specifically, the KLT allows a nonperiodic process to be expanded in a series of orthogonal functions. In the discrete case, the least MSE implies that KLT expansion minimizes the approximation error when $m < n$ basis vectors are used. The optimum properties are obtained by choosing as $m$ normalized eigenvectors those corresponding to the largest eigenvalues, i.e., components of maximum variance which are useful for discrimination. Such a process as described above is known as factor analysis or principal component analysis. Conversely, if we choose $m < n$ normalized vectors associated with the smallest eigenvalues, then the entropy is minimized and the process allows for cluster analysis. Finally, it can be shown that if $Y = TX$, then $\Sigma_Y = T\Sigma_xT' = \text{diag}(\lambda_1, \lambda_2, \ldots, \lambda_n)$; i.e., if we subject a signal $X$ to the KLT then the resulting signal is uncorrelated.

## 2. Fisher Discriminant

One main problem in pattern recognition is that of classifying one object as belonging to one among a number of given classes; linear (decision) discriminant functions $d(\mathbf{X})$ are widely used for such a purpose. Given some object and $\mathbf{X}$ being its set of features (it is desirable that the set is optimal in the KLT sense) then $d(\mathbf{X}) = W^{\mathrm{T}}\mathbf{X} + \omega_0$. This is equivalent to projecting $\mathbf{X}$ on the $W$ direction and comparing the result with the threshold $\omega_0$; in practical terms, it means to reduce the dimensionality of $\mathbf{X}$ by projecting it onto a line, and then finding that orientation for which the projected samples are well separated. The Fisher discriminant is aimed at achieving just such a result and it is also optimal in the MSE sense. Specifically, one tries to maximize the ratio of between-class scatter to that of within-class scatter. Formally, the criterion $J(W)$ is given as

$$J(W) = (\tilde{m}_1 - \tilde{m}_2)^2/(\tilde{\sigma}_1^2 + \tilde{\sigma}_2^2)$$
$$= W^{\mathrm{T}}S_{\mathrm{B}}W/(W^{\mathrm{T}}S_{\mathrm{W}}W)$$

where $\tilde{m}_i = W^{\mathrm{T}}m_i$, $\tilde{\sigma}_i^2 = W^{\mathrm{T}}\Sigma_i W$, and $S_{\mathrm{B}}$ and $S_{\mathrm{W}}$ are the between and within-class scatter matrices, respectively.

The generalized eigenvalue problem $S_{\mathrm{B}}W = \lambda S_{\mathrm{W}}W$ can be solved for $W = S_{\mathrm{W}}^{-1}(m_1 - m_2)$ if $S_{\mathrm{W}}$ is not singular. In general $S_{\mathrm{W}} = \frac{1}{2}(\Sigma_1 + \Sigma_2)$. For equal covariances $W = \Sigma^{-1}(m_1 - m_2)$. The threshold $\omega_0$ is given (Fukunaga, 1972)

$$\omega_0 = \frac{(m_2 - m_1)^T(\frac{1}{2}\Sigma_1 + \frac{1}{2}\Sigma_2)^{-1}(\sigma_1^2 m_2 + \sigma_2^2 m_1)}{\sigma_1^2 + \sigma_2^2}$$

If the probability distributions are multivariate $N(m_i, 1)$, then the threshold used to discriminate signals is the mean of the projected signals.

### F.  Formal Languages

Languages are defined in terms of grammar (syntax) and strings (patterns) are found to belong to a given language if they can be parsed successfully according to the corresponding grammar. There is more to a language than syntax, evidence to that being the fact that ungrammatical (noisy) sentences can still be recognized, while sometimes grammatically correct sentences make no sense. To deal with such issues one must consider the possibility of stochastic grammars as well as semantics (meaning), respectively. Still, it is the grammar which can capture the invariant structure of a particular class of objects. A grammar $G$ is formally defined as a set $G = \{V, T, P, S\}$, where $V$ and $T$ are finite sets of variables and terminals, respectively, $(V \cap T = \varnothing)$, $P$ is the production set (rules defining legal derivations within the grammar), and $S$ is a start symbol. Based on $P$, the grammars can be defined/classified as:

(1) Regular grammars (RG), where $P = \{A \to \omega B \mid \omega\}$, where $A$ and $B$ are variables and $\omega$ is a (possible empty) string of terminals. RG are equivalent to finite automata (FA), and they can capture the transformations among a set of entities. They are efficient, but they do not allow embedding, as it might be required for processing of /(parentheses)/nested expressions/patterns.

(2) Context-free grammars (CFG), where

$$P = \{A \to \alpha;\ A \in V,\ \alpha \in (V \cup T)^*\}.$$

CFG are equivalent to pushdown automate (PDA)/stacks.

(3) Context sensitive grammars (CSG), where $P = \{\alpha \to \beta;\ |\beta| \geq |\alpha|$ and the normal form is $\alpha_1 A \alpha_2 \to \alpha_1 \beta \alpha_2;\ \beta$ is not the empty string, and therefore the replacement is legal only in the context $\alpha_1 - \alpha_2\}$. CSG are equivalent to linear bounded automata (LBA) and they allow recognition in those situations where context is paramount for interpretation.

(4) Phrase structure grammar (PSG), where $P = \{\alpha \to \beta;\ \alpha$ and $\beta$ arbitrary strings, and $\alpha$ is not the empty string$\}$. PSG are accepted by Turing machines and are said to be RE (recursively enumerable).

The hierarchy theorem due to Chomsky refers to the above grammars— RE, CSG, CFG, and RG—as being characteristic of languages of type $0, 1, 2$, and 3, respectively. Except for the empty string, the type-$i$ language properly includes the type-$(i + 1)$ for $i = 0, 1, 2$. The more restricted the language, the more difficult to capture a particular internal structure, but the easier its implementation.

## IV. Geometrical Invariance

We consider in this section ways to achieve invariance to 2D geometrical distortions like LT (linear transformations) given as scale, rotation, and shift (translation). Pseudoinvariance to perspective distortions (slant and tilt) is dealt with through deprojections. The approach followed is deterministic; statistical methods are considered in the next section.

### A. Invariance to Linear Transformations

#### 1. Fourier Transform

The bread-and-butter approach for dealing with geometrical distortions is based on the Fourier transform (FT). We consider a 2D image characterized by $f(x, y)$, the picture value function. The corresponding (cyclic) FT is given by $F(u, v)$, where

$$F(u, v) = \int\int_{-\infty}^{\infty} f(x, y) \exp[-j2\pi(ux + vy)] \, dx \, dy$$

The discrete version of the FT is given by the discrete Fourier transform (DFT), which takes a periodic, discrete spatial image into another discrete and periodic frequency representation. Formally, the DFT is given as

$$F(k, l) = \frac{1}{MN} \sum_{n=0}^{N-1} \sum_{m=0}^{M-1} f(n, m) \exp\left[-j2\pi\left(\frac{kn}{N} + \frac{lm}{M}\right)\right]$$

where $0 \le k \le N - 1$, and $0 \le l \le M - 1$.

It is easy to show that the magnitude of both the FT and DFT are shift invariant, i.e., invariant to a shift in position due to translation. Specifically, let $F(u, v)$ be the FT of $f(x, y)$; i.e., $F(u, v) = \mathscr{F}\{f(x, y)\}$. Assume that $f(x, y)$ is shifted by $\mathbf{d} = (x_0, y_0)$. Then, the FT of $f(x - x_0, y - y_0)$ is given as

$$F_1(u, v) = \mathscr{F}\{f(x - x_0, y - y_0)\}$$

$$= \int\int_{-\infty}^{\infty} f(x - x_0, y - y_0) \exp[-j2\pi(ux + vy)] \, dx \, dy$$

Changing variables, $a = x - x_0$ and $b = y - y_0$, and it follows that

$$F_1(u, v) = \exp[-j2\pi(ux_0 + vy_0)] \int\int_{-\infty}^{\infty} f(a, b) \exp[-j2\pi(au + bv)] \, da \, db$$

$$= \exp[-j2\pi(ux_0 + vy_0)] F(u, v)$$

Finally,

$$|F_1(u, v)| = |\exp[-j2\pi(ux_0 + vy_0)]| |F(u, v)| = |F(u, v)|$$

In a similar way one can show the same property regarding the DFT. Specifically, let $F(k, l)$ be the DFT of the image function $f(m, n)$, i.e.,

$$F(k, l) = \frac{1}{MN} \sum_{n=0}^{N-1} \sum_{m=0}^{M-1} f(n, m) \exp\left[-j2\pi\left(\frac{kn}{N} + \frac{lm}{M}\right)\right]$$

Assume again that $f(n, m)$ has been shifted by $\mathbf{d} = (n_0, m_0)$. The corresponding DFT of $f(n - n_0, m - m_0)$ is given by $F_1(k, l)$ as

$$F_1(k, l) = \frac{1}{MN} \sum_{n=0}^{N-1} \sum_{m=0}^{M-1} f(n - n_0, m - m_0) \exp\left[-j2\pi\left(\frac{kn}{N} + \frac{lm}{M}\right)\right]$$

Substitute $q = n - n_0$ and $r = m - m_0$. Then,

$$F_1(k, l) = \frac{1}{MN} \sum_{q=-n_0}^{N-1-n_0} \sum_{p=-m_0}^{M-1-m_0} f(q, p) \exp\left[ -j2\pi \left( \frac{k(q + n_0)}{N} + \frac{l(p + m_0)}{M} \right) \right]$$

$$= \left\{ \exp\left[ -j2\pi \left( \frac{kn_0}{N} + \frac{lm_0}{M} \right) \right] \bigg/ MN \right\} \sum_{q=-n_0}^{N-1-n_0} \sum_{p=-m_0}^{M-1-m_0} f(q, p)$$

$$\times \exp\left[ -j2\pi \left( \frac{kq}{N} + \frac{lp}{M} \right) \right]$$

$$= \exp\left[ -j2\pi \left( \frac{kn_0}{N} + \frac{lm_0}{M} \right) \right] F(k, l)$$

Finally, one can readily obtain that

$$|F_1(k, l)| = \left| \exp\left[ -j2\pi \left( \frac{kn_0}{N} + \frac{lm_0}{M} \right) \right] \right| |F(k, l)| = |F(k, l)|$$

An important concept related to the FT is the similarity theorem. It states (Castleman, 1978) that

$$\mathscr{F}\{f(a_1 x + b_1 y, a_2 x + b_2 y)\} \leftrightarrow (A_1 B_2 - A_2 B_1) F(A_1 u + A_2 v, B_1 u + B_2 v)$$

The theorem can be shown as following from

$$\{f(a_1 x + b_1 y, a_2 x + b_2 y)\} = \int\!\!\!\int_{-\infty}^{\infty} f(a_1 x + b_1 y, a_2 x + b_2 y)$$

$$\times \exp[-j2\pi(ux + vy)] \, dx \, dy$$

Let $w = a_1 x + b_1 y$ and $z = a_2 x + b_2 y$. Then $x = A_1 w + B_1 z$ and $y = A_2 w + B_2 z$, where

$$A_1 = \frac{b_2}{a_1 b_2 - a_2 b_1}; \qquad B_1 = \frac{-b_1}{a_1 b_2 - a_2 b_1}$$

$$A_2 = \frac{-a_2}{a_1 b_2 - a_2 b_1}; \qquad B_2 = \frac{a_1}{a_1 b_2 - a_2 b_1}$$

Then

$$\mathscr{F}\{f(a_1 x + b_1 y, a_2 x + b_2 y)\} = \int\!\!\!\int_{-\infty}^{\infty} f(w, z) \exp\{-j2\pi[(A_1 u + A_2 v)w$$

$$+ (B_1 u + B_2 v)z]\}(A_1 B_2 + A_2 B_1) \, dz \, dw$$

$$= (A_1 B_2 - A_2 B_1) F(A_1 u + A_2 V, B_1 u + B_2 v)$$

Finally

$$\mathscr{F}\{f(a_1 x + b_1 y, a_2 x + b_2 y)\} \leftrightarrow (A_1 B_2 - A_2 B_1)F(A_1 u + A_2 V, B_1 u + B_2 v)$$

In other words, the similarity theorem relates the FT of a transformed function to its original FT. An interesting application of this concept is the central slice theorem used in CT (computer tomography). Specifically, consider the projection $g_Y(x)$ of the function $f(x, y)$ on the $x$ axis,

$$g_Y(x) = \int_{-\infty}^{\infty} f(x, y)\, dx$$

The corresponding one-dimensional FT of the projection is given by

$$G_Y(u) = \mathscr{F}\{g_Y(x)\} = \int_{-\infty}^{\infty} g_Y(x)\exp(-j2\pi ux)\, dx$$

$$= \iint_{-\infty}^{\infty} f(x, y)\exp(-j2\pi ux)\, dx\, dy = F(u, 0)$$

From the similarity theorem it follows that rotation of $f(x, y)$ through an angle $\theta$ yields the original spectrum rotated by the same amount, if we let

$$a_1 = \cos\theta; \qquad b_1 = \sin\theta; \qquad \theta_2 = -\sin\theta; \qquad b_2 = \cos\theta$$

so that

$$A_1 = \cos\theta; \qquad A_2 = \sin\theta; \qquad B_1 = -\sin\theta; \qquad B_2 = \cos\theta$$

Finally, $G_Y(u) = F(u, 0)$ combines with the rotation property to imply that the one-dimensional FT of $f(x, y)$ projected onto a line at an angle $\theta$ with the $x$ axis is just as if $F(u, v)$ were evaluated along a line at an angle $\theta$ with the $u$ axis. Therefore, one has a collection of FT evaluated along $\{\theta_i\}$ radial orientations, and if the sampling is fine enough, one can obtain (reconstruct) the original image as

$$f(x, y) = \mathscr{F}^{-1}[F(u, v)]$$

Another transformation one might consider is reflection. Based on the similarity theorem,

$$\mathscr{F}\{f(ax, by)\} = \frac{1}{|ab|}F\left(\frac{u}{a}, \frac{v}{b}\right)$$

where $F(u, v) = \mathscr{F}\{f(x, y)\}$. If one assumes that $a = b = -1$, then it follows that $\{f(-x, -y)\} = F(-u, -v)$, i.e., the FT of a function reflected through the origin is given by $F(-u, -v)$. If we were to reflect this image only with respect to the $y$ axis, then

$$\mathcal{F}\{f(-x, y)\} = F(-u, v)$$

Note also that the FT of the FT yields the reflected function. Specifically

$$\mathcal{F}\{\mathcal{F}\{f(x)\}\} = \int_{-\infty}^{\infty} F(u) \exp(-j2\pi ux)\, dx = f(-x)$$

Next we assume two-dimensional image functions $f(x, y)$ characterized by circular symmetry, i.e., $f(x, y) = f_R(r)$, where $r^2 = x^2 + y^2$. The corresponding FT is given by (Castleman, 1978)

$$\mathcal{F}\{f(x, y)\} = \iint_{-\infty}^{\infty} f(x, y) \exp[-j2\pi(ux + vy)]\, dx\, dy$$

$$= \int_0^{\infty} \int_0^{2\pi} f_R(r) \exp[-j2\pi qr \cos(\theta - \phi)]\, r\, dr\, d\theta$$

where $x + jy = r \exp(j\theta)$ and $u + jv = q \exp(j\phi)$. Then

$$\mathcal{F}\{f(x, y)\} = \int_0^{\infty} f_R(r) \left( \int_0^{2\pi} \exp(-j2\pi qr \cos\theta)\, d\theta \right) r\, dr$$

The zero-order Bessel function of the first kind is

$$J_0(x) = \frac{1}{2\pi} \int_0^{2\pi} \exp(-jz \cos\theta)\, d\theta$$

Therefore, one obtains by substitution of $J_0(z)$

$$\mathcal{F}\{f(x, y)\} = 2\pi \int_0^{\infty} f_R(r) J_0(2\pi qr) r\, dr$$

and one observes that the FT of a circularly symmetric function is a function only of a single radial frequency variable $q$, i.e., $F(u, v) = F_R(q)$, where $q^2 = u^2 + v^2$. The above pair of transforms is known as the Hankel transform of zero order, i.e., $\langle f_R(r), F_R(q) \rangle$, where

$$F_R(q) = 2\pi \int_0^{\infty} f_R(r) J_0(2\pi qr) r\, dr$$

$$f_R(r) = 2\pi \int_0^{\infty} F_R(q) J_0(2\pi qr) q\, dq$$

The above discussion showed that circular symmetry is invariant; i.e., it is preserved by the FT, and that the FT of a circularly symmetric function can be considered as a one-dimensional function of a radial variable only.

There is an additional important property relating the spacing and orientation of the spatial domain to that of the corresponding FT. Specifically, if there is a structure of parallel spatial lines, spaced by $\Delta$ and passing through the origin with a slope $\alpha$, then the Fourier spectrum is spaced $1/\Delta$ and is of slope $-1/\alpha$. Such a relationship is widely used for texture classification, discrimination, and segmentation. Formally, if $f(x, y) = \delta(y - \alpha x)$, i.e., if $f(x, y)$ is a line passing through the origin at a slope $\alpha$, then the FT of $f(x, y)$ is

$$F(u, v) = \int\int_{-\infty}^{\infty} \delta(y - \alpha x) \exp[-j2\pi(ux + vy)] \, dx \, dy$$

$$= \int_{-\infty}^{\infty} \exp[-j2\pi(ux + \alpha vx)] \, dx = \delta(u + \alpha v)$$

i.e., the FT is also a line passing through the origin but at a slope of $-1/\alpha$, and the spatial line and the FT line are perpendicular to one another, $\alpha(-1/\alpha) = -1$. Next assume parallel lines spaced by $\Delta$, i.e.,

$$f(x, y) = \Sigma_{-\infty}^{\infty} \delta(x - n\Delta)$$

(the orientation is parallel to the $y$ axis, but the result can be extended to any orientation $\alpha$). The corresponding FT is

$$F(u, v) = \int\int_{-\infty}^{\infty} \Sigma_{-\infty}^{\infty} \delta(x - n\Delta) \exp[-j2\pi(ux + vy)] \, dx \, dy$$

where $f(x, y)$ is periodic and can thus be replaced by its Fourier series representation

$$f(x, y) = \sum_{-\infty}^{\infty} c_n \exp\left(-j\frac{2\pi nx}{\Delta}\right), \qquad c_n = \int_{-\infty}^{\infty} f(x, y) \exp\left(j\frac{2\pi nx}{\Delta}\right) = \frac{1}{\Delta}$$

$$F(u, v) = \int\int_{-\infty}^{\infty} \sum_{-\infty}^{\infty} \frac{1}{\Delta} \exp\left(-j\frac{2\pi nx}{\Delta}\right) \exp[-j2\pi(ux + vy)] \, dx \, dy$$

$$= \int_{-\infty}^{\infty} \sum_{-\infty}^{\infty} \frac{1}{\Delta} \delta\left(u - \frac{n}{\Delta}\right) \exp(-j2\pi vy) \, dy$$

$$= \frac{1}{\Delta} \sum_{-\infty}^{\infty} \delta\left(u - \frac{n}{\Delta}\right) \delta(v)$$

The above expression describes a line along the $u$ axis sampled at $1/\Delta$; i.e., the

FT of parallel lines spaced at $\Delta$ is a sampled line perpendicular to those lines and sampled at the inverse of the distance separating them; i.e., the spacing is given by $1/\Delta$.

## 2. Mellin Transform

We introduce next the Mellin transform (MT) (Bracewell, 1978), which either implicitly or explicitly is a major component in most applications of invariant pattern recognition (Zwicke, 1983). Formally, MT is given by

$$M(u, v) = \int\int_0^\infty f(x, y) x^{-ju-1} y^{-jv-1} \, dx \, dy$$

where $j^2 = -1$. This discrete version of the MT is given as

$$M(k \Delta u, l \Delta v) = \sum_{n=1}^{N} \sum_{m=1}^{M} f(n \Delta x, m \Delta y)(n \Delta x)^{-jk \Delta u - 1} (m \Delta y)^{-jl \Delta v - 1} \Delta x \, \Delta y$$

The following relationship holds between FT and MT if we substitute $x = Qe^q$ and $y = Pe^p$,

$$M(u, v) = Q^{-ju} P^{-jv} \int\int_{-\infty}^{\infty} f(Qe^q, Pe^p) \exp[-j(qu + pv)] \, dq \, dp$$

If one notes that $|Q^{-ju}| = |P^{-jv}| = 1$, then the magnitude of the MT is the same as the magnitude of the FT evaluated for an exponentially sampled function. Note also that such an exponential sampling is similar to the human retina sampling over a polar grid (high resolution in the "central" fovea and low resolution in the preattentive "periphery").

Invariance properties to distortions like scaling can be easily proved. Assume, for the continuous case, that $g(x, y)$ is the scaled version of the original image function $f(x, y)$, i.e., $g(x, y) = f(k_1 x, k_2 y)$. If we denote the MT of $g(x, y)$ and $f(x, y)$ by $G(u, v)$ and $M(u, v)$, respectively, then

$$G(u, v) = Q^{-ju} P^{-jv} \int\int_{-\infty}^{\infty} f(k_1 Qe^q, k_2 Pe^p) \exp[-j(qu + pv)] \, dq \, dp$$

$$= Q^{-ju} P^{-jv} \int\int_{-\infty}^{\infty} f(Qe^{q+\ln k_1}, Pe^{p+\ln k_2}) \exp[-j(qu + pv)] \, dq \, dp$$

Substitute $t = q + \ln k_1$, and $z = p + \ln k_2$

$$G(u, v) = Q^{-ju} P^{-jv} \int\limits_{-\infty}^{\infty}\!\!\int f(Qe^t, Pe^z) \exp\{-j[(t - \ln k_1)^u + (z - \ln k_2)^v]\}\, dt\, dz$$

$$= k_1^{-ju} k_2^{-jv} M(u, v)$$

Therefore $|G(u, v)| = |M(u, v)|$. A similar proof holds for the discrete case. Specifically, assume that $g(n\,\Delta x, m\,\Delta y) = f(np_1\,\Delta x, mp_2\,\Delta y)$. Then, if the discrete MT are $G(k\,\Delta u, l\,\Delta v)$ and $M(k\,\Delta u, l\,\Delta v)$, respectively, one can obtain

$$G(k\,\Delta u, l\,\Delta v) = \sum_{n=1}^{N} \sum_{m=1}^{N} f(np_1\,\Delta x, mp_2\,\Delta y)(n\,\Delta x)^{-jk\,\Delta u - 1}(m\,\Delta y)^{-jl\,\Delta v - 1}\,\Delta x\,\Delta y$$

Substitute $p_1\,\Delta x = \Delta t$ and $p_2\,\Delta y = \Delta z$

$$G(k\,\Delta u, l\,\Delta v) = \sum_{n=1}^{N} \sum_{m=1}^{M} f(n\,\Delta t, m\,\Delta z)(n\,\Delta t)^{-jk\,\Delta u - 1}\,\Delta t$$

$$\times (m\,\Delta z)^{-jl\,\Delta v - 1}\,\Delta z^{(p_1 - 1) - jk\,\Delta u - 1}(p_1)^{-1}(p_2^{-1})^{-jl\,\Delta v - 1}(p_2)^{-1}$$

$$= p_1^{jk\,\Delta u} p_2^{jl\,\Delta v} \sum_{n=1}^{N} \sum_{m=1}^{M} f(n\,\Delta t, m\,\Delta z)(n\,\Delta t)^{-jk\,\Delta u - 1}(m\,\Delta z)^{-jl\,\Delta v - 1}\,\Delta t\,\Delta z$$

$$= p_1^{-jk\,\Delta u} p_2^{-jl\,\Delta v} M(k\,\Delta u, l\,\Delta v)$$

Therefore we obtain again that $|G(k\,\Delta u, l\,\Delta v)| = |M(k\,\Delta u, l\,\Delta v)|$; i.e., the magnitudes are the same disregarding the change in scale.

### 3. Conformal Mapping

One useful technique used to achieve invariance to rotation and scale changes is the complex-logarithmic (CL) conformal mapping. Specifically, assume that Cartesian plane points are given by $(x, y) = (\text{Re}(z), \text{Im}(z))$, where $z = x + jy$. Thus we can write $z = r \exp(j\theta)$, where $r = |z| = (x^2 + y^2)^{1/2}$ and $\theta = \arg(z) = \arctan(y/x)$. Now the CL mapping is simply the conformal mapping of points $z$ onto points $w$ defined by

$$w = \ln(z) = \ln[r \exp(j\theta)] = \ln r + j\theta$$

Therefore, points in the target domain are given by $(\ln r, \theta) = (\text{Re}(w), \text{Im}(w))$, and logarithmically spaced concentric rings and radials of uniform angular spacing are mapped into uniformly spaced straight lines. More generally, after CL mapping, rotation and scaling about the origin in the Cartesian domain correspond to simple linear shifts in the $\theta \pmod{2\pi}$ and $\ln r$ directions, respectively. [It is worth noting that the CL mapping has also been conjectured for the brain/cortex function by recent neurophysiological research

(Schwartz, 1977).] Correlation techniques can then be used to detect invariance between objects (to scale and rotation) within a linear shift.

The ideas suggested by the Mellin transform and the conformal mapping are incorporated into a system designed by Casasent and Psaltis (1977). The system can recognize an object invariant to LT (linear transformations), and step by step it works as follows:

(1) Evaluate the magnitude of the FT corresponding to the image function $f(x, y)$, i.e., $|F(u, v)| = |\mathscr{F}\{f(x, y)\}|$;

(2) Convert $|F(u, v)|$ to polar coordinates $F_1(r, \theta)$;

(3) Derive $F_1(\exp \zeta, \theta)$, where $\zeta = \ln r$;

(4) The magnitude of the FT of $F_1(\exp \zeta, \theta)$ yields the PRSI (position, rotation, scale invariant) function $M_1(\omega_p, \omega_0)$.

Step (1) removes translation distortion because the magnitude of the FT is shift invariant. Step (2) changes scale and rotation distortions into linear shifts along the $\ln r - \theta$ new system of coordinates. Steps (3) and (4) are equivalent to an MT done with respect to both axes; i.e., linear shifts are recognized as scale ($k$) and rotation ($\phi$) changes.

Another attempt to build a system for invariant recognition is the ORLEC system (Moss et al., 1983). The system consists of two parts:

(1) Camera subsystem for preprocessing; it centers and normalizes the image; subsequently, the image is transformed into polar form.

(2) Pattern recognition subsystem; it compares the polar image to a set of prestored prototypes. The technique involves projecting the image points onto a given direction and determining a set of (wedge and ring) profile signatures which are invariant to scale and rotation changes, respectively.

Messner and Szu (1985) have recently suggested an image processing architecture for real-time generation of scale- and rotation-invariant patterns; the approach is based on the CL mapping as well.

We consider next a different type of representation, one which is based on nonlinear transforms of the original image. One example is the Wigner distribution, which is a simultaneous spatial/spatial-frequency (s/sf) representation.

## B. The Wigner Distribution and Conjoint Spatial/ Spatial-Frequency Representations

There is signal processing which is easier to implement in the spatial domain while some other processing finds a more congenial solution in the frequency domain (e.g., convolution). Furthermore, one can also consider the case where a conjoint representation, like the Gabor and Wigner distribution, might be the best solution (Jacobson and Wechsler, 1985a).

Motivation for conjoint image representations comes also from biological vision. Such representations are useful since they can decouple image patterns that have common functional support regions in either the spatial or spatial-frequency domains. One can also exploit the same *separability* in order to decouple multiplicative image components (e.g., reflectance and illumination) by transforming the log function. Finally, one can maximize the benefits of conjoint representations by coupling them with existing methods that use geometric domain mappings, like the complex-log conformal mapping, to achieve recognition that is *invariant* to LT, or more generally to arbitrary changes in camera aspect via deprojection methods. Such invariant methods use one or more *foveated* joint representations to yield an efficient encoding that nonetheless provides high acuity along the camera line of sight. Another important dimension along which image representation should be considered is that of resolution. Jacobson and Wechsler (1985a) address this issue by defining a new measure of energy spread called *REPentropy*. REPentropy is defined in terms of *one* (joint) signal representation, not two, as is the case with *SIGentropy*, the standard way of measuring the uncertainty of a 2D signal. (SIGentropy is defined as the product of the effective spatial extent and effective bandwidth.) The representation with the lowest REPentropy has the highest conjoint resolution since, on the average, its energy is relatively more concentrated in tight clumps throughout the (space, spatial-frequency) = $(\mathbf{x}, \mathbf{u})$ space. The Wigner distribution (WD) is a conjoint pseudoenergy representation whose rapidly growing emergence comes from the fact that virtually all popular conjoint representations are equivalent to smoothed derivations of the WD. Given an arbitrary $N$-dimensional signal, $f(\mathbf{x})$, the WD, $J_f(W)(\mathbf{x}, \mathbf{u})$ of $f(\mathbf{x})$, is defined by

$$J_f(W)(\mathbf{x}, \mathbf{u}) = \int_{-\infty}^{\infty} f\left(x + \frac{\alpha}{2}\right) f^*\left(\mathbf{x} - \frac{\alpha}{2}\right) \exp[-j(\mathbf{u} \cdot \boldsymbol{\alpha})] \, d\boldsymbol{\alpha}$$

The WD and its approximations such as the composite pesudo-Wigner distribution (CPWD) can provide higher conjoint resolution than the two other image representations currently considered: DOG (Difference of Gaussians) and GABOR. For reasons of computational efficiency and modularity, one can derive the CPWD by using sequential computation such as DOG → GABOR → CPWD (Jacobson and Wechsler, 1985a). An experimental subsystem called FOVEA implements the above ideas regarding high-resolution conjoint image representations and provides invariance to LT and perspective distortions via CL mappings and deprojection (Jacobson and Wechsler,

1985b), respectively. FOVEA employs complete, canonical, object-centered representations at the interface with visual memory.

## C. Orthogonal Series Expansion

There is much research regarding signal representation in terms of orthogonal functions. The coefficients characterizing such an expansion enjoy some nice properties with respect to invariance, and one can make use of them for classification purposes. We restrict ourselves in the following to Fourier series expansion only. Other orthogonal series expansions are discussed at great length by Ahmed and Rao (1975).

## 1. Fourier Descriptors

One popular method derives Fourier descriptors based on objects' closed boundaries. One can look at the closed boundary via its "intrinsic function" $K(l)$, where $l$ is the distance (clockwise) from a preselected starting point along the curve, and $K(l)$ is the boundary curvature (Rosenfeld and Kak, 1982). [For line segments $K(l) = 0$, while circular arcs have constant curvature $K(l) = a$, where $a$ is the radius of curvature.] $K(l)$ is periodic, the period $L$ is the arc length (clockwise) from the starting point and back, and $K(l) = K(l + nL)$. $K(l)$ being periodic, one can consider its Fourier orthogonal series expansion, and choose the first $N$ coefficients as the classification features; the features are called Fourier descriptors. $K(l)$ can be plagued by sharp angles which become instantaneous impulse discontinuities. [Think of the square, where $K(l)$ is zero everywhere except the corners, where it is infinite.] Without loss of information the above problem can be overcome by defining the slope intrinsic function $G(l)$, where $G(l) = \int K(l) \, dl$, and the discontinuities are now step functions. $G(l)$ is not periodic, and it can be easily derived that cumulative effects yield $G(l + nL) = G(l) + 2\pi n$. The rising component can be eliminated, and then the resulting function $\hat{G}(l) = G(l) - 2\pi l/L$ is periodic, i.e., $\hat{G}(l + nL) = \hat{G}(l)$. The average value of the slope intrinsic function depends on what we choose as the starting point. If $\bar{G}(l)$ is the average, then

$$\bar{G}(l) = \hat{G}(l) - \frac{1}{L} \int_0^L \hat{G}(l) \, dl$$

Now, the signal average affects only the first term (dc) of the Fourier series, i.e., the $C_0$ term. By discarding $C_0$ we basically discard information related to the average and make the recognition invariant to rotation and choice of starting point. Finally, the Fourier series is given as

$$\hat{G}(l) = \sum_{-\infty}^{\infty} C_n \exp(2\pi n j l/L)$$

where $j^2 = -1$ and the coefficients $C_n$ are found as

$$C_n = \frac{1}{L} \int_0^L \hat{G}(l) \exp(-2\pi njl/L)\,dl$$

Note that the above coefficients are normalized by $(1/L)$, which renders the coefficients scale invariant. ($C_0$ was set to zero for rotation invariance.) Finally, one can note that the procedure is based on boundary information only, and that boundary location within the field of view is irrelevant. Thus we have also translation (shift) invariance, and therefore, we can achieve invariance to LT (linear transformation) if Fourier descriptors, as derived above, are used for shape classification through the use of traditional statistical methods. The above discussion applies as well to the discrete case, when one can make use of the chain code representation (0 through 7 directions rather than 0 through $2\pi$ for the continuous case). The modified slope intrinsic function is given by $\hat{G}(l) = G(l) - 8l/L$. This is a periodic function, and there exists a corresponding Fourier series expansion $\hat{G}(l)$ defined only for integer values of $l$. The discrete coefficients are given by

$$C_n = \frac{1}{L} \sum \hat{G}(l) \exp(-2\pi njl/L)$$

Since the coefficients are complex, their magnitudes $|C_n| = (a_n^2 + b_n^2)^2$ are used as features for statistical classification. Note also that the step size for the chain code is not uniform. Odd number directions stand for distances of 1.414 rather than the grid size of 1.0 for even directions. The above argument should be part of any attempt to accurately estimate the Fourier descriptors, or a subset of them. For an application of such descriptors see Mitchell and Grogan (1984). Another example of an orthogonal series expansion and the use of series coefficients is the Karhunen–Loève transform, which is discussed in the section dedicated to statistical invariance (Section III,E).

### D. Moments

Another widely used method to achieve invariant pattern recognition is that of the moments. Assuming a two-dimensional continuous function $f(x, y)$, we define the moment of order $(p + q)$ by the relation

$$m_{pq} = \int\limits_{-\infty}^{\infty}\int x^p y^q f(x, y)\,dx\,dy, \qquad p, q = 0, 1, 2, \ldots$$

The central moments can be expressed as:

$$\mu_{pq} = \int\limits_{-\infty}^{\infty}\int (x - \bar{x})^p(y - \bar{y})^q f(x, y)\, dx\, dy$$

where the average/mean/centroid/center of gravity is given by $(\bar{x}, \bar{y}) = (m_{10}/m_{00}, m_{01}/m_{00})$. Note that $m_{00}$ is the total function mass, i.e.,

$$m_{00} = \int\limits_{-\infty}^{\infty}\int f(x, y)\, dx\, dy$$

If a discrete image is considered, we have

$$\mu_{pq} = \sum_x \sum_y (x - \bar{x})^p(y - \bar{y})^q f(x, y)$$

Again, note that for a binary blob $m_{00}$ represents its area. A set of normalized central moments can be defined (Hu, 1962):

$$\eta_{pq} = \frac{\mu_{pq}}{\mu_{00}^\gamma}, \qquad \gamma = \frac{p + q}{2} + 1$$

The above normalized central moments can be used to define a set of seven moments invariant to LT. The set is given below.

$$M_1 = \eta_{20} + \eta_{02}$$
$$M_2 = (\eta_{20} - \eta_{02})^2 + 4\eta_{11}^2$$
$$M_3 = (\eta_{30} - 3\eta_{12})^2 + (3\eta_{21} - \eta_{03})^2$$
$$M_4 = (\eta_{30} + \eta_{12})^2 + (\eta_{21} + \eta_{03})^2$$
$$M_5 = (\eta_{30} - 3\eta_{12})(\eta_{30} + \eta_{12})[(\eta_{30} + \eta_{12})^2 - 3(\eta_{21} + \eta_{03})^2]$$
$$+ (3\eta_{21} - \eta_{03})(\eta_{21} + \eta_{03})[3(\eta_{30} + \eta_{12})^2 - (\eta_{21} + \eta_{03})^2]$$
$$M_6 = (\eta_{20} - \eta_{02})[(\eta_{30} + \eta_{12})^2 - (\eta_{21} + \eta_{03})^2]$$
$$+ 4\eta_{11}(\eta_{30} + \eta_{12})(\eta_{21} + \eta_{03})$$
$$M_7 = (3\eta_{21} - \eta_{03})(\eta_{30} + \eta_{12})[(\eta_{30} + \eta_{12})^2 - 3(\eta_{21} + \eta_{03})^2]$$
$$- (\eta_{30} - 3\eta_{12})(\eta_{12} + \eta_{03})[3(\eta_{30} + \eta_{12})^2 - (\eta_{21} + \eta_{03})^2]$$

It is easy to show as an example that $\eta_{pq}$ are invariant to scale change.

Assuming that $\alpha$ is the scale change, it follows that $x' = \alpha x$, $y' = \alpha y$, and

$$\eta'_{pq} = \frac{\mu^{pq}}{(\mu'_{00})^{(p+q)/2+1}} = \frac{\mu^{pq}\alpha^{p+q+2}}{(\alpha^2\mu_{00})^{(p+q)/2+1}} = \frac{\mu^{pq}}{(\mu_{00})^{(p+q)/2+1}} = \eta_{pq}$$

where

$$\mu'_{00} = \int\!\!\!\int_{-\infty}^{\infty} dx'\, dy' = \alpha^2 \int\!\!\!\int_{-\infty}^{\infty} dx\, dy = \alpha^2\mu_{00}$$

Note that the digital nature of data introduces errors which sometimes might mask the invariance. The functions $M_1$ through $M_6$ are also invariant under reflection, while $M_7$ changes sign under reflection.

Another approach in using the concept of moments is to consider them in polar coordinates, i.e., radially and angularly. The radial and angular moments are defined (Reddi, 1981) as

$$\psi(k, p, q, g) = \int_0^\infty \int_{-\pi}^{+\pi} r^k g(r, \theta) \cos^p \theta \sin^q \theta \, d\theta \, dr$$

One can note that

$$\mu_{pq} = \int_{-\infty}^{\infty} \int_{-\infty}^{\infty} f(x, y) x^p y^q \, dx \, dy$$

$$= \int_0^\infty \int_{-\pi}^{\pi} r^{p+q+1} \cos^p \theta \sin^q \theta \, g(r, \theta) \, d\theta \, dr$$

$$= \psi(p + q + 1, p, q, g)$$

The seven invariant functions defined above can then be expressed in terms of radial and angular moments as follows: Defining

$$\psi_r(k, g) = \int_0^\infty r_k g(r, \theta) \, dr, \qquad \psi_\theta(g) = \int_{-\pi}^{\pi} g(r, \theta) \, d\theta$$

$$M_1 = \psi_r(3, \psi_\theta(g))$$

$$M_2 = |\psi_r(3, \psi_\theta(ge^{j2\theta}))|^2$$

$$M_3 = |\psi_r(4, \psi_\theta(ge^{j3\theta}))|^2$$

$$M_4 = |\psi_r(4, \psi_\theta(ge^{j\theta}))|^2$$

$$M_5 = \mathrm{Re}[\psi_r(4, \psi_\theta(ge^{j3\theta}))\psi_r^3(4, \psi_\theta(ge^{-j\theta}))]$$

$$M_6 = \mathrm{Re}[\psi_r(3, \psi_\theta(ge^{j2\theta}))\psi_r^2(4, \psi_\theta(ge^{-j\theta}))]$$

$$M_7 = \mathrm{Im}[\psi_r(4, \psi_\theta(ge^{j3\theta}))\psi_r^3(4, \psi_\theta(ge^{-j\theta}))]$$

Invariance to rotation can be shown as follows: if $\theta \rightarrow \theta + \alpha$, then

$$M_5 = \text{Re}[\psi_r(4, \psi_\theta(ge^{j3(\theta+\alpha)}))\psi_r^3(4, \psi_\theta(ge^{-j(\theta+\alpha)}))]$$

$$= \text{Re}[\exp(j3\alpha)\psi_r(4, \psi_\theta(ge^{j3\theta}))\exp(-j3\alpha)\psi_r^3(4, \psi_\theta(ge^{-j\theta}))]$$

$$= \text{Re}[\psi_r(4, \psi_\theta(ge^{j3\theta}))\psi_r^3(4, \psi_0(ge^{-j\theta}))]$$

Changing $\theta$ to $-\theta$, for reflection, would leave the first 6 functions unchanged, but $M_7$ would change sign.

Advantages of the radial and angular moments are:

(1)   There is no restriction on positive integral powers of $r$ in generating the invariants.

(2)   The weighting function is not restricted to $r^k$ but could be exponentials and similar functions of $r$. (Such weighting could be better for target identification with low SNR.) Finally, by proper weighting, the functions could be made scale invariant.

The moments can also be used to define a natural (invariant) system of coordinates determined by the moments of inertia (Wechsler, 1979). Assume, as before, that the centroid is of coordinates $(\bar{x}, \bar{y})$. Then, the moment of inertia of $f(x, y)$ about the line $y = x \tan \theta$ [i.e., a line through the origin/centroid, $(x, y) = (x - \bar{x}, y - \bar{y})$]

$$M_\theta = \sum\sum (x \sin \theta - y \cos \theta)^2 f(x, y)$$

Let $\theta_0$ be the angle for which $M_\theta$ is minimal. If there is a unique $\theta_0$, then the line $y = x \tan \theta_0$ is called a principal axis of inertia with respect to $f(x, y)$. Define

$$M_{20} = \sum\sum x^2 f(x, y)$$

$$M_{11} = \sum\sum xy f(x, y)$$

$$M_{02} = \sum\sum y^2 f(x, y)$$

Then

$$M_\theta = M_{20} \sin^2 \theta - 2M_{11} \sin \theta \cos \theta + M_{02} \cos^2 \theta$$

and the minimum for $M_\theta$ is achieved when $\partial M_\theta / \partial \theta = 0$. It follows that $\theta_0$ is a solution of the following equation

$$\tan 2\theta_0 = 2M_{11}/(M_{20} - M_{02})$$

There are two solutions for $\theta_0$, $90°$ apart, and therefore one principal axis is perpendicular to the other. The moment of inertia is maximal with respect to one principal axis, and it is minimal with respect to the other axis. Define

$$\alpha = \partial^2 M_\theta / \partial \theta^2 = 2(M_{20} - M_{02}) \cos 2\theta_0 + 4M_{11} \sin 2\theta_0$$

and the moment of inertia will be minimal if $\alpha \geq 0$. Once the principal axes are determined, the $x$ axis of the desired system of coordinates is chosen according to the requirement $M_{20} > M_{02}$ and the positive direction of the $x$ axis is chosen according to whether $M_{30} = \sum x^3 f(x, y)$ is positive or negative. Note that if discrete images are used, the moment of inertia about the $x$ axis is defined (Freeman, 1974) as

$$M_2 = \sum_{i=1}^{n} \tfrac{1}{3} a_{i_x} (y_{i-1}^3 + \tfrac{3}{2} a_{i_y} y_{i-1}^2 + a_{i_y}^2 y_{i-1} + \tfrac{1}{4} a_{i_y})$$

Another process aimed at determining a natural system of coordinates is the KL transform, to be discussed within the framework of statistical invariance.

## V. Statistical Invariance

We consider in this section the possibility of performing invariant pattern recognition on images for whom the variance (distortion), which impedes the recognition process, has been reduced through statistical means.

### A. Wiener Filter

Smoothing (both spatially and/or temporally) a set of $M$ images increases the SNR (signal-to-noise ratio) by $M$. The assumption is that each image $I_i(x, y)$, $i = 1, \ldots, M$, is given as $I_i = S + N_i$, where $S$ is the original signal, $N_i$ uncorrelated random noise of zero mean value, and the signal-to-noise ratio is given by $S^2/E(N^2)$.

Castleman (1978) shows how smoothing and optimal filter (Wiener) theory can be used to enhance the boundaries of blood vessels for angiography. Specifically, the power spectrum of the signal is given as

$$P_s(s) = |\mathscr{F}\{S(\mathbf{x})\}|^2 \simeq \left|\mathscr{F}\left\{\frac{1}{N}\sum f_i(\mathbf{x})\right\}\right|^2$$

where there are $n$ lines scanned for the corresponding picture value function. The noise spectrum is estimated as

$$P_N(s) \simeq \frac{1}{N}\sum |\mathscr{F}\{f_i(x) - S(x)\}|^2$$

The Wiener estimator transfer function is given by $G_0(s) = P_s(s) + P_N(s)$. (This last expression is an alternative way of looking at the Wiener filter in the context of linear systems; compare it with the filter derived based on statistical considerations in the review section.) The obtained filter $G_0(s)$ is smoothed into

$G_0(s)$ and the impulse response $\tilde{g}(x)$ is readily obtained. To find accurate vessel boundaries one now convolves the original scan line $f_i(x)$ with $\tilde{g}'(x)$, the derivative of the impulse response.

## B. Autoregressive Models

An autoregressive (AR) model for two-dimensional shape classification is described by Dubois and Glanz (1986). The AR is a statistical model which attempts to capture the structure of the object's shape by expressing each pixel on the boundary as a weighted linear combination of a prespecified number ($m$) of preceding points, where $m$ is the order of the model. The closed boundary is represented as a sequence of radial distances from the shape centroid to the boundary. The spacing among the radii is $\psi = 2\pi t/N$, $t = 1, 2, \ldots, N$, and a corresponding periodic function $r(t)$ is obtained. Special extensions are required if the shape is not convex of if the radii meet the boundary at more than one point. (The one-to-many function needs unwrapping.) Following Kashyap and Cellappa (1981), the AR model for the periodic sequence $r(t)$ is given by

$$r_t = \alpha + \sum_{j=1}^{m} \theta_j r_{t-j} + \sqrt{\beta} w_t, \qquad t = 1, \ldots, N$$

where $\theta_j$ are the unknown lag coefficients, $\alpha, \sqrt{\beta}$ constants to be estimated, and $\{w_t\}$ a random sequence of independent zero-mean samples with unit variance. Solving for the unknowns one obtains

$$\alpha = \bar{r}\left(1 - \sum_{j=1}^{m} \theta_j\right)$$

$$\beta = \frac{1}{N} \sum_{t=1}^{N} \left(r_t - \alpha - \sum_{j=1}^{m} \theta_j r_{t-j}\right)^2$$

$\alpha$ is proportional to the mean radius vector length $\bar{r}$, and as such it is a descriptor of the shape size, $\beta$ is the residual variance, and $\alpha/\sqrt{\beta}$ gives a measure of the shape signal-to-noise ratio. The MSE yields the following solution for the $\{\theta_j\}$:

$$
\begin{bmatrix} \theta_1 \\ \theta_2 \\ \vdots \\ \theta_m \\ \alpha \end{bmatrix} =
\begin{bmatrix} \sum r_{t-1}^2 & \cdots & \sum r_{t-1} r_{t-m} & \sum r_{t-1} \\ & & & \\ \sum r_{t-m}^2 & & & \sum r_{t-m} \\ \sum r_{t-1} & \sum r_{t-m} & & N \end{bmatrix}
\begin{bmatrix} \sum r_{t-1} r_t \\ \\ \sum r_{t-m} r_t \\ \sum r_t \end{bmatrix}
$$

It is the set $\{\theta_j\}$ which is used to describe the shape and classify objects by providing procedures like the feature weighted method, or the rotated coordinated system method, and the features are $\{\theta_j\}$, as found for a given AR model of order $m$. If the sampling is fine enough, the set $\{\theta_j\}$ is invariant to translation, rotation, and the starting point used to follow the boundary, assertions which are readily intuitive. It can be also shown that $\theta$ is independent of the size of the object. Therefore, the AR model if used to encode the boundary can make the recognition procedure invariant to LT. [Compare this method with the one described in Section IV,C,1 (Fourier descriptors).]

### C. Viterbi Algorithm

OCR (optical character recognition) is a practical application for which the need to correct garbled text arises quite often. Single character recognition is prone to error, and contextual postprocessing might help. Markov chains, as described in the review of basic concepts, can encode (state) transitions between characters for a particular dictionary domain. Such statistics are gathered and then local processing (single character recognition) is combined with diagrams (probability of a transition between two consecutive characters), yielding a global measure of goodness of fit. The Viterbi algorithm (Forney, 1973) is a recursive optimal solution to the task of estimating the state (character) sequence of a discrete-time finite-state (alphabet) Markov chains embedded in memoryless noise. The algorithm is "analog" to FDP (forward dynamic programming) and is described next.

Assume that $\boldsymbol{\alpha} = (\alpha_1, \alpha_2, \ldots, \alpha_n)$ is the (local) measurement vector corresponding to (letter) position $i$ (within a word of length $n + 1$ and of known first letter $x_0$), and that $P(\boldsymbol{\alpha}) = (p(\alpha_1), \ldots, p(\alpha_M))$ is the a priori letter-class probability vector, i.e., the probability that a given observation $\alpha_i$ could be induced by some letter class $l \in \{1, 2, \ldots, M\}$, where $M$ is the size of the alphabet. The local letter-class assignments can be combined based on known probabilities for Markov-type transitions of a (subset) of the (English) dictionary. Specifically, one then looks for the maximum a posteriori probability (MAP) estimation, i.e., the best global solution. The Viterbi algorithm (VA), like the FDP, transforms the MAP estimation into seeking a shortest-path solution. One can write

$$P(\mathbf{X}, \boldsymbol{\alpha}) = P(\mathbf{X} \mid \boldsymbol{\alpha}) P(\boldsymbol{\alpha}) = P(\mathbf{X}) P(\boldsymbol{\alpha} \mid \mathbf{X})$$

$$= \prod_{k=0}^{n-1} P(x_{k+1} \mid x_k) \prod_{k=1}^{n} P(\alpha_k \mid x_k)$$

where $\mathbf{X}$ is the letter-class-assignment vector, $\mathbf{X} = (x_0, x_1, \ldots, x_n)$, and $P(\boldsymbol{\alpha})$ is the *a priori* probability vector. It can be shown that to maximize $P(\mathbf{X}, \boldsymbol{\alpha})$ is equivalent to finding a shortest path through a tree built as follows:

(1) Each node corresponds to a letter-class at step/position $k$ through the word (letter-sequence).

(2) Branches $\beta_k$ correspond to transitions between letter-states $x_k$ and $x_{k+1}$.

The tree starts with $k = 0$ (first letter—$x_0$) and ends with $k = n$ (last letter— $x_n$). Then, if the length of branch $\beta_k$ is defined as

$$\lambda(\beta_k) = -\ln P(x_{k+1} | x_k) - \ln P(\alpha_{k+1} | x_{k+1})$$

the total length for traversing the tree through some path is given as

$$\sum_{k=0}^{n-1} \lambda(\beta_k) = -\ln P(\mathbf{X}, \boldsymbol{\alpha})$$

and it is obvious that MAP is equivalent to finding the shortest path. The shortest-path problem can be efficiently solved if one observes that for any $k > 0$ and $x_k$ there is just one best minimal path, called a "survivor." The shortest complete path must always begin with one of those survivors, and recursively one extends, step by step, each survivor at step $k$ into a new survivor at step $k + 1$.

## D. Eigenvalue Methods

The KL transform, despite being a very expensive computational tool, suggests a number of ways for performing invariant pattern recognition. First, we remind the reader that the choice of basis vectors is always made such that they point in the direction of maximum variance, as given by the eigenvectors of the covariance matrix. If an object is rotated by an angle $\theta$, such that $\mathbf{y} = A\mathbf{x}$, where

$$A = \begin{pmatrix} \cos\theta & \sin\theta \\ -\sin\theta & \cos\theta \end{pmatrix}$$

then the eigenvectors are rotated by the same angle $\theta$, and registration/ alignment within a standard coordinate system is possible. If $\mathbf{x}$ is normalized and becomes of zero mean, then invariance to translation is achieved as well. while scaling by a factor of $k$ can be accounted by considering the factor of $k^2$ introduced into the covariance matrix $\sum$. The invariant aspect of KLT as discussed above and MSE approximation if truncation is considered (for data compression) are used by Marinovic and Eichman (1985) in conjunction with the Wigner distribution (WD) for invariant shape analysis. Specifically, the WD allows one to perform feature extraction and pattern classification in a conjoint space/spatial-frequency (s/sf) domain and, therefore, to enjoy the benefits of simultaneous s/sf representations. Assuming 2D (shape) patterns given in terms of their 1D (one-dimensional) parametric boundary representation, i.e., a sequence of $r(t)$ radial distances measured from the centroid to each of $N$ equidistant points on the boundary (if the shape is not convex, reservations similar to the ones made in the AR model apply), the corresponding WD is defined as

$$W_r(t, m) = \frac{1}{T} \int_{-T/2}^{T/2} r\left(t + \frac{\alpha}{2}\right) r^*\left(t - \frac{\alpha}{2}\right) \exp\left(-j\frac{2\pi m\alpha}{T}\right) d\alpha$$

where $T$ is the spatial period, and $0 \leq t < T$. If generalized singular value decomposition (KLT) is performed, the resulting diagonal eigenvalue matrix is given as $D = \text{diag}(\lambda_1, \lambda_2, \ldots, \lambda_N)$. If all but $k$ singular values are close to zero, truncation to $k$ terms is feasible. The KLT is performed using a sampled approximation $W(n, m)$ of the WD, where $n, m = 1, \ldots, N$. {Remember that the WD is a 2D representation (s/sf) of a 1D signal $[r(t)]$.} Then the generalized eigenvalue analysis problem can be stated as

$$W(n, m) = UDV' = \sum_{i=1}^{N} \lambda_i u_i(n) v'_i(m), \qquad \|W\|^2 = \sum_{i=1}^{N} \lambda_i$$

where $\{\lambda_i\}_{i=1}^{N}$ are the positive square roots of the square matrix $WW'$. Columns of $U$ (or $V$) are orthonormal eigenvectors $u_i$ (or $v_i$) of $WW'$ (or $W'W$). Finally, a set of $k$ descriptors obtained as described above was successfully used to discriminate among shapes like letters or airplane contours. One has to note that the invariance mentioned earlier is a kind of pseudo-invariance due to the nonlinear nature of selecting the first $k$ eigenvalues among $N$.

### E. Integral Geometry

Statistical invariance can be achieved not only through digital processing but also through optical pattern recognition. One example is the generalized chord transform (GCT) suggested by Casasent and Chang (1983). The GCT is a nice combination of concepts such as:

(1) Statistics on the chords of a pattern, a concept coming from integral geometry (Moore, 1972); and

(2) Polar representation and wedge and ring detectors, insensitive to scale and rotation. The polar concept is related to conformal mappings, and the detectors we saw implemented in OLREC (Moss et al., 1983).

Specifically, if the boundary of an object is given by $b(x, y) = 1$, the chord distribution (of length $r$ and angle $\theta$) is given by $h(r, \theta)$. Formally, a chord exists between two points $(x, y)$ and $(x + r\cos\theta, y + r\sin\theta)$ if $g(x, y, r, \theta) = b(x, y)b(x + r\cos\theta, y + r\sin\theta) = 1$. The statistical distribution $h(r, \theta)$ can be defined as

$$h(r, \theta) = \iint g(x, y, r, \theta)\, dx\, dy = \iint b(x, y)b(x + r\cos\theta, y + r\sin\theta)\, dx\, dy$$

Substitute $(\xi, \eta)$ for $(r\cos\theta, r\sin\theta)$ and then one obtains

$$h(\xi, \eta) = \iint b(x, y)b(x + \xi, y + \eta)\, dx\, dy = b(x, y) \otimes b(x, y)$$

i.e., the distribution $h(\xi, \eta)$ is the autocorrelation function of the boundary. If instead of restricting ourselves to the boundary only, we substitute any picture value function $f$ for the binary valued function $b$, one obtains the GCT as

$$h_G(\xi, \eta) = \iint f(x, y)f(x + \xi, y + \eta)\, dx\, dy$$

Both for computational reasons (data compression) and invariance, ring and wedge detectors are defined as, respectively,

$$h_G(r) = \int h_G(r\cos\theta, r\sin\theta)r\, d\theta$$

$$h_G(\theta) = \int h_G(r\cos\theta, r\sin\theta)r\, dr$$

One sees that the ring and wedge detection features are independent of the orientation and scale changes in the chord distribution. The Fisher discriminant can be used for classifying objects based on the chord distribution induced features.

## F. Matched Filters

The review of basic concepts introduced the Wiener filter as the optimal filter for recovering an unknown signal embedded in noise, i.e., approximating it in the MSE sense. There are situations, as in object pattern recognition, where all one is interested in is to determine the presence or absence of a given object at a given location or moment in time, i.e., locating a known signal in noisy background. The matched filter (MF) is used for just such a purpose. If the function we are looking for is $f_i(t)$, for class $i$, then the impulse response for the case of white noise is merely a reflected and shifted version of the signal itself, i.e., $g_i(t) = f_i(t_0 - t)$. The MF is then a cross-correlator, and its output is "high" at $t_0$ when $f_i(t)$ is present and "small" when it is absent. Specifically, if the available signal and noise are given by $m(t)$ and $n(t)$, respectively, then

$$m(t) = f_i(t) + n(t)$$

$$u(t) = f_i(t) \otimes g_i(t) = \int_{-\infty}^{\infty} f_i(\tau) f_i(t_0 - t + \tau) \, d\tau = R_m(t_0 - t)$$

$$v(t) = n(t) \otimes g_i(t) = \int_{-\infty}^{\infty} n(\tau) f_i(t_0 - t + \tau) \, d\tau = R_{mn}(t_0 - t)$$

The output $y(t) = u(t) + v(t) = R_m(t_0 - t) + R_{mn}(t_0 - t)$ has an autocorrelation component only when the class object $i$ is present and always a cross-correlation component. [Assuming correlation between noise and class $i$ is small, i.e., $R_{mn} \simeq 0$ and that $R_m(\tau)$ peaks at $\tau = 0$ so the MF output is large at $t = t_0$ as desired.) Both the SDF (synthetic discriminant function) and the GMF (generalized matched filter), to be presented next, try to capture the structure of an object by using the concept of the MF. The MF to be used is obtained through averaging over "representative" images defining an object and/or its stable states (i.e., object appearances which are more or less the same, i.e., they are stable over a given range of viewing conditions).

## 1. Synthetic Discriminant Functions

Hester and Casasent (1981) use the conceptual model of the SDF as the basis for developing synthetic discriminant functions (SDF), which are supposed to maintain reasonable recognition capability regardless of intensity and geometrical distortions. Assume one object is given by its training set as $\{f_n(x)\}$, and that each member can be expanded in terms of orthonormal basis functions $\phi_m(x)$ as

$$f(x) = \sum_{m=1}^{M} a_m \phi_m(x)$$

Then each input $f(x)$ can be expressed as a vector $\hat{f} = \langle a_1, a_2, \ldots, a_M \rangle$. Using geometrical considerations the energies $E_f$ and $E_{fg}$ (cross-correlation) are given as

$$E_f = \int f^2(x) \, dx = \sum_m a_m^2 = \hat{f} \cdot \hat{f} = R_s$$

$$E_{fg} = \hat{f} \cdot \hat{g} = \sum_m a_m b_m = f \otimes g \bigg|_{\tau=0} = R_{fg}(0) = R_p$$

where $R_s$ and $R_p$ stand for a hypersphere of radius $R_s$ and a hyperplane ($g$ is fixed), respectively. To perform classification one can then require finding a fixed reference function $\hat{g}$ such that objects belong to class $i$ if and only if $\hat{f} \cdot g_i \leq R_{p_i}$. This is equivalent with determining an M(S)F (matched spatial filter). Alternatively, one can require that objects belong to class $i$ if $\hat{f} \cdot \hat{f} \leq R_{s_i}$, i.e., be within an equienergy surface sphere of radius $s_i$. The orthonormal set $\{\phi_m\}$ and the desired filter are yet to be determined. One can write

$$\hat{f}_n = \sum_m a_{nm} \hat{\phi}_m, \qquad \hat{g} = \sum_m b_m \hat{\phi}_m$$

and the constraint needed for classification is

$$\hat{f}_n \cdot \hat{g} = R_{f_n} \hat{g}(0) = \sum_m a_{nm} b_m = R_p = \text{constant}$$

Finding the optimal $g$ one has to find both $b_m$ and $\hat{\phi}_m$. The algorithm to do just that is:

   (1)   Form the autocorrelation matrix $\mathbf{R}_{f_i f_j}(0) = \hat{f}_i \cdot \hat{f}_j$;
   (2)   Diagonalize [Gram–Schmidt (GS) or KLT] $\mathbf{R}$ and determine $\hat{\phi}_m$;
   (3)   Determine $a_{nm}$ coefficients from $a_{nm} = \hat{f}_n \cdot \hat{\phi}_m$;
   (4)   Use some $R_p$ as a constant, possibly $R_p = 1$, and from $\hat{f}_n \cdot \hat{g} = R_p$ determine $b_m$.

The algorithm presented above then provides the optimum filter $\hat{g}$. How can such an approach deal with additive (zero-mean) noise? The noisy input is $\hat{f}' = \hat{f} + \hat{n}$. Since $E[\langle n \rangle] = 0$, its average energy $E_n$ is given by

$$E_n = \langle \hat{n} \cdot \hat{n} \rangle = \sum_n \sigma_{nn}^2$$

and can thus be bound by the hypersphere centered at the origin and of radius $\rho = k \langle \hat{n} \cdot \hat{n} \rangle^{1/2}$, where $k$ is a function of $\{\sigma_{nn}^2\}$. Then, our correlation threshold criterion is modified as

$$\hat{f}'_n \cdot \hat{g} = R_p \pm \rho$$

i.e., $\hat{f}'$ is bound to lie within a sphere of radius $\rho$. The method presented above has been implemented such that the training set is given in terms of stable states like $f_1$, right side (RS); $f_2$, front side (FS); and $f_3$, rear (R). In other words, SDF attempts to create a match filter which is an average over all possible object appearances, both stable state and their translated versions (which are given as part of the training set as $\{f_1^r, f_2^j, f_3^k\}$

## 2. Generalized Matched Filters

   Caulfied and Weinberg (1982) follow a similar line of thought to the SDF of Hester and Casasent (1981) and suggest the use of generalized matched filters (GMF). They view the $M \times N$ samples of the Fourier transform of the input as an $MN$-component feature vector and use the Fisher discriminant to find that linear combination which best separates in the MSE sense the objects of interest from other patterns and/or noise. An important contribution is made regarding the complexity of the computation involved, i.e., solving large eigenvalue problems. Specifically, an assumption is made that $G$, the vector derived from the FT (using an arbitrary raster scan like left-to-right, top-down), is such that its individual components have no cross-correlation and hence that $B$ (between-class scatter matrix) and $W$ (within-class scatter matrix) are diagonal. As was the case with SDF, the GMF achieves invariances by clustering an object and its different appearances (according to viewpoints) into just one class, which then has to be discriminated/separated from other classes of objects, (white) noise being one among several alternative classes. Note that recognition achieves as a byproduct what a matched filter is supposed to, i.e., an indication regarding the presence/absence of a given object.

## VI. RESTORATION

One of the models suggested earlier for machine vision systems included a preprocessing step aimed at restoring the image quality. Such quality can be degraded by factors such as motion, uneven illumination, or noise due to the sensors. Restoration of (digital) imagery is a full-fledged subject, and it is treated in detail by Andrews and Hunt (1977). We treat the subject in the context of recovering intrinsic scene characteristics as defined by Tenenbaum and Barrow (1978). Specifically, one wants to parallel early (low-level) visual processing and to describe a scene in terms of intrinsic (veridical) characteristics—such as reflectance, optical flow (motion), range, and surface orientation—for each pixel in the image. Following Tenenbaum and Barrow, support for such an approach comes from: (1) usefulness of such characteristics for higher-level scene analysis; and (2) human ability to determine those characteristics, regardless of viewing conditions or familiarity with the scene. The difficulty for recovering such intrinsic properties stems from the fact that the information is confounded in the original brightness–intensity image: a single intensity value encodes all the intrinsic characteristics of the corresponding pixel. Recovery depends then on exploiting constraints, derived from assumptions about the nature of the scene and the physics of the imaging process.

### A. Homomorphic Filtering

Spatial filtering is one technique used in the process of recovering intrinsic characteristics, and through it image enhancement can be achieved as well. De filtering (high-pass filter) lowers the saturation due to large white/black blobs and by rescaling image intensities allows for finer details to be perceived. This last step is implemented by processing the image histogram, where the histogram is defined as a graph relating an image property and the corresponding number of pixels exhibiting it. Another attractive technique is that of homomorphic filtering, based on the multiplicative law of the image formation process. The intensity/brightness as recorded by the image formation process is given as $I(x, y) = R(x, y)\tilde{I}(x, y)$, where $R$ and $\tilde{I}$ stand for reflectance and the illumination falling on the surface, respectively. The intrinsic characteristic one is interested to derive is that of $R(x, y)$, whose perceptual correlate is called lightness. The illumination component $\tilde{I}(x, y)$ is composed usually of spatial low frequencies characteristic to gradual variations, whereas the reflectance component is rich in detail and subject to sudden variations at the boundary between different objects. If one were to take the logarithm then

$$\log I(x, y) = \log R(x, y) + \log \tilde{I}(x, y)$$

and the multiplicative law becomes additive. A dc filter can then suppress the illumination component and by performing exponentiation a reflectance intrinsic image can be recovered. One interesting question is related to lightness constancy, where lightness of a surface (white, gray, black) is largely independent of its illumination $\tilde{I}(x, y)$. What we measure is the brightness $I(x, y)$, while the illumination component $\tilde{I}(x, y)$ is usually unknown. Based on homomorphic filtering, Horn (1974) designed a computational procedure for recovering lightness. The algorithm starts by detecting strong edges in the image intensity $I(x, y)$ and relating them to changes in lightness $R(x, y)$ only. The step is implemented (via lateral inhibition) through convolution. Following thresholding [to keep only strong edges and to discard the weak ones due only to variation in the illumination component $\tilde{I}(x, y)$], the lightness (reflectance) image is reconstructed (via lateral facilitation) through deconvolution. Illusions like Craik–Cornsweet–O'Brien (Frisby, 1980) can be explained by such a process as given above and were known to the 11th-century Chinese manufacturers of Ting white porcelain (incisions with a sharp inner edge and a graded outer edge create the illusion of adjacent surfaces as being of different lightnesses, despite their being made of the same material).

### B. Simulated Annealing

Simulated annealing is a statistical mechanics method discussed by Kirpatrick *et al.* (1983) for solving large optimization problems. [It is interesting to note that some of the research on neural networks (see Section

VII,D) is based on just such methods (Hopfield, 1982; Hinton *et al.*, 1984).] Annealing, according to Webster's Dictionary, means to heat and then cool (as is the case with steel or glass) for softening and making something less brittle. The alternating process of heating and then cooling attempts in a similar fashion to relaxation (Rosenfeld *et al.*, 1976) to estimate the steady state of a particular system or image. As a consequence, the internal structure of a particular image can be recovered despite its being corrupted by possible noise. (The neural networks benefit from such an approach by learning an appropriate response configuration for a given input or settling into an appropriate steady-state response.)

Carnevali *et al.* (1985) used simulated annealing for image processing tasks related to restoration. Specifically, one can view restoration as an attempt to minimize some function $E(x_i)$, where the energy $E$ of the system is given in terms of $x_i$, the parameters to be estimated. The minimum energy yields a steady (ground) state; such a state can be reached by simulating the cooldown of a physical system starting at some temperature $T$ and going down to $T = 0$. The cooling must be slow enough in order to avoid thermodynamically metastable states, i.e., local minima (see the analogy to hill climbing methods).

The Metropolis *et al.* (1953) method for simulated annealing models the states of a physical system using the Boltzmann distribution $f = \exp(-E/T)$. A random configuration $\{x_i\}$ and a small random perturbation $\{\Delta x_i\}$ of it are chosen. The energy change $\Delta E = E(x_i + \Delta x_i) - E(x_i)$ is calculated; if $\Delta E < 0$, then the perturbation is accepted; otherwise it is accepted with probability $\exp(-\Delta E/T)$. The optimization can be extended to the case when the perturbations are accepted only if a set of given constraints is satisfied, i.e., the problem being solved is that of constrained minimization. (See the similarity to Lagrange multipliers.)

One use for simulated annealing was for parameter estimation (Carnevali *et al.*, 1985). Specifically, a binary image assumes a noiseless configuration of 50 rectangles. Degradation of the image is then achieved by adding random noise to it; i.e., pixels have their binary value changed according to some probability $r$. The task is that of estimating the location of the original, noisefree rectangles. The energy $E$ to be minimized is made up of two components:

(1)  The number of pixels on which the original and estimated image, $I$ and $I^*$, respectively, differ. (Note that the image model is known and the parameters to be estimated are the number of rectangles and their location.)

(2)  The Lagrange-type multiplier $k$ is introduced via $\alpha$, the number of rectangles. The total number of parameters is then $4\alpha + 1$, and the energy $E$ to be minimized is given as $E = k\alpha + \beta$, where $k$ is a penalty-like price to be paid

for too many rectangles, while $\beta$ is the discrepancy between the original and estimated image. (When $k$ is made large, small rectangles are considered as noise and discarded if the corresponding change in energy $\Delta E$ resulting from adding such rectangles is positive.) The perturbations considered for the above problem were add/delete rectangle, stretch, and/or split rectangle. The results, even for low SNR, were fairly good.

Another problem treated by simulated annealing is that of smoothing an image. The energy $E$ to be minimized is a function of $R$, the roughness of the smoothed image, and $D$, the discrepancy between the smoothed and the original image. Specifically, $E = kD + R$, where $k$, the Lagrange-type multiplier, trades off between roughness and fidelity of restoration. For $D$, one can use either the $L^1$ or the $L^2$ distance between the original $a$ image and its smoothed version $b$ [for binary $a$ and $b$, $|a - b| = (a - b)^2$], while for $R$, one can use the Laplacian (Rosenfeld and Kak, 1982). Minimizing $E$ is then equivalent to finding the ground state of an Ising ferromagnetic system imbedded in an external field (Carnevali et al., 1985). [The minimal ground-state configuration has to account for two competing forces on each particle with a magnetic moment (spin): (1) interaction between the neighbors which tends to align the spins (up or down); and (2) tendency to align the spin with the external field.] One can easily see that the external field is analogous to fidelity while neighborhood alignment is analogous to roughness. The experiments which were carried out showed that for low SNR, simulated annealing yields better results than standard smoothing algorithms. The main drawback of simulated annealing is its large computational requirements.

### C. Motion Deblur

Another example of restoration toward recovery of an intrinsic characteristic is that of motion deblur. Preprocessing is aimed at providing a good quality image for further processing and recognition. If an object moves faster than the exposure time of the camera, then the image is blurred. We show next how the effects of motion, in the case of linear translation, can be removed. [Linear translation could be assumed based on a particular optical flow (Ballard and Brown, 1982) field.] Assume that the object to be imaged is given by $f(x, y)$, the true picture value function, and that $T$ is the exposure or effective integration time of the camera. Then what the camera records is

$$h(x, y) = \int_{-T/2}^{T/2} f(x - \alpha(t), y - \beta(t)) \, dt$$

if $\alpha(t)$ and $\beta(t)$ stand for motion in the $x$ and $y$ directions, respectively. By taking the Fourier transforms one obtains

$$H(u, v) = \int\!\!\!\int_{-\infty}^{\infty} h(x, y) \exp[-j2\pi(ux + vy)] \, dx \, dy$$

$$= \int_{-T/2}^{T/2} \int\!\!\!\int_{-\infty}^{\infty} f(x - \alpha(t), y - \beta(t)) \exp[-j2\pi(ux + vy)] \, dx \, dy \, dt$$

Perform transformations as required by the substitutions $\{\xi = x - \alpha(t)$ or $x = \xi + \alpha(t)\}$ and $\{\eta = y - \beta(t)$ or $y = \eta + \beta(t)\}$ and then obtain

$$H(u, v) = \int_{-T/2}^{T/2} \int\!\!\!\int_{-\infty}^{\infty} f(\xi, \eta) \exp[-j2\pi(u\xi + v\eta)] \, d\xi \, d\eta$$

$$\times \exp\{-j2\pi[u\alpha(t) + v\beta(t)]\} \, dt$$

or, equivalently,

$$H(u, v) = F(u, v)G(u, v)$$

where

$$G(u, v) = \int_{-T/2}^{T/2} \exp\{-j2\pi[u\alpha(t) + v\beta(t)]\} \, dt$$

In other words, we showed that the effects of linear motion can be modeled as a position-invariant imaging system. If the motion is uniform in the horizontal direction, then

$$\{\alpha(t) = vt, \beta(t) = 0\}$$

and

$$G(u, v) = \int_{-T/2}^{T/2} \exp(-j2\pi vut) \, dt = \frac{\sin(\pi v Tu)}{\pi v u}$$

$$= T \operatorname{sinc}(\pi v Tu)$$

Once both $G$ and $H$ are known, the original picture value function can be recovered (in functional form) as

$$f = \mathscr{F}^{-1}\{H/G\}$$

## VII. Structural Methods

The methods presented in the preceding sections attempt to perform invariant pattern recognition by extracting a set of measurements from an image which has been rendered "invariant" via some functional transformation. Digitization and computationally induced errors make such methods sensitive to noise (uneven illumination, sensor fault, or background clutter). The structural methods, to be presented in this section, are complementary to methods based solely on measurements, and could make the process of pattern recognition more robust with respect to the invariance aspect. Structural methods, as the name implies, are aimed at deriving the relationship holding between internal components making up an object. We review next structural concepts drawn from topology, syntactic pattern recognition, AI, and distributed associative memories.

### A. Topology

Topology is useful for deriving global descriptions of an image which are invariant to distortions like shrinking, stretching, and/or rotation. Basically, topological features do not depend upon metric properties. Assuming that $C$ is the number of connected components in a given image and $H$ is the number of holes, one can define a topological property like the Euler number as $E = C - H$. Furthermore, for a polygonal approximation, assuming that $V$, $Q$, and $F$ are the number of vertices, edges, and faces, respectively, the Euler formula states that

$$E = V - Q + F = C - H$$

The Euler number is implicitly available as one of the discriminant features for the first generation of commercial binary vision systems. The same number could be also used in a sequential tree classifier for discriminating among digits and/or letters, and clustering them later into subsets. As an example, $A$, 6, and 9 have the Euler number of 0 while $B$ and 8 have $E = -1$. The reader should be aware of how connectivity is defined (4-, 6-, or 8-neighbor) because it could affect the derivation of both holes and connected components in pathological cases. Toriwaki et al. (1982) extend the above concept to the case of 3D binary images. (3D image analysis is relevant for CT-derived medical imagery.) Rather than dealing with pixels $(x, y)$, they are concerned with voxels $(x, y, z)$, and several types of connectivity/neighborhoods are introduced. Topological properties can be easily derived if an object is given by its outline encoded as a chain code. Two chains can be checked for congruency, disregarding the starting point of the chain by simple string matching. The residue of a chain $C$, denoted by $R(C)$, is the chain of minimum length from the start to the end point of $C$ and which has its links arranged in ascending value. Chain rotation, expansion, and contraction can be derived through straightforward algorithms. It should be pointed out that chain codes are by definition invariant to translation, and as a consequence invariance to linear transformation (LT) can then be achieved. Similarity (in shape and orientation) between two contours can be found using the chain correlation function and filtering (smoothing) of a chain might precede the correlation. Layout issues and the bounding rectangle problem can be approached also based on a chain code representation. Structural analysis using topology is in its infancy and deserves more attention. Interaction with mathematics where topology is a major area of research could be fruitful.

### B. Syntactic Pattern Recognition

Syntactic pattern recognition (Fu, 1982) aims at modeling the boundary of a particular shape as a string made up of primitive elements ("letters/words"). Such primitive elements are usually defined based upon splitting the boundary at high curvature points. Next, the elements are combined according to a grammar which is characteristic to a given class of object. Conversely, given a string made up of some elemental components, a parser will work on it assessing if it belongs to some corresponding language (grammar). The basic question is then to define the grammar $G = (V, T, P, S)$ and to build the parser. It is the grammar which hopefully will capture the structure of the object and achieve invariant pattern recognition. We review next two examples of how syntactic pattern recognition can be applied to chromosome analysis and (mechanical) industrial part recognition, respectively. It was Ledley (1964) who first approached the chromosome identification problem. The grammar $G$ was of a context-free type and the primitives were: $\{\widehat{a}, b|, \breve{c}, d, \breve{e}\}$. The nonterminals were $A$ through $F$, standing for armpair, bottom, side, arm, rightpart, and leftpart, respectively. There is a starting symbol $S$ and two additional pseudo-start symbols $S_1$ and $S_2$, which stand for specific subclasses of chromosomes: submedian and telocentric, respectively. Once a string representation for a given chromosome is obtained, a parser will look for class identity (1 or 2) or reject (if the string is not legal according to the grammar definition). To gain some flavor regarding the grammar, we give some of its productions and discuss their meaning. $S \rightarrow S_1 \,|\, S_2$ is obvious. $S_1 \rightarrow AA$ means that a ⟨submedian chromosome⟩ is made up of two ⟨armpair⟩ (BNR—Backus normal form—notation is used to describe the grammar). $A \rightarrow CA$ is a self-referencing rule (characteristic to CFG) and means that an armpair is defined as a side followed by an armpair. Definitely, there are rules which allow creation of terminal elements like $A \rightarrow (F)(D) \rightarrow (DC)(D) \rightarrow aca$, i.e., a leftarm

corresponds to $\underbrace{\overset{\frown}{a} \overset{\frown}{c} \overset{\frown}{a}}$ .

It should be clear that defining an appropriate grammar is not a trivial task. The task of defining substrings as appropriate primitive terminals, which is quite difficult by itself, is complicated further by errors in outlining the boundary, which can lead to missing and/or extra terminals. As a consequence, error-correcting grammars might be required. The designer can determine statistically the likelihood and nature of possible errors, and the result is then a (stochastic) (error-correcting) grammar. Furthermore, to define an appropriate grammar significant learning might be required. Learning a grammar, except for trivial cases, is quite cumbersome, and it is usually an empirical rather than algorithmic method. Similar ideas to the ones used for chromosome analysis were applied to forensic applications like fingerprint recognition and human face identification. To improve on human face identification in particular, and syntactic pattern recognition in general, additional information in the form of filtering constraints can be used. Such additional constraints, in the form of semantic contextual information, is similar in its use to contextual cues used by listeners to disambiguate (speech-discourse) casual conversation. In formal terms this means going from CFG (context-free grammars) to CSG (context-sensitive grammars), and paying the price in the form of a more difficult (computationally/storagewise) parser. The system developed by Vamos (1977) for industrial objects and machine part recognition is another example of using syntactic pattern recognition techniques. A context-free grammar is used as well but this time in the form of the Chomsky-normal form. (Any CFG without the empty string can be generated by a grammar $G$ in which all the productions are of the form $A \rightarrow BC$ or $A \rightarrow a$. Here, $A$, $B$, $C$ are nonterminal variables, and $a$ is a terminal.) The reason for using the Chomsky-normal form is that rules containing only nonterminal symbols on the right side correspond to preprocessors used for feature extraction of primitive components. As such they are the same for all object classes to be defined by corresponding grammars $G_i$. The rules involving nonterminals only define the internal structure of each class $i$ and stand for the grammars $G_i$. The recognition process is further enhanced by the availability of relations like concatenation, loops, parallelism, layout (left, right, below, over, inside), attributes for primitives (numerical values like slopes, end-point coordinates, radii), and relevance grades relating components and their importance to identification. Similar to sequential tree classifiers, relevance grades help in choosing the shortest path in the recognition process. As can be seen, more specific knowledge is made available to the system, and as a consequence the system performance is enhanced both in terms of reliability and cost. Such a system is already half-way between pure syntactic pattern recognition and expert systems for computer vision, where AI techniques are widely used.

## C. The Artificial Intelligence Approach

The artificial intelligence (AI) approach to invariant pattern recognition is the next evolutionary trend following syntactical pattern recognition. It is the knowledge base (KB) of the (expert) system (ES) which drives both the image formation (sensing) and image interpretation processes. (How such ES for CV work was shown generically in the review of basic concepts.) As was the case with the relevance grades introduced by Vamos for machine part recognition, AI systems derive their efficiency from their ability to determine critical aspects of the problem being investigated and then dedicating most of their computational resources toward solving them. The difficulties, which ES for CV have to cope with, stem from poor image quality (noise), aspect appearance in 3D space (i.e., distortions due to LT and perspective with respect to some standard stable state) and the stable states themselves (those aspects of a 3D structure which look more or less the same for a given range of viewing angles). The radar target classification (RTC) (McCune and Drazovich, 1983) system is one attempt in dealing with the problems listed above. It employs several layers of processing techniques characterized by an increased level of compactness with respect to the representation being used. The pairs of (processing techniques, data representation) are (signal processing, raw radar signals), (image processing, 2D radar image), (PR, image features), and (KB scene interpretation, semantic/contextual description of a scene). The suggested ES reasons about signal (radar) images by interpreting the rules available in the production-rule (PR) part of the system. The conditional part of the rules has to be matched against the working memory containing up-to-date information regarding the scene being analyzed. Contextual information (heuristics, facts) is part of the inference engine and helps in choosing a subset of hypotheses (data interpretation). Example of such metaknowledge used in ship classification can be shipping lanes, weather, ocean geometry, library of ships (model and tactics), intelligence reports, and so on. [This, by the way, is not too different from the approach followed in the HASP/SIAP expert system (Nii et al., 1982) for identification and tracking of ships using ocean sonar signals.] Basically, what RTC attempts to do is to extract features from the image and associate them with high-level models of the possible ship classes stored in the system's KB. Once such high-level models are triggered, ship components are predicted, and their appearance in the image is suggested. [This is equivalent to the concept of deprojection suggested in the

ACRONYM system (Brooks, 1981; Binford, 1982) whereby image-derived features suggest the stable state and its 3D aspect while image transformations "deproject" the stored-standard-canonical model into such a position and then attempt to match it at fine resolution with the raw image.] Therefore, invariance is captured in such a system by (1) well-structured models, (2) available geometrical transformations⁺ which can try for registration between the model and possible but still unknown object appearances. Last, but not least, the same way a human interpreter will perform, the system can use contextual (metaknowledge) information to patch up error and gloss over details which might be added/missing by error. In other words, the system can see and/or disregard data based on beliefs (*a priori* knowledge). [Remember the adage, "One sees what he/she wants (is prepared) to see"!] The whole area of KB systems is in its very early stages. However, the approach holds much promise and significant research is currently being carried out.

## D. Neural Networks

We mentioned earlier that the most promising strategy for AI-related goals (and invariant pattern recognition is one of the most significant ones) is that of synergism between disciplines like neurophysiology, psychophysics, cognitive sciences, physics, and signal analysis. A firmly grounded concept suggested by such synergism is that of distributed associative memories (DAM) (Kohonen, 1984). Conceptually, DAMs are related to the GMF (generalized match filters) and SDF (synthetic discriminant functions). Like them, the DAMs attempt to capture a distributed representation which somehow averages over variational input representations belonging to the same class. As was mentioned earlier, low-level vision is characterized by data-independent and extensive parallel processing. The representation derived is then matched against learned, canonical prestored images. Neurophysiology and memory research on one side, as well as the requirement for recognizing noisy, partially occluded objects, suggest the use of DAM. The DAM allows for memory to be "reconstructive"; i.e., the matching process between incoming and stored data representations yields the entire output vector (or a closed approximation of it) even if the input is noisy, partially present, or if there is noise in the memory matrix itself (computer components, like neurons, are not always reliable, and they might die as well). The DAM is like a "holographic memory," and its basic principle of operation for the autoassociative case is given below:

$$M = FF^+$$

where $M$ is the optimum DAM, and $F^+$ is the pseudoinverse of the input matrix $F$. If one can assume linear independence between the input images (taken as vectors) which build up the matrix $F$, then

$$M = F(F^T F)^{-1} F^T$$

The recall operation for a given test image $t$ yields a retrieved approximation given by

$$\hat{f} = Mt$$

The memory matrix $M$ is orthogonal to the space spanned by the input image representations. The retrieved data $\hat{f}$ obeys the relationship

$$t = \hat{f} + \tilde{f}$$

where $\hat{f}$ plays the role of an optimal autoassociative recollection/linear regression in terms of $M$ and $\tilde{f}$, the residual is orthogonal to the space spanned by $M$, The $\tilde{f}$ component is what is new in $t$ with respect to the stored data in $M$ and as such it represents novelty. (The smallest the $\tilde{f}$ component is, the more invariant the recognition results are.) the $\hat{f}$ component filters out that part of the image representation that can be explained in terms of the stored data. It is the definition of $t$ in terms of $(\hat{f}, \tilde{f})$ which can be then used to facilitate learning/discovering new things. Finally, one has to ask how can a system as we suggest learn the attitude in space of an object perceived for the first time? Toward that end, ecological optics in the form of 3D cues (like shape from shading, texture, motion, and/or monocular cues) could provide for both the derivation of the scene's layout *and* environment rather than viewer-centered perception of objects. The information such derived allows then the canonical representation stored in the DAM to be associated with a particular 3D geometry. One additional point has to be made out. The DAM as presented here capture the (internal) structure in a statistical sense only, as the GMF and SDF do. However, the concept of mental transformations (Shepard and Metzerl, 1981) and that of spatial (polar) mappings in the primary sensory projection areas (Schwartz, 1977) suggest that the complex-log mapping, discussed earlier in our survey, should be made complementary to DAMs in order to achieve invariance to linear transformations. We are at present carrying out research based on just such an approach.

Finally, still within the framework of structural methods and a synergistic philosophy, we consider neural networks. We note first the work of Edelman and Reeke (1982) regarding higher-brain function as applied to invariant pattern recognition, because it suggests an interesting path to follow. The approach is firmly grounded in molecular biology and evolutionary theory and it is also suggested by psychologists like Hebb. Specifically, Edelman and Reeke claim that the brain is a selective system more akin in its working to evolution itself than to computation. The brain has been compared, throughout history, to the machines characteristic of those times. It is only recently that the view of the brain as an information processing machine prevailed over behaviorism (Skinner) and direct perception (Gibson). Still, assuming that the brain as an information processing unit view is adopted, there is, to quote Edelman, "The famous ghost that haunts all considerations of the brain, namely the homunculus, who, in fact, is telling whom what to do? Who is writing the program?" [Those issues are also addressed by Searle (1985) in his polemics with AI regarding consciousness and awareness.] Following Edelman, the last question does not have to be asked if one adheres to the notion that the brain, in its higher function at least, is a Darwinian system. Specifically, each brain is a selective system in which a large pre-existing set of variants of neural networks, formed during embryonic life in the absence of any contemporary information that might be necessary for adult functions, are later selected for and against during the worldly stay of the animal. The issue of invariant pattern recognition is reflected in optical illusions which illustrate the difference between the phenomenological (perceptual) and physical (world) order. To reconcile such differences, and to make perception invariant, the organism has to *adapt* its internal classification schemes. A selective, self-adapting system will then operate based on the following requirements. First, there is some kind of variable repertoire upon which selection can act. Second, there is some kind of reasonably frequent encounter between the selective conditions (images to be perceived) and those elements in the repertoire that are to be selected (memory representations). Finally, there is some kind of differential amplification, i.e., the alteration of the connection strengths of the synapses of any particular network which happens to fit a particular input pattern. The last step, like {DAM, GMF, SDF}, attempts to build that network which will be invariant in its recognition capability for a specific object. The model presented above also suggests the possibility of a degenerate set, i.e., a nonisomorphic set of isofunctional structures, such that the recognition is good over the ensemble of possibilities to be encountered in the future. Therefore, invariance is achieved through several distributed model representations. Edelman and Reeke implemented DARWIN II, which is based on just such a group selection theory as presented above. There are two subnetworks, DARWIN and WALLACE, respectively, in DARWIN II. The function of DARWIN is to make a representative transformation of local features of an object in network terms. In contrast, the function of WALLACE is to give an more or less invariant response of its groups to all objects of a given class having similar global features. The net effect is that two networks with two different functions (representative transformations for individuals and invariant responses for classes) have been linked to create a new function, that of an associative memory (AM). It is difficult to assess the theory based on

DARWIN II's performance, but both philosophically and conceptually the theory is very appealing, and it deserves further and *significant* attention. Another example of a neural network model for associative memories has been recently suggested by Fukushima (1984). Specifically, the model consists of a hierarchical multilayered network to which efferent (top-down) connections are added, as to provide positive feedback loops in pairs with afferent (bottom-up) connections. As a result, both pattern recognition and associative recall progress simultaneously in the pyramid-like hierarchically structured network. Converging afferent paths allow for local features, extracted from the original input, to be integrated into more global features. (There are specific features extraction cells, and the corresponding receptive field are allowed to overlap and gradually cover more of the original input, i.e., globality of analysis is achieved by reducing the resolution as one approaches the top of the pyramid.) Finally, the last layer, the classification one, decides about the object identity. For the purpose of associative recall, the integrated information is passed over to the lower cells by a diverging efferent path, and yields the capability of regenerating at the lowest level the original input object despite noise, overlapping or missing components. Neural networks were tried on very simple test data and therefore it is difficult to assess their power. Nevertheless, it is worth mentioning them because of several factors:

(1)   Researchers eventually got over some of the limitations of the original PERCEPTRON (Minsky and Papert, 1968) and tried to cope with inherently complex constraints, like connectivity.

(2)   The basic mechanism, that of DAM, has a strong theoretical foundation and has much promise.

(3)   The model allows for self-organization, i.e., unsupervised training. Therefore, it answers one of the major claims made against AI, that of placing the intelligence in the program tutor rather than instilling it into the system for further use.

### E.  The Environmental Approach and Ecological Optics

Our surroundings, in the form of the physical environment (ground, horizon) can provide a useful frame of reference for determining the structure of a given object. Gibson (1950) showed how rich and useful environmental information can be; ecological optics is a term introduced later on by Gibson to coin an approach where stimulus/environmental information is considered crucial to perception. The validity and implications of Gibson's approach (misnamed as direct perception) are beyond our immediate goals herein; for a detailed discussion see Ullman (1980). There are, however, a number of strong points in the environmental approach on which builders of emerging (robotic) vision systems will agree:

(1)  Interdependence between action and perception; as an example, see active versus passive stereo systems;

(2)  Richness of the visual stimulus in the optical array;

(3)  The need for appropriate frames of references within which a systemic approach for object recognition and location can be undertaken. The advantages of having available a viewer-independent representation are obvious. Palmer (1983) makes a strong point about properties of objects and events being perceived relative to stable reference standards that unify the various parts of the visual field. Furthermore, the reference orientation is environmental and/or gravitational rather than retinal, even though there is a powerful tendency toward establishing the orientation of a reference frame along an axis of reflectional symmetry, if one exists. One could add that experimental evidence points out the strong relationship between the systemic/internal structure of an object, as determined by its axes of symmetry, *and* the environment; (vertical axes, ground,...) frames of reference, for object recognition or motion interpretation, are derived based on just such a relationship and can be computed using vector analysis on the available axes.

(4)  Most of computer vision is concerned with image transformation and the attempt to capture invariant object aspects despite such transformations. Palmer (1983) suggested that grouping (gestalt) phenomena and "good" figures can be explained in terms of minimal transformations.

Spatial perception is facilitated by our ability to determine the spatial layout of the environment (Sedgwick, 1983). If one deals with a rich environment rather than poor displays, the local depth cues are less important, and paramount to perception are the following scales of space as defined by Haber (1983):

(1)  Stationary perspective (texture gradients, vanishing points);

(2)  Motion perspective; and

(3)  Binocular disparity.

The above scales interface and together they cover the whole visual field, through the use of different frames of reference. The stationary scale covers the ground, away from our feet. Motion refers to our point of fixation, and binocular disparity refers to the point of (convergent) fixation (zero disparity).

Achievement of constancy (invariance) is fundamental to all scales of space processing.

Let us consider first the stationary scale. The visual/terrestrial environment is continuous, more or less horizontal, and the objects of interest rest on the ground surface. As we are going to see soon, scale information as given by perspective is the intrinsic data which can be derived, and based on it the range can be estimated if and when needed. Such an approach is characteristic of object/environment rather than viewer-centered representations. One should remember that in viewer-centered representations, like Marr's 2.5D sketch, range (depth) is derived first and (relative) scale can be found later on. One of the stationary scales is related to texture density gradients as a function of the distance along the ground. Specifically, assume that texture elements ($t_e$) are regularly spaced over the ground and then one can show that there is a constant scale invariance under perspective. Following Purdy (1960), define the direction of slant as being perpendicular to the zero-gradient direction. The amount of slant along the slant direction can be found using the following definitions: the texture element ($t_e$) is of size $s \times s$; the observer ($O$) is at a height $h$ from the ground surface and a distance $d$ along the ground from $t_e$. The line of regard of length $r$ from $O$ meets $t_e$ at (optical slant) angle $R$ and makes an angle $F$ with the horizontal. The angular width (visual angle) is $L$, and the angular depth subtended by $t_e$ is $M$. One can then define the texture size $W = ML$, and the corresponding texture density as $N = 1/W$. The relative rate of change in texture size, as the line of regard is swept along the direction of slant, is given as $G_W = (dW/dF)W$. Computation then yields $G_W = 3/\tan R$, where $\tan R = h/d$ and $d/h = G_W/3$. Size and distance are specified relative to a scale factor as given by the texture element $t_e$. Further computation reveals that $s/r = 2\tan(L/2)$ and that for small angles given in radians $s/r \simeq L$. Relative size is given as $r_1/r_2 \simeq L_2/L_1$. Absolute range data can be found if the texture scale ($s$) is known. Furthermore, sizes of all objects resting on the ground are specified, to within a scale factor, by their projective relationships. The perspective scale is environment-centered and it is given in terms of vanishing points (the location where parallel lines converge) and horizon (a line through vanishing points). Sizes of all objects resting on the ground are specified, to within a scale factor, by their projective relations with the terrestrial horizon.

The motion scale can provide valuable information as well. Expansion patterns (of optical flow) are characteristic of approaching objects, while the concentricity of the optical flow is related to the path of the approaching object.

We conclude the discussion on the environmental approach by pointing out (Gilchrist, 1977) that spatial layout (depth) recovery precedes lightness computation. Such information is relevant to ordering the functional modules of any machine vision system; furthermore, it raises a question mark related to the indiscriminate use of edge detection as a first processing step.

### F. Bin Picking

We have presented, so far, different computational methods for achieving invariance in pattern recognition. We next consider the bin-picking problem as an application area where invariant recognition is called for. Industrial automation, both for assembly and quality inspection reasons, requires the capability to recognize both the identity and the attitude in space of objects despite image variability (noise, uneven illumination, geometric distortions, and occlusion). Invariant recognition despite such variability is generically known as the bin-picking problem. A survey of solutions suggested to solve the problem for flat 2D patterns is presented, and an assessment of the methods is made as well. Assessment is made in terms of computational requirements (memory and speed of computation), performance, invariance, and system approach.

The case was made earlier that once a computational task is defined, one has to choose appropriate image representations and the algorithms suitable to operate on such representations. We detail next some of the main approaches and discuss their limitations. All methods are based on derived image representations and their counterparts, learned models prestored in memory, *and* the attempt to match among them.

### 1. The Feature-Based Approach

Most of the commercially available machine vision systems (Crowley, 1984) are restricted to binary vision and/or silhouette defined objects. Connectivity analysis to define blobs and/or edge detection to determine the outline of the object are used. Information about the surface of the object, like texture, is discarded *and* the environment is assumed to be clean. Specifically, the illumination is even, no shadows are allowed, and objects do not touch each other. Feature extraction yields for blobs properties like perimeter ($P$), area ($A$), major and minor axes, minimum and maximum radii (from the center of gravity to the object's boundary), and compactness ($C = P^2/4\pi A$). Such properties are invariant to rotation and translation. Scale invariance is explicitly achieved through system calibration. Statistical classifiers use then such features for object recognition. The silhouette based systems use a polygonal approximation for the object's boundary. The segments used in approximation are like letters in some alphabet and their characteristics are in terms of length and orientation. The joins (angle) between such segments approximate grammatical constraints. Syntactical pattern recognition and/or graph search algorithms are used to identify the objects.

## 2. Intrinsic Functions and Correlations

Once occlusion is taken into consideration there is the need for sub-template matching. One particular object representation which facilitates such a task is the intrinsic $(\theta-s)$ (parametric) curve (Perkins, 1978), where the angle of the slope $(\theta)$ and the corresponding arc length $(s)$ along the boundary are used as a new system of coordinates. The $\theta-s$ representation allows descriptions of the form $\theta(s)$, where a change in orientation $(\alpha)$ in the $x-y$ space corresponds to simlply adding $(\alpha)$ to $\theta(s)$ in the $\theta-s$ space. The method is sensitive to the quantization noise introduced when determining the boundary curvature $\theta(s)$. Furthermore, one has yet to determine what the sub-templates are even though salient (significant) segments could be thought of.

## 3. The Hough Transform

The Hough transform is characteristic to connectionist models and gathers supporting evidence for object characterization and detection in some parametric space. Its use for the bin-picking problem has been suggested by Turney *et al.* (1985).

The Hough transform is implemented through the use of an accumulator array, and subtemplate $(\theta-s)$ matches for each of the mechanical parts to be recognized point to possible objects and their centroids. The goodness of the fit is used to increment the accumulator cells selectively. Specifically, if $B$ and $T$ stand for the object boundary and a particular template, respectively, then by defining $\tilde{T}$ as $T$ rotated by $180°$, it can be shown (Sklansky, 1978) that convolution between $B$ and $\tilde{T}$ is equivalent to cross-correlation between $B$ and $T$. In other words, the attempt to implement the Hough transform by looking for potential matches between the boundary $B$ and the rotated template $T$ is equivalent to match-filter detection. One can hardly over-emphasize the fact that most of object recognition and, therefore, bin-picking, is match filter (Section V,F) in one form or another. Occlusion is dealt with by determining the saliency of the subtemplates in terms of local properties. Orthogonality of the subtemplate–coordinate systems is achieved through factor analysis/minimization over the whole set of parts to be recognized.

Assume that $\gamma_{ij}$ is the matching coefficient (goodness of fit) between the $(\theta-s)$ representations corresponding to template $\tau_i$ and boundary segment $b_j$, where $h$ is the number of pixels, and $\alpha$ is the difference between the average slopes. Then

$$\gamma_{ij}^* = \min \gamma_{ij} = \sum_{l=1}^{h} \{\theta_{b_j}(s_p) - [\theta_{\tau_i}(s_p) + \alpha]\}^2$$

Assume next that templates $A$ and $B$ have $|t|$ and $|b|$ subtemplates, respectively, and that $|t| \, |b|$ matches are attempted. If the image is of the size $M \times N$, an accumulator array $a_s$ can be defined [where $s = (N - 1)*$ row index + column index] as

$$a_s = \sum_{i=1}^{|t|} \sum_{j=1}^{|b|} \delta_{ij}^s c_{ij} w_i$$

where $c_{ij} = 1/(1 + \gamma_{ij}^*)$ are the weighted contributions (by $w_i$) used to adjust the accumulator array. The above equation can be rewritten as $\mathbf{a} = D\mathbf{w}$, where $d_{si} = \Sigma_{j=1}^{|b|} \delta_{ij}^s c_{ij}$. The weights $w_i$ have to be chosen so as to minimize the correlation of template $A$ with boundary $B$. Specifically, one has to minimize $\Sigma a_s^2 = \mathbf{a}^T \mathbf{a} = \mathbf{w}^T D^T D \mathbf{w}$ subject to $\mathbf{1}^T \mathbf{w} = 1$ and $\mathbf{w} \geq 0$. The above discussion can then be generalized to a whole set of parts. Saliency of subtemplates through appropriate weighting factors improves upon the unrestricted use of the Hough transform. Drawbacks in using the Hough concept are mainly related to the memory requirements needed to implement big accumulator arrays and the computation itself, which can be quite wasteful due to the exhaustive nature of the algorithm. Some strategies suggested by Turney et al. (1985) to improve on the performance are listed below:

(1)   Eliminate segments (templates) from further consideration as soon as they are identified;

(2)   Coarse to fine resolution strategy;

(3)   Sequential recognition; and

(4)   Multiprocessor hardware architecture.

## 4. Match Filters

The solutions suggested for solving the bin-picking problem vary in their generality. Partial solutions can be considered as well, whereby one merely attempts to locate potential holdsites for enabling a robot end-effector to grasp and to hold a mechanical part. Recognition of the part, if and when needed, could follow. Dessimoz et al. (1984) suggest the use of match filters for exactly such a task as described above. Assume that the holdsites are described by local masks $p(i, j)$ and that the original image $(D)$ is given as $f(i, j)$. Potential holdsites are then located when the similarity between $f$ and $p$ is high, or equivalently for those locations where the error $E^2$ given as

$$E^2 = \sum_{(m,n) \in D} [f(i + m, j + n) - p(m, n)]^2$$

$$= \sum [f^2(i + m, j + n) + p^2(m, n) - 2f(i + m, j + n) + p(m, n)]$$

is minimized. It is clear that $E^2$ is minimized if the cross-correlation of $f$ and $p$ as given by $f(i + m, j + n)p(m, n)$ is maximized. If noisy data $[W(u, v)]$ are a possibility, the following equations hold

$$h(i, j) = p(-i, -j)$$

$$g(i, j) = \sum h(m, n) f(i - m, j - n) = h \otimes f$$

$$G(u, v) = H(u, v)F(u, v) = P^*(u, v)F(u, v)$$

$$H(u, v) = kP^*(u, v)/W(u, v)$$

If the energy (illumination) of the image is not uniform, then normalized cross-correlation should be used.

## 5. Probabilistic (Labeling) Relaxation

Bhanu and Faugeras (1984) suggest the use of probabilistic labeling. The approach is based on relaxation (Davis and Rosenfeld, 1981), and it is similar to the Viterbi algorithm discussed in Section V,C. The object is given as a polygonal approximation, and one attempts to label each segment as one of the predefined subtemplates. Optimality is achieved by seeking the best solution in a global sense. Computational requirements are high, as noted by Ayache and Faugeras (1986) because all the potential solutions are calculated and stored.

## 6. Graph Theory

Given one image representation and a corresponding memory model, one has to perform some kind of (sub)graph isomorphism, which is known to be NP complete. Different strategies can be used to speed up the computation. One can look for maximum cliques (Bolles and Cain, 1982), use some heuristic evaluation function (to prune out unlikely matches) within the framework of the $A^*$ algorithm (Ballard and Brown, 1982), and/or do clustering analysis in some feature space, maybe helped by the MST (minimum spanning tree) approach.

## 7. Geometry

Invariance to linear transformations can be achieved through the use of conformal mappings. We mentioned and discussed earlier (Section IV,A,3) the OLREC system (Moss et al., 1983), which is based upon CL mappings. Another idea whose origin comes from differential geometry is that of the Gaussian sphere and EGI (extended Gaussian sphere) (Horn and Ikeuchi, 1984). The EGI is equivalent to an orientation histogram corresponding to the normals ("needles") as estimated across a 3D structure. Potential problems related to the EGI are listed below:

(1) The "needle" map is obtained through the use of photometric stereo and is highly dependent on the illumination. Fluctuating illumination and shadows can easily distort the results. As Yang and Kak (1985) point out, the EGI method could benefit from the use of range instead of brightness as an intrinsic characteristic.

(2) High computational requirements and the difficulty to deal with occlusion. The generation of models in terms of their EGI is also quite difficult.

## 8. Deprojection and Recursive Estimation

Ayache and Faugeras (1986) suggest an approach called HYPER (hypotheses predicted and evaluated recursively). The method has an AI flavor by considering (suggesting) different hypotheses, making predictions, and then verifying them. The loop is traversed until enough confidence is gained. Estimation about the attitude of the object, assuming rigidity, is made recursively through the application of a Kalman filter. Specifically, the objects are given as polygonal approximations, including salient segments with which the match starts. The affine transformation which takes some model into the image is given as $T(\theta, k, t_x, t_y)$, where $\theta$, $k$, $t_x$, $t_y$ stand for rotation, scale, and translation. Estimating $T$ allows one to locate the attitude of the object in space. HYPER operates in an iterative fashion. At step (1), assuming $M_0$ was the most salient segment, the program chooses segment model $M_i$ which is the closest in physical location to $M_0$. $T_{i-1}$ applied to $M_i$ yields a deprojected segment $M_i^*$, and a similarity measure $s$ between $M_i^*$ and its closest image segment $S_j$ is calculated. If $s(M_i^*, S_j)$ is above some threshold the match is accepted; otherwise occlusion is assumed. The recursive estimation of $T$ is performed as follows. Define $\mathbf{a} = (k \cos \theta, k \sin \theta, t_x, t_y)^T$, and assume that segments are given by the coordinates of their middle points. Then, if $M_i^*(x_i^*, y_i^*)$, one can write, based upon $\mathbf{a}(i-1)$ and $M_i(x, y)$

$$\begin{bmatrix} x_i^* \\ y_i^* \end{bmatrix} = \begin{bmatrix} t_x \\ t_y \end{bmatrix} + k \begin{bmatrix} \cos \theta & -\sin \theta \\ \sin \theta & \cos \theta \end{bmatrix} \begin{bmatrix} x \\ y \end{bmatrix}$$

The potential match $S_j$ is given by $S_j(x_j', y_j')$, where

$$\begin{bmatrix} x_j' \\ y_j' \end{bmatrix} = \begin{bmatrix} x_i^* \\ y_i^* \end{bmatrix} + \begin{bmatrix} \varepsilon_{x_i} \\ \varepsilon_{y_i} \end{bmatrix}$$

where the error is $(\varepsilon_{x_i}, \varepsilon_{y_i})$. One can further write

$$y_i = C_i \mathbf{a} + \varepsilon_i$$

where $E(\varepsilon_i \varepsilon_i^{\mathsf{T}}) = W_i$ and

$$C_i = \begin{bmatrix} x_i & -y_i & 1 & 0 \\ y_i & x_i & 0 & 1 \end{bmatrix}$$

Then the Kalman filter equations needed to estimate $\mathbf{a}(i)$ are given as

$$\hat{\mathbf{a}}(i) = \hat{\mathbf{a}}(i-1) + K(i)[Y_i - C_i\hat{a}(i-1)]$$

$$K(i) = S(i-1)C_i^{\mathsf{T}}[W_i - C_iS(i-1)C_i^{\mathsf{T}}]^{-1}$$

$$S(i) = [I - K(i)C_i]S(i-1)$$

where $K(i)$ stands for gain and the variance-error matrix $S(i)$ is given by

$$S(i) = E[[\mathbf{a} - \hat{a}(i)][\mathbf{a} - \hat{\mathbf{a}}(i)]^{\mathsf{T}}]$$

One can initialize the filter by estimating the error between the model and the center of mass for the object image. Once the procedure described above runs to completion, i.e., after an attempt was made to match each model segment $M_i$ ($i = 0, \ldots, n-1$), a final estimate $\hat{\mathbf{a}}(n)$ is obtained. Using $\mathbf{a}(n)$ a new attempt is made to deproject the model segments $M_i$ and to find a match form them. A goodness-of-fit score and occlusion percentage are then used to accept/reject a potential match.

The methods discussed lack generality and depend heavily on *a priori* knowledge. The computational resources required render most of the methods impractical for real applications. Furthermore, the approaches presented so far are sensitive to noise, illumination, and/or occlusion. As of now, the bin-picking problem still waits for a general and robust solution. Partial solutions, like those attempting to identify potential holdsites, might prove to be useful short-term solutions.

## VIII. CONCLUSIONS

We have reviewed the problem of invariance in pattern recognition. The motivation for dealing with such an issue is similar to the main goal of artificial intelligence. Specifically, visual processes involve by their very nature complexity, uncertainty, ambiguity, lack of information, facts which can "distort" the phenomena ("imagery") under observation. However, following the human example, any artificial machine vision system should process information such that the result is invariant to the vagaries of the acquisition process.

We have considered the issue of invariant PR within the framework provided by AI in general, and computational vision in particular. We have reviewed basic concepts like Markov chains, graph theory (MST and cliques), signal analysis (FT and linear systems), statistical PR (KLT and Fisher discriminant), formal languages, and the way they can work toward capturing the invariant aspect of the image to be processed and recognized. Following the review, we considered both deterministic and statistical methods for performing invariant PR. Within the deterministic methods used to achieve invariance to distortions caused by linear transformations (LT), we looked at the FT, Mellin transform (MT), conformal mappings (CL), orthogonal series expansion, and the moments approach. The statistical methods attempt to capture an "average" aspect of the imagery to be dealt with. Such an approach is similar with a match filter, and examples of the method are the synthetic discriminant functions (SDF) and the generalized matched filters (GMF). We considered not only geometric distortions but distortions caused by the image formation process itself. To remove illumination components which are not representative of the image we looked at homomorphic filtering. Removing the effects of motion was considered as well. Finally, advanced/recent methods of dealing with invariance were considered. Specifically, we looked at those attempts aimed at capturing the structure of the pattern. Methods described included syntactic pattern recognition, AI, and neural networks. Our treatment is not exhaustive. We dealt with the important issue of 3D invariant PR only marginal through the use of deprojection methods. The field of invariant 3D is in its infancy and is treated elsewhere (Besl and Ramesh, 1985b).

One important warning regarding the methods described is appropriate at this point. Our treatment was analytical, and the methods work *well* on paper. We considered only the task and the appropriate representation and algorithms by which the task can be accomplished. Implementation is not trivial, and one has to be concerned with "digital" distortions caused by discrete grids, quantization and accuracy effects, and so on.

Before concluding our review we have to ask ourselves what looks most promising at the present time in terms of future research directions for invariant PR. We strongly believe that the following approaches are most likely to advance the field:

(1) A cognitive approach, where an adequate knowledge base, via deprojection, can account for geometrical distortions and thus both identify the object and its attitude in space. Such an approach, if implemented, would represent an expert system (ES) for computational vision (CV).

(2) The neural networks introduced earlier are an exciting possibility. Such networks go beyond simplistic perceptions, and in fact they provide us with a unique opportunity. Specifically, they allow for self-organizing systems, i.e., unsupervised learning. Furthermore, they suggest a selectionist rather

than instructionist approach to model the information processing task of any intelligent system.

(3)   Finally, it is unlikely that one approach by itself will provide all the answers. It is more likely that a synergistic approach will be more successful. The reasons are twofold: the analogies drawn from different fields and the powerful mix that such methods can provide if they are taken together. We saw how matched filters (signal analysis) and statistical PR can be combined into SDF. We also suggested the mix of neural network DAM (distributed associative methods) and statistical analysis. It seems that the recognition memory has to approximate in a conceptual way the equivalent of a match filter.

We are only beginning the arduous task of invariant/PR. The problem is exciting, and the next several years will undoubtedly provide us with exciting solutions as well.

ACKNOWLEDGMENTS

This work was supported in part by the National Science Foundation under Grant ECS-8310057. I would like to thank Saad Bedros for his contributions on geometric invariance.

REFERENCES

Ahmed, N., and Rao, K. R. (1975). "Orthogonal Transforms for Digital Signal Processing." Springer-Verlag, Berlin.
Andrews, H. C., and Hunt, B. R. (1977). "Digital Image Restoration. Prentice-Hall, New York.
Ayache, N. (1983). Ph.D. Thesis', L'Université de Paris-Sud Centre D'Orsay.
Ayache, N., and Faugeras, O. D. (1986). *IEEE Trans. Pattern Anal. Machine Intelligence* **8,** 44–54.
Ballard, P. H., and Brown, C. M. (1982). "Computer Vision." Prentice-Hall, New York.
Bamieh, B., and De Figueiredo, R. J. P. (1985). *Proc. IEEE Int. Conf. Robot. Automation.*
Besl, P., and Ramesh, J. (1985a). *Proc. Computer Vision Pattern Recognition Conf.*
Besl, P., and Ramesh, J. (1985b). *Comput. Surv.* **17,** 1.
Bhanu, B., and Faugeras, O. D. (1984). *IEEE Trans. Pattern Anal. Machine Intelligence* **6,** 137–155.
Binford, T. O. (1971). *IEEE Conf. Syst. Control, Miami.*
Binford, T. O. (1982). *Int. J. Robot. Res.* **1.**
Bolles, R. C., and Cain, R. A. (1982). *Proc. IEEE Comput. Soc. Conf. Pattern Recognition Image Process., Las Vegas* 498–503.
Bracewell, R. N. (1978). "The Fourier Transform and Its Application," 2nd Ed. McGraw-Hill, New York.
Brooks, R. (1981). *Artif. Intelligence* **17.**
Carnevali, P., L. Coletti, and Patarnello, S. (1985). *IBM J. Res. Dev.* **29,** 6.

Casasent, D., and Chang, W. T. (1983). *Appl. Opt.* **22,** 14.

Casasent, D., and Fetterly, (1984). *Proc. SPIE* **456.**

Casasent, D., and Psaltis, D. (1977). *Proc. IEEE* **65,** 1.

Castelman, J. R. (1978). "Digital Image Processing." Prentice-Hall, New York.

Caulfield, H. J., and Weinberg, M. H. (1982). *Appl. Opt.* **21,** 9.

Cooper, L. A. (1983). *In* "Physical and Biological Processing of Images" (O. J. Braddick and A. C. Sleigh, eds.). Springer-Verlag, Berlin.

Crowley, J. L. (1984). "Machine vision: Three generations of commercial systems." CMU-RI-TR-84-1, The Robotics Institute, Carnegie-Mellon University.

Davis, L. S., and Rosenfeld, A. (1981). *Artif. Intelligence* **17,** 1–3, 245–263.

Dessimoz, J. D., Birk, J. R., Kelley, R. B., Martins, H. A. S., and C. L. I. (1984). *IEEE Trans. Pattern Analy. Machine Intelligence* **6,** 686–697.

Dubois, S. R., and Glanz, F. H. (1986). *IEEE Trans. PMAI* **8,** 1.

Duda, R., and Hart, P. (1973), "Pattern Classification and Scene Analysis" Wiley, New York.

Edelman, G. M., and Mountcastle, V. B. (1982). "The Mindful Brain: Cortical Organization and the Group-Selective Theory of Higher Brain Function." MIT Press, Cambridge Mass.

Edelman, G. M., and Reeke, G. N. (1982). *Proc. Natl. Acad. Sci. U.S.A.* **79.**

Forney, D. G. (1973). *Proc. IEEE* **61,** 3.

Freeman, H. (1974). *Comput. Surv.* **6,** 1.

Frisby, J. P. (1980). "Seeing." Oxford Univ. Press, London.

Fu, K. S. (1982). "Syntactic Pattern Recognition and Applications." Prentice-Hall, New York.

Fukunaga, K. (1972) "Introduction to Statistical Pattern Recognition." Academic Press, New York.

Fukushima, K. (1984) *Biol. Cybern.* **50.**

Gevarter, W. B. (1985). Intelligent Machines." Prentice-Hall, New York.

Gibson, J. J. (1950). "The Perception of the Visual World. Houghton-Mifflin, Boston.

Gilchrist, A. L. (1977). *Science* **195,** 185–187.

Gonzales, R. C., and Wintz, P. (1979). "Digital Image Processing." Addison-Wesley, Reading, Mass.

Haber, R. N. (1983). *In* "Human and Machine Vision" (J. Beck, B. Hope, and A. Rosenfeld, (eds.). Academic Press, New York.

Haber, R. N. (1985). *Comput. Vision, Graphics, Image Process.* **31,** 3.

Hester, C., and Casasent, D. (1981). *Proc. SPIE Infrared Technol. Target Detect. Classif.* **302.**

Hinton, G. E., Sejnowski, T. J., and Ackley, D. H. (1984). Boltzmann machines: Constraint satisfaction networks that learn. TR-CS-84-119, Computer Science Dept., Carnegie-Mellon University.

Hopfield, J. J. (1982). *Proc. Natl. Acad. Sci. U.S.A.* **79.**

Horn, B. K. P. (1974). *Computer Graphics Image Process.* **3.**

Horn, B. K. P. (1984). *Proc. IEEE* **72.**

Horn, B. K. P., and Ikeuchi, K. (1984). *Sci. Am.* **Aug.,** 100–111.

Hu, M. K. (1982). *IEEE Trans. Inform. Theory* **8.**

Jacobson, L., and Wechsler, H. (1984). *IEEE Trans PAMI* **6,** 3.

Jacobson, L., and Wechsler, H. (1985a). *SPIE Conf. Intelligent Robots Comput. Vision Cambridge, MA.*

Jacobson, L., and Wechsler, H. (1985b). *Proc. Int. SPIE Conf. Comput. Vision Robots, 2nd, Cannes.*

Jacobson, J., Thomanschefsky, U., and Wechsler, H. (1984). *Proc. SPIE Conf. Intelligent Robots Comput. Vision, Cambridge, MA.*

Kashyap, R. L., and Chellapa, R. (1981). *IEEE Trans. Inform. Theory* **27.**

Kirpatrick, S., Gelatt, C. D., and Vecchi, M. P. (1983). *Science* **220,** 671.

Kohonen, T. (1984). Self-Organization and Associative Memory. Springer-Verlag, Berlin.

Ledley, R. S. (1964). *Science* **146**, 3461.

McCune, B. P., and Drazovich, R. D. (1983). *Defense Electron.* **Aug.**

Marcelja, S. (1980). *J. Opt. Soc. Am.* **70.**

Marinovic, N. M., and Eichmann, G. (1985). *Proc. SPIE* **579.**

Marr, D. (1982). "Vision." Freeman, San Francisco.

Messner, R. A., and Szu, H. H. (1985). *Comput. Vision Graphics Image Process.* **32**, 2.

Metropolis, N., Rosenbluth, A., Rosenbluth, M., Teller, A. and Teller, E. (1953). *J. Chem. Phys.* **21**, 1087.

Minsky, M., and Papert, S. (1968). "Perceptrons." MIT Press, Cambridge, Mass.

Mitchell, G. R., and Grogan, T. A. (1984). *Opt. Eng.* **23**, 5.

Moore, D. J. H. (1972). *IEEE Trans SMC* **2.**

Moore, D. J. H., and Parker, D. J. (1974). *Pattern Recognition* **6**, 2.

Moss, R. H., Robinson, C. M., and Poppelbaum, W. J. (1983). *Pattern Recognition* **16**, 6.

Nii, H. P., Feigenbaum, E. A., Anton, J. J., and Rockmore, A. J. (1982). *AI Mag.* **Spring.**

Palmer, S. E. (1983). *In* "Human and Machine Vision" (J. Beck, B. Hope, and A. Rosenfeld, eds.). Academic Press, New York.

Pavlidis, T. (1977). "Structural Pattern Recognition." Springer-Verlag, Berlin.

Perkins, W. A. (1978). *IEEE Trans. Comput.* **C-27**, 126–143.

Purdy, W. C. (1960). The hypothesis of psychophysical correspondence in space perception. GE Tech. Info. Series, R6OELC56.

Reddi, D. (1981). *IEEE Trans. PAMI* **3**, 2.

Rosenfeld, A., and Kak, A. (1982). "Digital Image Processing." Academic Press, New York.

Rosenfeld, A., Hummel, R. A., and Zucker, S. W. (1976). *IEEE Trans. SMC* **6**, 6.

Schwartz, E. L. (1977). *Biol. Cybern.* **25.**

Searle, J. (1985). "Mind, Brain & Science." Harvard Univ. Press, Cambridge, Mass.

Sedgwick, H. A. (1983). *In* "Human and Machine Vision" (J. Beck, B. Hope, and A. Rosenfeld, eds.). Academic Press, New York.

Sedgwick, H. A., and Levy, S. (1985). *Comput. Vision, Graphics, Image Process.* **31**, 2.

Shepard, R. N., and Metzerl, J. (1981). *Science* **171.**

Shirai, Y. (1979). *In* Computer Vision and Sensor-Based Robots" Plenum, New York.

Shirai, Y. (1982). *Computer* **15.**

Sklansky, J. (1978). *IEEE Trans Comput.* **C-27**, 923–926.

Tenenbaum, J. M., and Barrow, H. G. (1978). *In* "Computer Vision Systems" (A. Hanson, and E. Riseman, eds.). Academic Press, New York.

Toriwaki, J., Yokoi, Y., and Yonekura, Y. (1982). *Proc. Int. Conf. Pattern Recognition.*

Tou, J. J. (1980). *Int. J. Comput. Inf. Sci.* **9**, 1.

Turney, J. L., Mudge, T. N., and Voltz, R. A. (1985). *IEEE Trans Pattern Anal. Machine Intelligence* **7**, 410–421.

Ullman, S. (1980). *Behav. Brain Sci.* **3**, 373–415.

Vamos, T. (1977). *In* Syntactic Pattern Recognition Applications" (K. S. Fu, ed.). Springer-Verlag, Berlin.

Wechsler, H. (1979). *Comput. Graphics Image Process.* **9**, 3.

Wechsler, H. (1981). *Comput. Graphics Image Process.* **17**, 4.

Yang, H. S., and Kak, A. C. (1985). *Proc. IEEE Workshop Comput. Vision: Representation and Control, 3rd, Bellaire, Michigan.*

Zwicke, P. E., and Kiss, I. (1983). *IEEE Trans. PAMI* **5**, 2.

# Index